Structure and Dynamics
of
Molecular Systems - II

Structure and Dynamics of Molecular Systems – II

Edited by

R. Daudel

J.-P. Korb

J.-P. Lemaistre

and

J. Maruani

Centre de Mécanique Ondulatoire Appliquée,
CNRS, Paris, France

D. REIDEL PUBLISHING COMPANY

A MEMBER OF THE KLUWER ACADEMIC PUBLISHERS GROUP

DORDRECHT / BOSTON / LANCASTER / TOKYO

Library of Congress Cataloging in Publication Data

CIP

Structure and dynamics of molecular systems – II

"Invited lectures given at the international seminar of the same title held at the Centre de mécanique ondulatoire appliquée du Centre national de la recherche scientifique in Paris (France) from October 1983 to May 1985." Pref.

Includes bibliographies and indexes.

1. Molecular structure–Congresses. 2. Molecular dynamics–Congresses. I. Daudel, Raymond.

QD461.S924 1985 541.2'2 85–2010

ISBN-13: 978-94-010-8572-4 e-ISBN-13: 978-94-009-4662-0
DOI: 10.1007/978-94-009-4662-0

Published by D. Reidel Publishing Company,
P.O. Box 17, 3300 AA Dordrecht, Holland.

Sold and distributed in the U.S.A. and Canada
by Kluwer Academic Publishers,
190 Old Derby Street, Hingham, MA 02043, U.S.A.

In all other countries, sold and distributed
by Kluwer Academic Publishers Group,
P.O. Box 322, 3300 AH Dordrecht, Holland.

Softcover reprint of the hardcover 1st edition 1986

TABLE OF CONTENTS

PREFACE

This volume is the second of a set of two which contain 28 selected from the 40 invited lectures given at the international seminar of the same title held at the Centre de Mécanique Ondulatoire Appliquée du Centre National de la Recherche Scientifique in Paris (France) from October 1983 to May 1985. They are intended to provide a survey of topics of current interest relative to the structure and the dynamics of molecular systems. The papers have been selected on the basis of their relevance to the following four topics: i) molecular conformations and transformations; ii) molecular relaxation and motion; iii) charge, spin and momentum distributions and intermolecular interactions; iv) collective phenomena in condensed matter. The first volume deals mostly with the first two topics, the second volume mostly with the last two. The two volumes consist of an approximately equal number of self-contained, reference contributions covering recent achievements in active branches of molecular physics and physical chemistry.

The first two papers of the present volume deal with theoretical aspects of intermolecular interactions: the first paper with the physical origin of the so-called non-exchange molecular terms, a complete derivation of which is given using Rayleigh-Schrödinger second-order perturbation theory; the second paper with the symmetry analysis of the effects of interactions between rigid molecules and crystal environments, using the isodynamic-group theoretical approach devised by Altmann for non-rigid systems. The next two papers provide a survey of the information which can be derived from X-ray diffraction data on either charge densities or magnetic structures, both the charge and spin of the electron interacting with the electromagnetic radiation. The electronic structure and internal dynamics of isolated molecules can be further investigated by the use of X-ray or electron-beam Compton profiles and laser-pulse-induced non-linear oscillations, as shown in the next two papers.

The following three papers present spectroscopic, NMR and theoretical approaches to local and collective dynamic properties of polymer systems, in solution, gel, melt or glass form, with an emphasis on the scaling approach and the reptation model. The next two papers propose theoretical models for transition systems of particular interest: micro-emulsion structures with varying ratios of the main components and percolating clusters undergoing fractal to Euclidean cross-over. Finally, two papers present thorough experimental and theoretical surveys of some novel materials: superionic conductors (including the effect of anisotropic diffusion on nuclear magnetic resonance) and metal hydrides (particularly bond stability and superconductive behavior). Most of these papers contain an abundant bibliography which should help the interested reader to go deeper into the subject.

The editors wish to express their gratitude to the Direction de la Chimie of the C.N.R.S. for lending financial support to the international seminar.

THE EDITORS

MOLECULAR CHARGE DISTRIBUTIONS AND RESPONSE FUNCTIONS: MULTIPOLAR AND PENETRATION TERMS; APPLICATION TO THE THEORY OF INTERMOLECULAR INTERACTIONS

P. CLAVERIE
Dynamique des Interactions Moléculaires
Université Pierre et Marie Curie
Tour 22
4 place Jussieu
75230 Paris Cedex 05
France

ABSTRACT. The present work pertains to the perturbation-theoric approach of intermolecular interactions. We first recall briefly the general framework, namely the exchange perturbation theory. We then concentrate our attention upon the non-exchange terms, namely the so-called Rayleigh-Shrödinger (RS) terms, at 1st-order (electrostatic term) and 2nd-order (induction and dispersion). These terms exhibit both long-range (R^{-k}) and short-range (exp(-aR)) behaviours, and , in order to account properly for this feature, we avoid the usual multipole expansion of the intermolecular potential V, using instead its exact expression in terms of molecular charge density operators. In the case of the electrostatic term, we are thus able to split it into a purely long-range part ("multipolar") and a purely short-range one ("penetration" or "charge overlap"). The long-range part of the molecular electrostatic potential can be represented, with arbitrary high accuracy, through sets of multipoles located on a large enough set of centers (such as atoms, middles of bonds,....), namely a so-called multi-center multipole representation. As concerns the second-order terms (induction and dispersion), we use the Green's function formalism, and we are thus led to use "molecular susceptibility functions" (both static and frequency-dependent) which generalize the usual multipole polarizabilities, so as to account exactly for the short-range behaviour, and not for the long-range one only. By using these susceptibility functions (and, in the case of the induction term, the exact electrostatic molecular potential), we obtain the desired exact formulae for the induction and dispersion terms, for molecules of arbitrary size and shape.

R. Daudel et al. (eds.), Structure and Dynamics of Molecular Systems – II, 1–33.

1. THEORETICAL BACKGROUND: EXCHANGE PERTURBATION TREATMENTS FOR INTERMOLECULAR INTERACTIONS

In order to investigate and evaluate intermolecular interaction energies, it is obviously essential to decompose them inasmuch as possible into a sum of contributions with simple behavior as functions of intermolecular distances and relative orientations (see e.g. [1-18]). Such a task can be achieved by using perturbation theory. The total Hamiltonian H of the complex is written as

$$H = H_0 + V \tag{1.1}$$

with

$$H_0 = H^{(1)} + H^{(2)} \tag{1.2}$$

where $H^{(i)}$ denotes the Hamiltonian of the isolated molecule (i) (see e.g. [2], section II.A) and V denotes the intermolecular interaction potential:

$$V = \sum_{\nu^{(1)}} \sum_{\nu^{(2)}} \frac{Z_{\nu^{(1)}} Z_{\nu^{(2)}}}{|\vec{r}_{\nu^{(1)}} - \vec{r}_{\nu^{(2)}}|} - \sum_{\nu^{(1)}} \sum_{j^{(2)}} \frac{Z_{\nu^{(1)}}}{|\vec{r}_{\nu^{(1)}} - \vec{r}_{j^{(2)}}|} - \sum_{i^{(1)}} \sum_{\nu^{(2)}} \frac{Z_{\nu^{(2)}}}{|\vec{r}_{i^{(1)}} - \vec{r}_{\nu^{(2)}}|} + \sum_{i^{(1)}} \sum_{j^{(2)}} \frac{1}{|\vec{r}_{i^{(1)}} - \vec{r}_{j^{(2)}}|} \tag{1.3}$$

where $\nu^{(m)}$ and $i^{(m)}$ label the nuclei and the electrons respectively of molecule m (here m=1,2).

Treating V as a perturbation in the framework of usual Rayleigh-Schrödinger (RS) perturbation theory provides the correct asymptotic behavior of the interaction energy ΔE for large values of the intermolecular distance R [19,20], but fails at intermediate and short distances [21-24] [2, section II] [8, section IV.3] essentially because the requirement of antisymmetry (Pauli principle) pertaining to <u>inter</u>-molecular electron exchange is not properly taken into account. As a consequence, the RS treatment would converge towards a too low "mathematical" eigenstate (with partially bosonic character) instead of the genuine physical ground state [21] (at large distance, there is first an apparent convergence towards some energy value intermediate between those of the mathematical and physical eigenstates, followed by an <u>extremely slow</u> convergence towards the mathematical eigenstate [20]).

In order to overcome this difficulty, a number of so-called "symmetrized" or "exchange" perturbation treatments have been proposed (see the reviews [2, section II.B] [8, chaps. III-V] , [9] and references therein). Here we shall not dwell into the intricacies of the comparison of these treatments. We shall merely note that three of these treatments, namely SRS (symmetrized Rayleigh-Schrödinger), MS-MA (Murrel-Shaw, Musher-Amos) and JK-ISF (Jeziorski-Kolos or Intermediate Symmetry Forcing), give identical 1st- and 2nd- order perturbation terms, each one exhibiting correct asymptotic behavior for large distances[9]. Moreover these common 1st- and 2nd- order terms appear to give satisfactory results for rather different types of complexes, namely Helium-Helium [27] and Water-Water [28]. We shall therefore deal from now on with these two terms, and we introduce some notations in order to write them down.

The eigenfunctions of the Hamiltonian $H^{(m)}$ of molecule m (in the present work we limit ourselves to m=1,2) are denoted

$$|0^{(m)}\rangle,\ |1^{(m)}\rangle, \ldots, |i^{(m)}\rangle, \ldots \qquad (1.4)$$

with the corresponding eigenvalues

$$E_0^{(m)},\ E_1^{(m)}, \ldots\ldots, E_i^{(m)}, \ldots\ldots \qquad (1.5)$$

Then the eigenfunctions of $H_0 = H^{(1)} + H^{(2)}$ are merely the products:

$$|0^{(1)}0^{(2)}\rangle, \ldots, |0^{(1)}j^{(2)}\rangle, \ldots, |i^{(1)}0^{(2)}\rangle, \ldots, |i^{(1)}j^{(2)}\rangle, \ldots \qquad (1.6)$$

with the coresponding eigenvalues

$$(E_0^{(1)} + E_0^{(2)}), \ldots, (E_0^{(1)} + E_j^{(2)}), \ldots, (E_i^{(1)} + E_0^{(2)}), \ldots, (E_i^{(1)} + E_j^{(2)}), \ldots \qquad (1.7)$$

For the sake of brevity, we shall also use whenever possible a single-label notation for the eigenfunctions and eigenvalues:

$$|\varphi_0\rangle = |0^{(1)}0^{(2)}\rangle, |\varphi_1\rangle, \ldots\ldots, |\varphi_k\rangle, \ldots \qquad (1.8)$$

$$E_0^{(0)} = E^{(1)} + E^{(2)}, E_1^{(0)}, \ldots\ldots, E_k^{(0)}, \ldots\ldots \qquad (1.9)$$

Then the expressions of the 1st- and 2nd- order terms ε_1 and ε_2 are respectively [9,25,26]:

$$\varepsilon_1 = \langle \varphi_0 | V \mathcal{A} | \varphi_0 \rangle / \langle \varphi_0 | \mathcal{A} | \varphi_0 \rangle \qquad (1.10)$$

$$\varepsilon_2 = - \langle \varphi_0 | V \mathcal{R}_0 \mathcal{A} (V - \varepsilon_1) | \varphi_0 \rangle / \langle \varphi_0 | \mathcal{A} | \varphi_0 \rangle \qquad (1.11)$$

where $\mathcal{R}_0$ denotes the reduced resolvent of H_0 :

$$\mathcal{R}_0 = \sum_k{}' | \varphi_k \rangle \langle \varphi_k | / (E_k^{(0)} - E_0^{(0)}) \qquad (1.12)$$

(the prime in $\sum'$ means as usual that the value k=0 is excluded from the summation), and $\mathcal{A}$ denotes some antisymmetrization operator (since $\mathcal{A}$ appears in both the numerator and the denominator of eqs. (1.10) and (1.11), the normalization of $\mathcal{A}$ is immaterial). Since the functions $|i^{(1)}\rangle$ and $|j^{(2)}\rangle$ of the subsystems are already antisymmetric with respect to intra-molecular electron permutation it is sufficient to build $\mathcal{A}$ from inter-molecular electron permutations only, namely (see e.g. [2], appendix A):

$$\mathcal{A} = 1 - \mathcal{A}' = 1 - P_{(1)} + P_{(2)} - \ldots\ldots + (-1)^{n_{inf}} P_{(n_{inf})} \qquad (1.13)$$

where $P_{(1)} = \sum_i^{(1)} \sum_j^{(2)} P_{ij}$ denotes the sum of all permutations which exchange one electron (i) of molecule (1) and one electron (j) of molecule (2), and similar definitions hold for $P_{(2)}, P_{(3)}, \ldots$ (n_{inf} denotes the smallest of n_1 and n_2, the numbers of electrons of molecules 1 and 2 respectively).

Then, by using the decomposition $\mathcal{A} = 1 - \mathcal{A}'$, where $\mathcal{A}' = P_{(1)} - P_{(2)} + P_{(3)} - \ldots$, we can decompose each n-th order perturbation term into the usual n-th order RS perturbation term $\varepsilon_n^{(RS)}$ plus a so-called n-th order exchange term $\varepsilon_n^{(ex)}$ (see e.g. [2,3,8,9]):

$$\varepsilon_n = \varepsilon_n^{(RS)} + \varepsilon_n^{(ex)} \qquad (1.14)$$

with

$$\varepsilon_1^{(RS)} = \langle \varphi_0 | V | \varphi_0 \rangle \tag{1.15}$$

$$\varepsilon_1^{(ex)} = - \langle \varphi_0 | V (\mathcal{A}' - \langle \mathcal{A}' \rangle_0) | \varphi_0 \rangle / \langle \varphi_0 | \mathcal{A} | \varphi_0 \rangle \tag{1.16}$$

and

$$\varepsilon_2^{(RS)} = - \langle \varphi_0 | V \mathcal{R}_0 V | \varphi_0 \rangle \tag{1.17}$$

$$\varepsilon_2^{(ex)} = \langle \varphi_0 | V \mathcal{R}_0 (\mathcal{A}' - \langle \mathcal{A}' \rangle_0)(V - \varepsilon_1) | \varphi_0 \rangle / \langle \varphi_0 | \mathcal{A} | \varphi_0 \rangle \tag{1.18}$$

The exchange terms are purely short-range, namely they are exponentially decreasing functions of the intermolecular distance R: they indeed involve intermolecular integrals over molecular orbitals of overlap, hybrid and exchange type, all of which exhibit this exponential decrease at large R. By contrast, the Rayleigh-Schrödinger terms are of mixed type, i.e. they exhibit both a long-range behavior (decrease of R^{-n} type) and a short-range behavior (exponential decrease). This behavior is due to the exponentially decreasing part (the so-called penetration part) of the coulomb and nuclear integrals; this part is itself related with the fact that the electronic charge distribution extend to infinity with an exponential-like decrease, hence the name "charge-overlap effect" [29] which is also used for denoting this short-range behavior of the RS terms. It must be emphasized that, from the mathematical point of view, this short-range behavior of the RS terms manifests itself in a different way for the 1st- order term, on one hand, and for the higher-order terms, on the other hand: the 1st-order term may actually be expressed as the sum of a purely long-range part (E_1^{mult} multipolar part) and a purely short-range one (E_1^{pen} :penetration part), while such an additive decomposition is not possible for the higher-order terms: accordingly, less trivial representations must be developped for these terms (see e.g. [10-12]). The remaining part of this paper will precisely be devoted to the study of the two terms $\varepsilon_1^{(RS)}$ and $\varepsilon_2^{(RS)}$, noticeably by paying adequate attention to the "double behavior" (long- and short- range) of these terms: this is necessary since, at medium range (region of the Van der Waals minimum), considering the long-range part only is not enough: the short-range part must be properly represented, too (as concerns the purely short-range exchange parts, they will not be considered further in the present work: some further developments concerning the 1st-order exchange term may be found in [2] (section IV and appendices C,D) and [17,18]).

2. ELECTROSTATIC (1st-ORDER) TERM CHARGE DISTRIBUTIONS : MULTICENTER MULTIPOLAR REPRESENTATIONS

Let us insert in the expression (1.15) the explicit expression of $|\varphi_0\rangle$, namely $|0^{(1)}0^{(2)}\rangle$ according to eqs. (1.8):

$$\varepsilon_1^{(RS)} = \langle 0^{(1)}0^{(2)} | V | 0^{(1)}0^{(2)} \rangle \tag{2.1}$$

For a long time, the treatment of this term was based upon the so-called multipole expansion of the interaction potential V (see e.g. [8, Chap. VI], [9, Section 1.4] and references therein [30-38]). This approach consists in expressing V as a sum of interactions between the multipoles moments of each molecule, these multipoles being evaluated separately for each molecule (1 or 2) with respect to some origin (O_1 or O_2, respectively) attached to each molecule. This is obtained by using the so-called bipolar expansion (with respect to the "poles" O_1 and O_2, defined by the position vectors R_1 and R_2 respectively, of the various coulombic terms $1 / |\vec{r} - \vec{r}'|$ appearing in V, which amounts to express $1 / |\vec{r} - \vec{r}'|$ in terms of the ionverse powers R^{-k} of $R = |\vec{R_2} - \vec{R_1}|$ and the polar coordinates r_1, θ_1, φ_1 and r_2, θ_2, φ_2 of $(\vec{r} - \vec{R_1})$ and of $(\vec{r}' - \vec{R_2})$ respectively [31,32,35,38].

It must be strongly emphasized that this multipole expansion of V is exact only when the two charge distributions are entirely contained inside two non-overlapping spheres centered at O_1 and O_2 respectively. This cannot be rigorously realized for electronic charge distributions, since they extend to infinity, with an exponentially decreasing tail. As a consequence, the electrostatic interactions involving such large distributions exhibit, beyond their multipole part (R^{-k} terms), exponentially decreasing terms (penetration or charge-overlap terms): see e.g. refs. [38], [8, section VI.4] and references therein. As a simple example, we can consider the interaction between a hydrogenoid electronic charge distribution and a proton, i.e. a so-called nuclear attraction integral [39]:

$$\langle 1s_{\zeta}(\vec{r})\,|\,1/|\vec{R}-\vec{r}|\,|\,1s_{\zeta}(\vec{r})\rangle = \frac{1}{R}[1-(1+R')e^{-2R'}] \qquad (2.2)$$

where $R' = \zeta R$ and $1s_{\zeta}(\vec{r}) = 2\zeta^{3/2}(4\pi)^{-1/2}\exp(-\zeta r)$.
We immediately recognize in eq. (2.2) the multipole part 1/R and the penetration part $-[(1+R')/R]\exp(-2R')$. From the practical point of view, it appears interesting to put this penetration part together with 1st-order exchange term $\varepsilon_1^{(ex)}$, so as to get a total 1st-order short-range (exponentially decreasing) contribution, for which simplified formulae may be searched [2, section IV and appendix D], [17,18].

But, beyond the lack of the penetration terms (which can be remedied as just indicated), the usual multipole expansion of V presents a more serious defect, which results from the use of molecular multipole moments evaluated with respect to a single center: indeed, when the molecule has a markedly non spherical shape, there are regions of free space which are closer to the molecular center than the atoms lying farthest from this center, and accordingly the molecular electrostatic potential in such regions will be rather poorly represented by the usual one-center multipole exoansion [40]. This implies, for example, that the electrostatic interaction between flat molecules in stacking interactions cannot be represented with sufficient accuracy if we represent each molecular charge distribution through a set of one-center multipoles, whatever the choice of this center [40]. The remedy for this defect obviously consists in introducing several "centers of force" in order to account in a satisfactory way for the actual shape of the molecule, and this line of research has now been pursued by several authors [2, section V] [41-58] (we shall essentially follow here the treatment given in refs. [2] and [57]).

The first essential idea is to introduce instead of the (approximate) multipole expansion of V, the representation in terms of molecular charge density operators proposed by Longuet-Higgins [59], namely:

$$V = \iint \frac{\rho^{(1)}(\vec{r}^{\,1})\;\rho^{(2)}(\vec{r}^{\,2})}{|\vec{r}^{\,1}-\vec{r}^{\,2}|}\,d\vec{r}^{\,1}\,d\vec{r}^{\,2} \qquad (2.3)$$

where the molecular charge density operators $\rho^{(m)}(\vec{r})$ (m=1,2,.... labels the molecules) are expressed as follows :

$$\rho^{(m)}(\vec{r}) = \sum_{\nu}^{(m)} Z_\nu \, \delta(\vec{r} - \vec{R}_\nu) - \sum_{i}^{(m)} \delta(\vec{r} - \vec{r}_i) \quad (2.4)$$

where $\vec{R}_\nu$ and r_i denote the position vectors of the nuclei ν and electrons i, respectively. Note that $\rho(\vec{r})$ operates on functions of these variables $\vec{r}_i$ (and possibly $\vec{R}_\nu$), while $\vec{r}$ merely appears as a parameter. It must be emphasized that this expression (2.3) of V is rigorously exact.

By inserting in eq. (2.1) the expression (2.3) of V, we get [2, section III.B]:

$$\varepsilon_1^{(RS)} = \iint \frac{\langle 0^{(1)} 0^{(2)} | \rho^{(1)}(\vec{r}^1) \, \rho^{(2)}(\vec{r}^2) | 0^{(1)} 0^{(2)} \rangle}{|\vec{r}^1 - \vec{r}^2|} \, d\vec{r}^1 \, d\vec{r}^2$$

$$= \iint \frac{\langle 0^{(1)} | \rho^{(1)}(\vec{r}^1) | 0^{(1)} \rangle \langle 0^{(2)} | \rho^{(2)}(\vec{r}^2) | 0^{(2)} \rangle}{|\vec{r}^1 - \vec{r}^2|} \, d\vec{r}^1 \, d\vec{r}^2$$

$$= \iint \frac{f_{00}^{(1)}(\vec{r}^1) \; f_{00}^{(2)}(\vec{r}^2)}{|\vec{r}^1 - \vec{r}^2|} \, d\vec{r}^1 \, d\vec{r}^2 \quad (2.5)$$

where $f_{00}^{(m)}(\vec{r})$ denotes the charge density of the ground state of molecule (m):

$$f_{00}^{(m)}(\vec{r}) = \langle 0^{(m)} | \rho^{(m)}(\vec{r}) | 0^{(m)} \rangle \quad (2.6)$$

and it thus clearly appears from eq. (2.5) that $\varepsilon_1^{(RS)}$ actually is the electrostatic interaction between the exact charge distribution of the two molecules in their (unperturbed) ground states.

The next step is to break up the complete molecular charge distribution (which is usually too far from spherical shape) into smaller pieces, distributed over the molecule: due to the smaller size of these partial charge distributions, it will now be possible to represent each of them through a multipole expansion performed with respect to some center specifically suited to the partial charge distribution. Since such a center will, as a general rule, vary with the partial charge distribution under consideration, we are led quite naturally to the concept of a multi-center muktipolar representation [41-58].

In order to be more specific, we shall now consider the standard approximation [2, sections III.B, V.B] [57] where the nuclei are kept fixed (in their equilibrium configuration) and the electronic wave function is approximated by a Slater determinant (generalization to a multi-configurational wave function would be easy:

$$|0\rangle = |a_1(1).....a_n(n)| \;/\; n! \tag{2.7}$$

where $a_i = b_i\,\sigma_i$ denotes a spin-orbital, b_i denotes the orbital part and σ_i the spin part (α or β). With the electronic wavefunction (2.7), the charge distribution is easily found to take the form:

$$\rho_{00}(\vec{r}) = \sum_{\nu} Z_\nu\,\delta(\vec{r}-\vec{R}_\nu) - \sum_{i} |b_i(\vec{r})|^2 \tag{2.8}$$

where the sum $\sum_i$ runs over the spinorbitals, or

$$\rho_{00}(\vec{r}) = \sum_{\nu} Z_\nu\,\delta(\vec{r}-\vec{R}_\nu) - \sum_{I} n_I^{occ}\,|b_I(\vec{r})|^2 \tag{2.9}$$

where the sum $\sum_I$ now runs over the orbitals and n_I^{occ} denotes the occupation number of the orbital I.

The first term of the sum corresponds to the fixed nuclear charges Z_ν located at the points $\vec{R}_\nu$ and therefore make no problem. We can therefore concentrate our attention upon the second term, namely the continuous electronic charge density. According to eq. (2.9), this density already appears as a sum of molecular orbital densities $|b_I(\vec{r})|^2$, but this decomposition is not especially useful if the molecular orbitals are the canonical ones, since these are still delocalized over the whole molecule. The situation looks better if we consider localized molecular orbitals (bond orbitals, lone pair orbitals), since each partial density $|b_I(r)|^2$ is now of limited size and can therefore be appropriately represented by a (limited) multipole expansion defined with respect to some well-suited center, e.g. the center of charge of this partial distribution (see Lavery et al. [46]).

Here we shall follow a different way, namely we shall assume the LCAO framework

$$b_I = \sum_\alpha c_{I\alpha} \chi_\alpha \qquad (2.10)$$

where the χ_α's denote the atomic orbitals. Then, upon inserting eq. (2.10) into eq. (2;9), we get:

$$f_{00}(\vec{r}) = \sum_\nu Z_\nu \delta(\vec{r} - \vec{R}_\nu) - \sum_\alpha \sum_\beta p_{\alpha\beta} \chi_\alpha^*(\vec{r}) \chi_\beta(\vec{r}) \qquad (2.11)$$

where $p_{\alpha\beta}$ denotes the elements of the charge and bond order matrix pertaining to the atomic basis set under consideration:

$$p_{\alpha\beta} = \sum_I c_{I\alpha}^* c_{I\beta} \qquad (2.12)$$

We have now to deal with the "elementary charge distributions" $\chi_\alpha^*(\vec{r})\chi_\beta(\vec{r})$ Let us assume, that each basic atomic orbital is the product of some radial part R(r) by some spherical harmonic angular part $Y_\ell^m(\theta,\varphi)$

$$\chi(\vec{r}) = R(r)\, Y_\ell^m(\theta,\varphi) \qquad (2.13)$$

(1) *one-center elementary distributions*

Let us consider first the case where the two atomic orbitals χ_α, χ_β belong to the same atomic center. Then :

$$\chi_\alpha^*(\vec{r})\, \chi_\beta(\vec{r}) = R_\alpha(r)\, R_\beta(r)\, Y_{\ell_\alpha}^{-m_\alpha}(\theta,\varphi)\, Y_{\ell_\beta}^{m_\beta}(\theta,\varphi) \qquad (2.14)$$

where r,θ,φ are the polar coordinates with respect to the common center. Then, according to the formula giving the product of spherical harmonics (see e.g. [60] or [61]) the angular part of the one-center distribution (2.14) reduces to a finite sum of spherical harmonics:

$$Y_{\ell_\alpha}^{-m_\alpha}(\theta,\varphi)\, Y_{\ell_\beta}^{m_\beta}(\theta,\varphi) = \sum_{\ell}\sum_{m=-\ell}^{\ell} C_{\ell_\alpha\,\ell_\beta\,\ell}^{-m_\alpha\, m_\beta\, m}\; Y_\ell^m(\theta,\varphi) \quad (2.15)$$

with the two conditions

$$|\ell_\alpha - \ell_\beta| < \ell < \ell_\alpha + \ell_\beta \quad (2.16)$$

$$-m_\alpha + m_\beta + m = 0 \quad (2.17)$$

Now, the so-called irreducible multipole moments Q_l^m of a charge distribution $\rho(r,\theta,\varphi)$ are given by the scalar product with the basic function $r^\ell Y_\ell^m(\theta,\varphi)$ (see e.g. [35], chap.12, eqs. (12.1)-(12.18b)):

$$Q_\ell^m = \iiint \rho(r,\theta,\varphi)\, r^\ell\, Y_\ell^m(\theta,\varphi)\, r^2 \sin\theta\, dr\, d\theta\, d\varphi \quad (2.18)$$

Then, as a consequence of eqs. (2.14)-(2.18), the charge distribution $\chi_\alpha^* \chi_\beta$ gives rise to a finite number of multipole moments Q_l with $m = m_\alpha - m_\beta$ and $|\ell_\alpha - \ell_\beta| < \ell < \ell_\alpha + \ell_\beta$. Let us give a few examples:

(a) χ_α and χ_β are s orbitals ($\ell_\alpha = \ell_\beta = 0$):
then $l=0$, i.e. there is only one multipole moment, namely the total charge

$$\int \chi_\alpha^*(\vec{r})\, \chi_\beta(\vec{r})\, d\vec{r} = \langle \chi_\alpha | \chi_\beta \rangle \quad \text{(the overlap integral).}$$

(b) χ_α is an s orbital ($\ell_\alpha = 0$) and χ_β is a p orbital ($\ell_\beta = 1$): then $\ell = 1$, i.e. we have a dipole moment only.

(c) χ_α and χ_β are p orbitals ($\ell_\alpha = \ell_\beta = 1$): then $0 < \ell < 2$; in actual fact we find quite generally a non-zero quadrupole moment ($l=2$), and a charge ($\ell = 0$) equal to the overlap integral $\langle \chi_\alpha | \chi_\beta \rangle$.

(2) *two-center elementary distributions*

The situation is now less simple unless we use special atomic orbitals such as the gaussian ones (see below). We can perform a multipole expansion with respect to some suitable center, e.g. the middle of the segment joining the centers pertaining to the orbitals χ_α and χ_β . This procedure was actually used in the so-called Overlap Multipole expansion (OMTP) [41-45]. By contrast with the previous case (one-center distributions), there is now, in principle, an infinite number of non-zero multipoles. A favourable factor, however, is due to the localized character of the basic atomic orbitals: accordingly, the charge distributions $\chi^*_\alpha \chi_\beta$ exhibit non-negligible values only if their atomic centers are not too far apart (and noticeably for chemically bonded atoms), and this means that these dominant two-center distribution should be adequately represented by a rather limited number of multipoles.

As mentioned above, the case of gaussian basic orbitals allows for a remarkable simplification [49-55] [57]. Indeed, the product of two gaussian

$$g_\alpha(\vec{r}) = \exp(-\zeta_\alpha r_\alpha^2) \qquad \text{and} \qquad g_\beta(\vec{r}) = \exp(-\zeta_\beta r_\beta^2) \tag{2.19}$$

where

$$r_\alpha = |\vec{r} - \vec{R}_\alpha| \qquad \text{and} \qquad r_\beta = |\vec{r} - \vec{R}_\beta| \tag{2.20}$$

is nothing but a single gaussian (see e.g. [62]):

$$g_\alpha(\vec{r})\, g_\beta(\vec{r}) = K_{\alpha\beta} \exp[-(\zeta_\alpha + \zeta_\beta) r^2] \tag{2.21}$$

where

$$r = |\vec{r} - \vec{R}_G| \quad \text{and} \quad \vec{R}_G = (\zeta_\alpha \vec{R}_\alpha + \zeta_\beta \vec{R}_\beta)/(\zeta_\alpha + \zeta_\beta) \tag{2.22}$$

i.e. R_G is the barycenter of the centers R_α and R_β , with the respective weights ζ_α and ζ_β , and $K_{\alpha\beta}$ is a constant:

$$K_{\alpha\beta} = \exp[-(R_\beta - R_\alpha)^2 \zeta_\alpha \zeta_\beta / (\zeta_\alpha + \zeta_\beta)] \tag{2.23}$$

In the case of orbitals g_α and g_β with more complicated angular parts (p,d,....types), the product $g_\alpha g_\beta$ reduces to a <u>sum</u> of gaussians with various symmetry types, all of them centered at the barycenter $\vec{R}_G$ (see [57] for further details). Thus, in the case of a basis of gaussian orbitals, each elementary distributions generates a <u>finite</u> number of multipole moments, centered at some point depending upon the distribution under consideration.

We thus have a "multi-centered multipole representation" which represents exactly the long-range part of the electrostatic molecular potential: the missing part is the purely short-range "penetration" or "charge overlap" part.

The price to pay for such an exact representation of the long-range part of the molecular electrostatic potential is, of course, the large number of centers involved (namely the atoms and all their barycenters generated according to eq. (2.22)): note, in particular, that this number of centers increases with the size of the basis set. Thus, it is of great practical interest to devise systematic simplification procedures in order to reduce the number of these centers. Such a procedure has been developped by Vigné-Maeder and Claverie [57,58], and we only summarize here its main features.

First, some final set of centers is chosen for building the desired approximate representation. This final set needs not to be a subset of the set of numerous centers which are associated with some exact multi-centered multipole expansion as defined above: for example as a rather standard choice, we may take the atoms and and the middles of the chemical bonds. Then, for each center to be supppressed, the corresponding multipoles are decomposed over the centers of the "final set" according to the following rule: we take among the final centers those two (say P_1 and P_2) which lie closest to the center P to be suppressed, then the (finite) set ($\mathcal{M}$) of the multipoles centered at P is divided into two parts $\lambda_1(\mathcal{M})$ and $\lambda_2(\mathcal{M})$, with $\lambda_1+\lambda_2=1$ (as concerns the choice of λ_1 and λ_2 , see below), and finally each set of multipoles $\lambda_i(\mathcal{M})$ is replaced by its multipole expansion (limited to some order) around the center P_i (i=1,2).

As concerns the choice of the weights λ_1 and λ_2 , there is some degree of arbitrariness; a reasonable requirement is of course that $\lambda_i \to 1$ when $P \to P_i$. In ref. [57], a simple rule fulfilling this requirement has been proposed, namely taking λ_i proportional to $1/PP_i$, i.e. :

$$\lambda_1 = \frac{1/PP_1}{1/PP_1 + 1/PP_2} = \frac{PP_2}{PP_1 + PP_2} \tag{2.24}$$

$$\lambda_2 = \frac{1/PP_2}{1/PP_1 + 1/PP_2} = \frac{PP_1}{PP_1 + PP_2} \tag{2.25}$$

This completes the brief description of the systematic reduction procedure of the number of centers for an approximate multi-center multipole representation (by the way, this procedure may also be used for generating molecular charge distributions by combining the charge distributions of smaller molecular fragments [63,64]). Thus, it may be concluded that the representation of the long-range part of the electrostatic energy $\mathcal{E}_1^{(RS)}$ is now well mastered, since we are able to define (at least for gaussian basis sets) an exact (multi-centered multipolar) representation and to build from it, in a systematic way, a series of approximations with decreasing accuracy and increasing simplicity, so as to meet in an optimal way the specific needs of practical applications.

3. INDUCTION AND DISPERSION (2nd ORDER) TERMS: RESPONSE FUNCTIONS

3.1 Some Basic Theoretical Elements

We now want to consider the Rayleigh-Schrödinger 2nd-order perturbation term, expressed, as usual, as ths sum of two induction terms and the dipersion term:

$$\mathcal{E}_2^{(RS)} = E_{ind}^{(1)} + E_{ind}^{(2)} + E_{disp} \tag{3.1}$$

$$E_{ind}^{(1)} = -\sum_{i^{(1)}}' |\langle 0^{(1)} 0^{(2)} | V | i^{(1)} 0^{(2)} \rangle|^2 / \Delta E_i^{(1)} \tag{3.2}$$

(induction energy of molecule (1))

$$E_{ind}^{(2)} = -\sum_{j^{(2)}}' |\langle 0^{(1)} 0^{(2)} | V | 0^{(1)} j^{(2)} \rangle|^2 / \Delta E_j^{(2)} \tag{3.3}$$

(induction energy of molecule (2))

$$E_{disp} = -\sum_{i^{(1)}}' \sum_{j^{(2)}}' |\langle 0^{(1)} 0^{(2)} | V | i^{(1)} j^{(2)} \rangle|^2 / (\Delta E_i^{(1)} + \Delta E_j^{(2)}) \tag{3.4}$$

(dispersion energy).

In these formulae $\Delta E_i^{(1)} = E_i^{(1)} - E_0^{(1)}$ and $\Delta E_j^{(2)} = E_j^{(2)} - E_0^{(2)}$ denote the excitation energies of molecules (1) and (2) respectively and the symbol $\sum'$ means that the ground state ($i^{(1)}=0^{(1)}$ or $j^{(2)}=0^{(2)}$) is excluded from the summation. The three terms of this decomposition correspond to the various kinds of excited states of the complex that may be distinguished, namely excitation of molecule (1) only ($|\, i^{(1)} 0^{(2)} \rangle$), excitation of molecule (2) only ($|\, 0^{(1)} j^{(2)} \rangle$), and finally of both molecules ($| i^{(1)} j^{(2)} \rangle$). It is appropriate to deal separately with each of these terms.

Each of the matrix elements of V involved in the numerators of eqs.(3.2)-(3.4) can be expressed , like $\varepsilon_1^{(RS)} = \langle 0^{(1)} 0^{(2)} \,|\, V \,|\, 0^{(1)} 0^{(2)} \rangle$, as an electrostatic interaction involving so-called state transition charge distributions [2, section III.B], by using the Longuet-Higgins [59] expression of V (eq.2.3); for example

$$\langle 0^{(1)} 0^{(2)} \,|\, V \,|\, i^{(1)} j^{(2)} \rangle = \iint \frac{\rho_{0i}^{(1)}(\vec{r}^{\,1}) \; \rho_{0j}^{(2)}(\vec{r}^{\,2})}{|\vec{r}^{\,1} - \vec{r}^{\,2}|} \, d\vec{r}^{\,1} \, d\vec{r}^{\,2} \qquad (3.5)$$

appears as the interaction between the transition distributions $\rho_{0i}^{(1)}$ and $\rho_{0j}^{(2)}$

$$\rho_{0i}^{(1)}(\vec{r}^{\,1}) = \langle 0^{(1)} | \rho^{(1)}(\vec{r}^{\,1}) | i^{(1)} \rangle \qquad \rho_{0j}^{(2)}(\vec{r}^{\,2}) = \langle 0^{(2)} | \rho^{(2)}(\vec{r}^{\,2}) | j^{(2)} \rangle \qquad (3.6)$$

Accordingly, each matrix element such as (3.5) could be expressed as a sum of a purely long-range (multipolar) and a purely short-range (penetration or charge overlap) part. But when we consider the whole series (3.2)-(3.4) instead of the individual terms, an essential difference appears, namely the series of the long-range contributions, on one hand, and of the short-range ones, on the other hand, are not separately convergent: each of them is merely an asymptotic series [9, section 1.4] [65,66]. This may be related with the fact that the short-range part of the matrix elements of V is not uniformly bounded with respect to the long-range part when $i^{(1)}$ and $j^{(2)}$ are higher and higher excited states, since they become less and less localized. For example, the 1/R expansion (multipolar part) of the induction energy of a Hydrogen atom polarized by a proton (at distance R) is [65,66,67] :

$$\Delta E_{ind}^{(H^+\to H)} \sim -\sum_{n=1}^{\infty} \frac{1}{R^{2n}} \frac{(2n+2)!}{2^{2n+2}} \frac{n+2}{n(n+1)} \qquad (3.7)$$

and this is obviously a divergent asymptotic series (for mathematical details, see e.g. Erdelyi [68] or Bender and Orszag [69]). In actual fact, the full expression of this induction energy has also been derived [67] and it involves terms such as e^{-2t} Ei(2t) and e^{2t} Ei(-2t), where t= R and Ei(x) denotes the exponential integral. Now this function exhibits an essential singularity at infinity, therefore E_{ind} has only an asymptotic expansion in terms of powers of R^{-1}, and this is the source of the expansion (3.7).

Although no rigorous proof is available for more general cases, we must expect that this behaviour actually holds quite generally: it is impossible to express the 2nd-order terms (induction and dispersion) as the sum of a purely long-range part ($\sum_n c_n/R^n$) and a purely short-range part (exponentially decreasing formula exp(-aR)): both types of behaviours remain entangled. A possible practical solution to this problem consists in introducing so-called "damping functions" $g_n(R)$ in the multipole expansion, namely (see e.g., [10,11]):

$$\Delta E = \sum_{n=1}^{\infty} c_n\, g_n(R)\,/\,R^n \qquad (3.8)$$

where $g_n(R) \to 0$ as $R\to 0$ and $g_n(R) \to 1$ as $R \to \infty$,in a non-uniform way with respect to n (it may be appropriate to introduce the functions $\chi_n(R) = 1 - g_n(R)$, which vary from 1 to 0 in an exponentially decreasing way when R varies from 0 to ∞).

In order to get such a representation of the 2nd-order terms, it is of course necessary to avoid the multipole expansion of V, since this could only provide the (asymptotic divergent) multipolar expansion (in terms of powers of R^{-1}). Treatments fulfilling this requirement have been until now limited to interactions between spherically symmetric systems such as H....H^+,H....H, He....He [12], [67], [70-71].

The treatment described below intends to be fully general. It is based upon three main ideas (see the brief summary [16]):

(1) the exact representation of the interaction potential V in terms of the molecular charge density operators [59], that we already used for treating the electrostatic energy (see eqs. (2.3), (2.4)). It must be emphasized here that, at the expense of introducing of having introduced the integration over $\vec{r}^1$ and $\vec{r}^2$, we obtain in the numerator of eq.(2.3) a separation into a product of (charge density) operators pertaining separately to each of the interacting molecules. This feature is essential in order to ultimately succeed in expressing the interaction energy in terms of physical properties of the isolated molecules;

(2) the use of Green's functions instead of "sums of eigenstates" for expressing the solution of perturbation equations (see refs. [73-79] for other examples of application of Green's function approach to the theory of intermolecular interactions). This approach can take davantage of various efficient procedures for the solution of inhomogeneous (partial) differential equations (such as the perturbation equations, precisely): see Dalgarno and Lewis [65] and Hirschfelder et al. [80] (section III.A).
One of the most powerful procedures for solving an inhomogeneous equation of the form

$$(H_0 - E)\psi = g \tag{3.9}$$

consists in introducing a new unknown function defined by

$$\psi = f\,\varphi_0 \tag{3.10}$$

where φ_0 denotes the ground state eigenfunction of H_0, with eigenvalue E_0. Then (3.9) can be rewritten [65,80] as:

$$(H_0 - E_0) f \varphi_0 + (E_0 - E) f \varphi_0 = g \tag{3.11}$$

and we have

$$(H_0 - E_0) f \varphi_0 = H_0 f \varphi_0 - f E_0 \varphi_0 =$$

$$= H_0 f \varphi_0 - f H_0 \varphi_0 = [H_0, f] \varphi_0 \qquad (3.12)$$

hence the transformed equation for f:

$$[H_0, f]\varphi_0 + (E_0 - E) f \varphi_0 = 0 \qquad (3.13)$$

This equation for f turns out to be markedly easier to solve than the original one (3.9) in important cases [65,80].

The case where the right-hand side g of eq.(3.9) is a δ-distribution precisely gives us the Green's function. We indeed use here for this function G(R , R' ; E) the standard definition, namely the kernel of the resolvent operator (H_0 - E) i.e. the solution of the equation

$$(H_0 - E)\, G(\vec{R}, \vec{R}'; E) = \delta(\vec{R} - \vec{R}') \qquad (3.14)$$

The formal expression of G in terms of the eigenfunctions and eigenvalues E_i of H_0 is:

$$G(\vec{R}, \vec{R}'; E) = \sum_i \frac{\varphi_i(\vec{R})\, \varphi_i^*(\vec{R}')}{E_i - E} \qquad (3.15)$$

but we also need a compact expression which should be obtained by solving (3.9). The solution of any inhomogeneous equation of the form (3.9) may be expressed in a straightforward way in terms of G(R , R' ; E):

$$\psi(\vec{R}) = \int G(\vec{R}, \vec{R}'; E)\, g(\vec{R}')\, d\vec{R}' \qquad (3.16)$$

In order to solve perturbation equations, it is interesting to introduce also <u>reduced</u> Green's functions, where one (or possibly several) term(s) of the formal expansion (3.10) are suppressed, e.g.

$$\begin{aligned} G_0(\vec{R}, \vec{R}'; E) &= \sum_{i \neq 0} \frac{\varphi_i(\vec{R})\, \varphi_i^*(\vec{R}')}{E_i - E} \\ &= G(\vec{R}, \vec{R}'; E) - \frac{\varphi_0(\vec{R})\, \varphi_0^*(\vec{R}')}{E_0 - E} \end{aligned} \qquad (3.17)$$

indeed, by making $E = E_0$, we get :

$$G_0(\vec{R}, \vec{R}'; E_0) = \sum_{i \neq 0} \frac{\varphi_i(\vec{R})\, \varphi_i^*(\vec{R}')}{E_i - E_0} \qquad (3.18)$$

namely the so-called <u>reduced resolvent</u> which enables us to solve the perturbation equations pertaining to the unperturbed eigenstate φ_0 of H_0;

(3) for the dispersion term specifically, we shall make use of the well-known identity:

$$\frac{1}{a+b} = \frac{2}{\pi} \int_0^{+\infty} \frac{a}{a^2 + \xi^2} \frac{b}{b^2 + \xi^2}\, d\xi \qquad (3.19)$$

with $a = \Delta E_i^{(1)}$ and $b = \Delta E_j^{(2)}$: indeed, at the expense of introducing the integration with respect to, we thus transform the factor $1 / (\Delta E_i^{(1)} + \Delta E_j^{(2)})$ appearing in the expression (3.4) of E_{disp} in such a way as to realize a "<u>separation</u>" into a <u>product</u> of quantities pertaining to each molecule separately:

$$\frac{1}{\Delta E_i^{(1)} + \Delta E_j^{(2)}} = \frac{2}{\pi} \int_0^{+\infty} \frac{\Delta E_i^{(1)}}{(\Delta E_i^{(1)})^2 + \xi^2} \frac{\Delta E_j^{(2)}}{(\Delta E_j^{(2)})^2 + \xi^2}\, d\xi \qquad (3.20)$$

We recognize here the same feature that we already emphasized under heading (1) above concerning the representation of the interaction potential V: such a separation opens the way towrds expressing the interaction energy in terms of physical properties pertaining to the isolated molecules.

We shall now apply the three ingredients just described to the treatment of the induction and dispersion terms.

3.2 Induction Terms $E_{ind}^{(1)}$ and $E_{ind}^{(2)}$

We shall consider for definiteness the induction term $E_{ind}^{(1)}$ for molecule (1) (the treatment of the induction term $E_{ind}^{(2)}$ for molecule (2) would be completely similar).

$$E_{ind}^{(1)} = -\sum_{i^{(1)}}' \frac{\langle 0^{(1)}0^{(2)}|V|i^{(1)}0^{(2)}\rangle\langle i^{(1)}0^{(2)}|V|0^{(1)}0^{(2)}\rangle}{\Delta E_i^{(1)}} \tag{3.21}$$

where, according to the expression (2.3) of V:

$$\langle 0^{(1)}0^{(2)}|V|i^{(1)}0^{(2)}\rangle = \iint \frac{\langle 0^{(1)}|\rho^{(1)}(\vec{R}_1)|i^{(1)}\rangle \rho_{00}^{(2)}(\vec{R}_2)}{|\vec{R}_1-\vec{R}_2|} d\vec{R}_1 d\vec{R}_2 \tag{3.22}$$

where

$$\rho_{00}^{(2)}(\vec{R}_2) = \langle 0^{(2)}|\rho^{(2)}(\vec{R}_2)|0^{(2)}\rangle \tag{3.23}$$

denotes the charge distribution pertaining to the ground state of molecule (2) (for convenience, we shall from now on use $\vec{R}_1$ and $\vec{R}_2$ instead of $\vec{r}^1$ and $\vec{r}^2$ for the dummy integration coordinates over space, since no confusion with nuclear coordinates can appear). We have a similar expression for the other matrix element of V, namely

$$\langle i^{(1)}0^{(2)}|V|0^{(1)}0^{(2)}\rangle = \iint \frac{\langle i^{(1)}|\rho^{(1)}(\vec{R}'_1)|0^{(1)}\rangle \rho_{00}^{(2)}(\vec{R}'_2)}{|\vec{R}'_1-\vec{R}'_2|} d\vec{R}'_1 d\vec{R}'_2 \tag{3.24}$$

hence, upon inserting (3.22) and (3.24) in (3.21):

$$E_{ind}^{(1)} = -\iint \frac{d\vec{R}_1 d\vec{R}_2}{|\vec{R}_1-\vec{R}_2|}\iint \frac{d\vec{R}'_1 d\vec{R}'_2}{|\vec{R}'_1-\vec{R}'_2|}\rho_{00}^{(2)}(\vec{R}_2)\rho_{00}^{(2)}(\vec{R}'_2) \times$$
$$\times\left[\sum_{i^{(1)}}' \langle 0^{(1)}|\rho^{(1)}(\vec{R}_1)|i^{(1)}\rangle\langle i^{(1)}|\rho^{(1)}(\vec{R}'_1)|0^{(1)}\rangle/\Delta E_i^{(1)}\right] \tag{3.25}$$

Now, we recognize in eq.(3.25) the expression (3.18) of the reduced Green's function pertaining to the ground state $\varphi_0 = |0^{(1)}0^{(2)}\rangle$ of H_0, namely:

$$\sum_{i^{(1)}}' |i^{(1)}(\vec{r}_1,\dots,\vec{r}_{n_1})\rangle\langle i^{(1)}(\vec{r}'_1,\dots,\vec{r}'_{n_1})|/\Delta E_i^{(1)}$$
$$= G_0^{(1)}(\vec{r}_1,\dots,\vec{r}_{n_1};\vec{r}'_1,\dots,\vec{r}'_{n_1};E_0) \tag{3.26}$$

Since $|0^{(1)}\rangle$ and $|i^{(1)}\rangle$ are orthogonal, the nuclear part of $\rho^{(1)}$ does not contribute to the matrix elements $\langle 0^{(1)}|\rho^{(1)}|i^{(1)}\rangle$, and the bracketed expression in (3.25) thus becomes:

$$\begin{aligned}[\quad] &= \int d\vec{r}_1 \dots d\vec{r}_{n_1} \int d\vec{r}_1{}' \dots d\vec{r}_{n_1}{}' \quad O^{(1)} \; (\vec{r}_1, \dots, \vec{r}_{n_1}) \times \\ &\times \left(- \sum_{k=1}^{n_1} \delta\,(\vec{R}_1 - \vec{r}_k)\right) G_0^{(1)}(\vec{r}_1, \dots, \vec{r}_{n_1}; \vec{r}_1{}', \dots, \vec{r}_{n_1}{}'; E_0) \times \\ &\times \left(- \sum_{\ell=1}^{n_1} \delta\,(\vec{R}_1{}' - \vec{r}_\ell{}')\right) \; O^{(1)*}(\vec{r}_1{}', \dots, \vec{r}_{n_1}{}') \qquad (3.27)\end{aligned}$$

where the sums $\sum_k$ and $\sum_\ell$ run over the electrons of molecule (1). Then, by using the property of antisymmetry of $|O^{(1)}(\vec{r}_1, \dots, r_{n_1})\rangle, |O^{(1)}(\vec{r}_1{}', \dots, \vec{r}_{n_1}{}')\rangle$ and of $G_0(\vec{r}_1, \dots; \vec{r}_1{}', \dots; E_0)$ with respect to electron permutation inside each of the two sets $\{\vec{r}_1, \dots, \vec{r}_{n_1}\}$ and $\{\vec{r}_1{}', \dots, \vec{r}_{n_1}{}'\}$, we get in (3.27) the same contribution from each product $\delta(\vec{R}_1 - \vec{r}_k)\,\delta(\vec{R}_1{}' - r_\ell{}')$, hence we have n_1^2 times the contribution corresponding to k=1, ℓ=1, say:

$$\begin{aligned}[\quad] &= n_1^2 \int d\vec{r}_2 \dots d\vec{r}_{n_1} \int d\vec{r}_2{}' \dots d\vec{r}_{n_1}{}' \quad O^{(1)}(\vec{R}_1, \vec{r}_2, \dots, \vec{r}_{n_1}) \\ &\times \; G_0(\vec{R}_1, \vec{r}_2, \dots, \vec{r}_{n_1}; \vec{R}_1{}', \vec{r}_2{}', \dots, \vec{r}_{n_1}{}'; E_0) \; O^{(1)*}(\vec{R}_1{}', \vec{r}_2{}', \dots, \vec{r}_{n_1}{}') \\ &= K^{(1)}(\vec{R}_1, \vec{R}_1{}'; 0) \qquad (3.28)\end{aligned}$$

Eq.(3.28) defines the function $K^{(1)}(\vec{R}, \vec{R}'; 0)$. This "response function" appears in the present treatment in the place of the various <u>static</u> multipole polarizabilities which appear in the standard treatments based upon the multipole expansion of the matrix elements of V. It may thus be called a "polarizability function" or "susceptibility function" for molecule (1). The fact that this function corresponds to <u>static</u> (i.e. <u>zero-frequency</u>) susceptibility is indicated by the third variable with the value 0 in our notation $K^{(1)}(\vec{R}, \vec{R}'; 0)$ (when dealing with the dispersion term below, we shall be led to introduce the values of the susceptibility function for arbitrary non-zero frequency). It is worth emphasizing that this susceptibility function depends on six variables $\vec{R}$, $\vec{R}'$ whatever the number of electrons of the molecule under consideration. The fact that the usual static (dipole) polarizability of a molecule may be represented in terms of bond

polarizabilities is an incitation to consider the possibility of such representations in terms of local contributions for the susceptibility function $K^{(1)}$,too.
We now insert $K^{(1)}$ into the expression (3.25) of E_{ind} ,and we get our final formula:

$$E_{ind}^{(1)} = -\iint \frac{d\vec{R}_1\, d\vec{R}_2}{|\vec{R}_1 - \vec{R}_2|} \iint \frac{d\vec{R}'_1\, d\vec{R}'_2}{|\vec{R}'_1 - \vec{R}'_2|}\, \rho_{00}^{(2)}(\vec{R}_2)\, \rho_{00}^{(2)}(\vec{R}'_2) \times \times K^{(1)}(\vec{R}_1, \vec{R}'_1; 0)$$

$$= -\int d\vec{R}_1 \int d\vec{R}'_1\, \mathcal{V}_{00}^{(2)}(\vec{R}_1)\, \mathcal{V}_{00}^{(2)}(\vec{R}'_1)\, K^{(1)}(\vec{R}_1, \vec{R}'_1; 0) \quad (3.27)$$

where

$$\mathcal{V}_{00}^{(2)}(\vec{R}) = \int \frac{\rho_{00}^{(2)}(\vec{R}_2)}{|\vec{R} - \vec{R}_2|}\, d\vec{R}_2 \quad (3.28)$$

is nothing but the electrostatic potential created at the point $\vec{R}$ by the charge distribution of molecule (2).
We would of course get a similar formula for the induction energy $E_{ind}^{(2)}$ of molecule (2), namely:

$$E_{ind}^{(2)} = -\int d\vec{R}_2 \int d\vec{R}'_2\, \mathcal{V}_{00}^{(1)}(\vec{R}_2)\, \mathcal{V}_{00}^{(1)}(\vec{R}'_2)\, K^{(2)}(\vec{R}_2, \vec{R}'_2; 0) \quad (3.29)$$

with obvious meaning for $\mathcal{V}_{00}^{(1)}$(the electrostatic potential created by molecule 1) and $K^{(2)}$ (the susceptibilty function of molecule 2).

This formula actually resembles the familiar formula for the induction energy when the response to an external field $\vec{\mathcal{E}}$ is represented through the point dipole polarizability tensor $\overline{\overline{\alpha}}$, namely $E_{ind} = -\frac{1}{2}\, \vec{\mathcal{E}}.\overline{\overline{\alpha}}.\vec{\mathcal{E}}$. This formula, as well as those involving higher order multipole polarizabilities would be recovered by introducing a suitable expansion of the susceptibility function $K^{(1)}$ (for an example of this kind of expansion, see Hameka [81]).

3.3 Dispersion Term E_{disp}

We start from the formula (3.4):

$$E_{disp} = -\sum_{i^{(1)}}{}' \sum_{j^{(2)}}{}' \frac{\langle 0^{(1)} 0^{(2)} | V | i^{(1)} j^{(2)} \rangle \langle i^{(1)} j^{(2)} | V | 0^{(1)} 0^{(2)} \rangle}{\Delta E_i^{(1)} + \Delta E_j^{(2)}} \qquad (3.30)$$

and, as we did previously for E_{ind}, we insert inside each matrix element of the numerator the expression (2.3) of V in terms of the charge density operators, and we replace the factor $1 / (\Delta E_i^{(1)} + \Delta E_j^{(2)})$ by its expression (3.20):

$$E_{disp} = -\sum_{i^{(1)}}{}' \sum_{j^{(2)}}{}' \iint \frac{\langle 0^{(1)} | \rho^{(1)}(\vec{R}_1) | i^{(1)} \rangle \langle 0^{(2)} | \rho^{(2)}(\vec{R}_2) | j^{(2)} \rangle}{|\vec{R}_1 - \vec{R}_2|} d\vec{R}_1 \, d\vec{R}_2$$

$$\times \iint \frac{\langle i^{(1)} | \rho^{(1)}(\vec{R}'_1) | 0^{(1)} \rangle \langle j^{(2)} | \rho^{(2)}(\vec{R}'_2) | 0^{(2)} \rangle}{|\vec{R}'_1 - \vec{R}'_2|} d\vec{R}'_1 \, d\vec{R}'_2$$

$$\times (2/\pi) \int_0^{+\infty} \frac{\Delta E_i^{(1)}}{(\Delta E_i^{(1)})^2 + \xi^2} \; \frac{\Delta E_j^{(2)}}{(\Delta E_j^{(2)})^2 + \xi^2} \, d\xi$$

$$= -(2/\pi) \int_0^{+\infty} d\xi \iint \frac{d\vec{R}_1 \, d\vec{R}_2}{|\vec{R}_1 - \vec{R}_2|} \; \frac{d\vec{R}'_1 \, d\vec{R}'_2}{|\vec{R}'_1 - \vec{R}'_2|} \times$$

$$\times \left[\sum_{i^{(1)}}{}' \langle 0^{(1)} | \rho^{(1)}(\vec{R}_1) | i^{(1)} \rangle \langle i^{(1)} | \rho^{(1)}(\vec{R}'_1) | 0^{(1)} \rangle \, \Delta E_i^{(1)} / [(\Delta E_i^{(1)})^2 + \xi^2] \right]$$

$$\times \left[\sum_{j^{(2)}} \langle 0^{(2)} | \rho^{(2)}(\vec{R}_2) | j^{(2)} \rangle \langle j^{(2)} | \rho^{(2)}(\vec{R}'_2) | 0^{(2)} \rangle \, \Delta E_j^{(2)} / [(\Delta E_j^{(2)})^2 + \xi^2] \right] \qquad (3.31)$$

The two bracketed expressions are response functions pertainig to molecules (1) and (2) respectively. We shall denote them $[\ \ ]^{(1)}$ and $[\ \ ]^{(2)}$ respectively. We immediately see that, for $\xi = 0$, these expressions become identical with the susceptibility functions $K^{(1)}(\vec{R}_1, \vec{R}'_1; 0)$ and $K^{(2)}(\vec{R}_2, \vec{R}'_2; 0)$ introduced in the previous subsection (eqs.(3.27) and (3.28)). By analogy with the notation $\alpha(i\xi)$ for the polarizabilities at imaginary frequency $i\xi$ (see e.g.[2] section V.E.3 and references therein,[82-86]), we therefore introduce the notation:

$$[\ \ ]^{(1)} = K^{(1)}(\vec{R}_1, \vec{R}'_1; i\xi) \qquad [\ \ ]^{(2)} = K^{(2)}(\vec{R}_2, \vec{R}'_2; i\xi) \qquad (3.32)$$

We shall consider for definiteness $K^{(1)}(\vec{R}_1, \vec{R}_1'; i\xi)$ (the treatment for $K^{(2)}$ would of course be completely similar). By using the formula:

$$\frac{a}{a^2+\xi^2} = \frac{1}{2}\left(\frac{1}{a-i\xi} + \frac{1}{a+i\xi}\right) \qquad (3.33)$$

with $a = \Delta E_i^{(1)}$, we can write:

$$K^{(1)}(\vec{R},\vec{R}'; i\xi) = \frac{1}{2}\left[u^{(1)}(\vec{R},\vec{R}'; i\xi) + u^{(1)}(\vec{R},\vec{R}'; -i\xi)\right] \qquad (3.34)$$

where $u^{(1)}(\vec{R},\vec{R}'; i\xi)$ is defined by:

$$u^{(1)}(\vec{R},\vec{R}'; i\xi) = \langle 0^{(1)} | \rho^{(1)}(\vec{R}) \left(\sum_{i^{(1)}}' \frac{|i^{(1)}\rangle\langle i^{(1)}|}{\Delta E_i^{(1)} - i\xi} \right) \rho^{(1)}(\vec{R}_1') | 0^{(1)} \rangle \qquad (3.35)$$

and

$$\sum_{i^{(1)}}' \frac{|i^{(1)}\rangle\langle i^{(1)}|}{\Delta E_i^{(1)} - i\xi} = \sum_{i^{(1)}}' \frac{|i^{(1)}\rangle\langle i^{(1)}|}{E_i^{(1)} - (E_0^{(1)} + i\xi)} \qquad (3.36)$$

is nothing but the general resolvent of H_0 for the value $E = E_0^{(1)} + i\xi$, deprived from the contribution of the ground state $|0\rangle$. Its kernel is the reduced Green's function $G_0^{(1)}(\vec{r}_1, \dots; \vec{r}_1', \dots; E)$ that we defined previously (see eq.(3.17), with $\vec{R}$ replaced by $\{\vec{r}_1, \dots, \vec{r}_n\}$ and $\vec{R}'$ replaced by $\{\vec{r}_1', \dots, \vec{r}_n'\}$). Then, by inserting the expression (2.4) of $\rho^{(1)}$ (in actual fact, only the electronic part contributes), and by using (as we did previously for $K^{(1)}(\vec{R}_1, \vec{R}_1'; 0)$ the antisymmetry of $|0^{(1)}\rangle$ and of G_0 with respect to electron exchange, we get

$$\begin{aligned} u^{(1)}(\vec{R},\vec{R}'; i\xi) &= n_1^2 \int d\vec{r}_2 \dots d\vec{r}_{n_1} \int d\vec{r}_2' \dots d\vec{r}_{n_1}' \times \\ &\times 0^{(1)}(\vec{R}, \vec{r}_2, \dots, \vec{r}_{n_1}) \; 0^{(1)*}(\vec{R}', \vec{r}_2', \dots, \vec{r}_{n_1}') \times \\ &\times G_0^{(1)}(\vec{R}, \vec{r}_2, \dots, \vec{r}_{n_1}; \vec{R}', \vec{r}_2', \dots, \vec{r}_{n_1}'; E_0^{(1)} + i\xi) \end{aligned} \qquad (3.37)$$

In the same way as $K^{(1)}(\vec{R}_1, \vec{R}_1'; 0)$ was a generalization of the static multipolar polarizabilities, the susceptibility function $K^{(1)}(\vec{R}_1, \vec{R}'_1; i\xi)$ is a generalization of the dynamic (frequency-dependent) polarizabilities. This susceptibility function includes exactly the penetration (or charge overlap) effects, thus going beyond any multipole approximation. We could use for it the name "dynamic susceptibility function".

Now, by inserting $K^{(1)}(\vec{R}_1, \vec{R}'_1; i\xi)$ and $K^{(2)}(\vec{R}_2, \vec{R}'_2; i\xi)$ instead of the brackets $[\ ]^{(1)}$ and $[\ ]^{(2)}$ in the expression (3.31) of E_{disp}, we get our final formula:

$$E_{disp} = -\frac{2}{\pi}\int_0^{+\infty} d\xi \iint \frac{d\vec{R}_1\, d\vec{R}_2}{\vec{R}_1 - \vec{R}_2} \iint \frac{d\vec{R}'_1\, d\vec{R}'_2}{\vec{R}'_1 - \vec{R}'_2} \times \times K^{(1)}(\vec{R}_1, \vec{R}'_1; i\xi)\; K^{(2)}(R_2, R'_2; i\xi) \qquad (3.38)$$

As we did in the case of the induction term, we can notice that this formula is clearly analogous to the well-known formula expressing the $1/R^6$ "dipole-dipole" part of the dispersion energy in terms of the dynamic polarizabilities at imaginary frequency (for example, in the case of the interaction between atoms, $E^{(6)}_{disp} = -\frac{3}{\pi}\frac{1}{R^6}\int_0^{+\infty} d\xi\, \alpha_1(i\xi)\alpha_2(i\xi)$, see e.g. [82-85]). This formula, and the others for higher multipole-multipole terms, would be recovered by using a multipole expansion for the electrostatic-like interactions terms between $K^{(1)}$ and $K^{(2)}$ corresponding to the integrations over $\vec{R}_1, \vec{R}_2$ and $\vec{R}'_1, \vec{R}'_2$ with the Coulomb kernels $1/|\vec{R}_1 - \vec{R}_2|$ and $1/|\vec{R}_1 - \vec{R}_2|$.

4. CONCLUSION

The Rayleigh-Schrödinger (RS) terms $\varepsilon_1^{(RS)}$, $\varepsilon_2^{(RS)}$ of the intermolecular interaction energy are very commonly reduced to their long-range terms, thereby neglecting their short-range (penetration or charge overlap) behaviour. This approach is likely to become unsufficient at medium-range (region of the Van der Waals minimum), and still more serious, a simply additive decomposition of each RS term into a purely long-range (multipolar) and a purely short-range part is not possible beyond the 1st-order term $\varepsilon_1^{(RS)}$. Accordingly, the present work has been devoted to a systematic treatment of the exact RS terms, involving both their long-range and short-range behaviours. An essential ingredient was the use of the exact representation of the intermolecular interaction potential V in terms of molecular charge density operators.

It is thus possible to express the electrostatic term $\varepsilon_1^{(RS)}$ as the sum of a long-range (multipolar) part and a short-range (penetration or charge overlap) part. The long-range part can be accurately described through finite sets of multipoles (charge, dipole, quadrupole in current actual practice) located at suitable centers (e.g. atoms and middles of chemical bonds): this is the so-called multi-center multipole representation.

As concerns the 2nd-order term $\varepsilon_2^{(RS)}$ (induction and dispersion), we were led to introduce, through the Green's function formalism, molecular "susceptibility functions" (both static and frequency dependent). These response functions generalize the usual multipole polarizabilities, and give the possibility of accounting correctly for the entangled long-range and short-range behaviours. More specifically, the induction terms are expressed in terms of the exact molecular electrostatic potentials (thus involving, in principle, both the long-range and short-range contributions) and of the static (zero-frequency) susceptibility function; on the other hand, the dispersion term is expressed in terms of the dynamic susceptibility functions (at imaginary frequency, to be precise).

It is appropriate to recall here that, according to the well-known fluctuation-dissipation theorems (see e.g. [84-86]), the response functions under consideration could be related with the charge fluctuations. Accordingly, the dispersion term could be called "charge fluctuation" term, as advocated by London [90] and later by Jehle [91].

We can also mention how connection may be established with the so-called "damping function" expansions [10-12], for interaction between spherically symmetric systems. In that case, the susceptibility function K (R , R';) depends only on the moduli R = | R |, R' = | R' | and the angle between R and R'. Then, following Hameka [81], we can introduce the expansion:

$$K(R, R', \gamma; i\xi) = K_\ell(R, R'; i\xi)\, P_\ell(\cos\gamma) \qquad (4.1)$$

where P_l denotes the Legendre Polynomial of order l. For the hydrogen atom, each K_ℓ "partly" factorizes, in the sense that

$$K_\ell(R, R'; i\xi) = k_\ell^{>}(R_>; i\xi)\, k_\ell^{<}(R_<; i\xi) \qquad (4.2)$$

where $R_> = \text{Max}(R,R')$ and $R_< = \text{Min}(R,R')$. This form makes rather easy the integrations over $\vec{R}_1, \vec{R}'_1$ and $\vec{R}_2, \vec{R}'_2$ in the formulae (3.27,3.29) and (3.28) of the previous section, and it should then be possible to recover the "damping function" expansion of the dispersion energy from the expansions (4.1).
For many-electron systems, upon accepting as an approximation to H_0 a Hartree or Hartree-Fock Hamiltonian, it is in principle possible, by following the procedure of Hirschfelder et al. [80], to reduce the search for the Green's function (and hence the response function) to a series of one-particle problems. Furthermore, by introducing localized molecular orbitals (instead of the delocalized canonical ones), it could be attempted to reduce further the many-center problems to one- and two-center problems (corresponding to orbitals localized on atoms or bonds respectively). The ultimate purpose of such a line of research would be, of course, to express (approximately) the response functions in terms of local (bond and atom) contributions, in the same way as this is customarily done for the dipole polarizability (see e.g. [4],[92]).

REFERENCES

[1] Claverie, P. (1976), Chap. 2, p. 127-152 in Localization and Delocalization in Quantum Chemistry, Vol. II (ed. by O. Chalvet, R. Daudel, S. Diner and J.P. Malrieu), Reidel, Dordrecht.

[2] Claverie, P. (1978), p. 69-305 in Intermolecular Interactions: from Diatomics to Biopolymers (ed. by B. Pullman), Wiley, New York.

[3] Claverie, P. (1982), p. 151-175 in Quantum Theory of Chemical Reactions, vol. III (ed by R. Daudel, A. Pullman, L. Salem and A. Veillard), Reidel, Dordrecht.

[4] Amos, A.T. and Crispin, R.J. (1976) Theoretical Chemistry (Advances and Perspectives) **2**, 1-66 (Academic Press, New York).

[5] Buckingham, A.D. (1978), Chap. 1, p. 1-68 in Intermolecular Interactions : from Diatomics to Biopolymers (ed. by B. Pullman), Wiley, New York.

[6] Kitaigorodsky, A.I. (1978), Chem. Soc. Rev. **7**, 133-163.

[7] Kaplan, I.G. and Rodimova, O.B. (1978), Soc. Phys.-Uspekki **21**, 918-943.

[8] Arrighini, P. (1981), Intermolecular Forces and their Evaluation by Pertubation Theory (Lecture Notes in Chemistry nr 25), Springer – Verlag, Berlin.

[9] Jeziorski, B. and Kolos, W. (1982), 1-46 in Molecular Interactions, Vol. III (ed. by H. Ratajczak and W.J. Orville-Thomas), Wiley, New-York.

[10] Hepburn, J., Scoles, G. and Penco, R. (1975), Chem. Phys. Lett. **36**, 451.

[11] Ahlrichs, R., Penco, R. and Scoles, G. (1977), Chem. Phys. **19**, 119.

[12] Koide, A. (1976), J. Phys. B **9**, 3173.

[13] Gresh, N., Claverie, P. and Pullman, A. (1979), Intern. J. Quantum Chem. Symp. **13**, 243.

[14] Gresh, N., Claverie, P. and Pullman, A. (1985), "Intermolecular Interactions. Elaboration on an Additive Procedure including an Explicit Charge-Transfer Contribution", Intern. J. Quantum Chem. (in press).

[15] Singh, U.C. and Kollman, P.A. (1985), J. Chem. Phys. **83**, 4033.

[16] Claverie, P. (1983), Intern. J. Quantum Chem. **23**, 1687.

[17] Bohm, H.J. and Ahlrichs, R. (1982), J. Chem. Phys. **77**, 2028.

[18] Hoinkis, J. , Ahlrichs, R. and Bohm, H.J. (1983), Intern. J. Quantum Chem. **23**, 821.

[19] Ahlrichs, R. (1976), Theoret. Chim. Acta (Berlin) **41**, 7.

[20] Chalasinski, B. , Jeziorski, B. and Szalewicz, K. (1977), Intern. J. Quantum Chem. **11**, 247.

[21] Claverie, P. (1971), Intern. J. Quantum Chem. **5**, 273.

[22] Kutzelnigg, W. (1980), J.Chem. Phys. **73**, 343.

[23] Jeziorski, B. , Schwalm, W.A. and Szalewicz, K. (1980), J. Chem. Phys. **73**, 6215.

[24] Chipman, D.M. and Hirschfelder, J.O. (1980), J.Chem. Phys. **73** , 5164.

[25] Jeziorski, B. and Kolos, W. (1977), Intern. J. Quantum Chem. **12**, Suppl. 1, 91.

[26] Jeziorski, B. , Szalewicz, K. and Chalasinski, B. (1978), Intern. J. Quantum Chem. **14**, 271.

[27] Szalewicz, K. and Jeziorski, B. (1979), Mol. Phys. **38**, 191.

[28] Jeziorski, B. and Van Hemert, M. (1976), Mol. Phys. **31**, 713.

[29] Bukta, J.F. and Meath, W.J. (1972), Intern. J. Quantum Chem. **6**, 1045.

[30] Carlson, B.C. and Rushbrooke, G.S. (1950), Proc. Cambr. Phil. Soc. **46**, 626.

[31] Buehler, R.J. and Hirschfelder, J.O. (1951), Phys. Rev. **83**, 628.

[32] Buehler, R.J. and Hirschfelder, J.O. (1952), Phys. Rev. **85**, 149.

[33] Rose, M.E. (1958), J. Math. and Phys. **37**, 215.

[34] Fontana, P.R. (1961), Phys. Rev. **123**, 1865.

[35] Hirschfelder, J.O., Curtis, C.F. and Bird, R.B. (1964), Molecular Theory of Gases and Liquids, (2nd edition) Wiley. New York.

[36] Steinborn, D. and Ruedenberg, K. (1973), Advances in Quantum Chemistry 7, 1 (ed. by P.O. Lowdin), Academic Press, New York.

[37] Wormer, P.E.S. (1975), Intermolecular Forces and the Group Theory of Many-Body Systems., Thesis, University of Nijmegen.

[38] Ng, K.C., Meath, W.J. and Allnatt, A.R. (1976), Mol. Phys. **32**, 177.

[39] Roothaan, C.C.J. (1951), J. Chem. Phys. **19** , 1445.

[40] Pollak, M. and Rein, R. (1967), J. Chem. Phys. **47**, 2045.

[41] Dreyfus, M. (1970), Etude non-empirique de la liaison Hydrogène dans les dimères de la formamide, Thèse (Doctorat de 3e cycle), University of Paris VI.

[42] Port, G.N.J. and Pullman, A. (1973), FEBS Lett. **31**, 70.

[43] Pullman, A. and Perahia, D. (1978), Theoret. Chim. Acta (Berlin) **48**, 29.

[44] Goldblum, A., Perahia, D. and Pullman, A. (1979), Intern. J. Quantum Chem. **15**, 121.

[45] Pullman, A., Zakrzewska, K. and Perahia, D. (1979), Intern. J. Quantum Chem. **16**, 395.

[46] Lavery, R., Etchebest, C. and Pullman, A. (1982), Chem. Phys. Lett. **85** , 266.

[47] Pack, G.R., Wang, H. and Rein, R. (1972), Chem. Phys. Lett. **17**, 381.

[48] Rein, R. (1973), Advances in Quantum Chemistry **7**, 335 (ed. by P.O. Lowdin), Academic Press, New York.

[49] Hall, G.G. (1973), Chem. Phys. Lett. **20**, 501.

[50] Tait, A.D. and Hall, G.G. (1973) , Theoret. Chem. Acta (Berlin) **31**, 311.

[51] Martin, D. and Hall, G.G. (1981), Theoret. Chem. Acta (Berlin) **59**, 281.

[52] Hall, G.G. (1983), Theoret. Chem. Acta (Berlin) **63**, 357.

[53] Shipman, L.L. (1975), Chem. Phys. Lett. **31**, 361.

[54] Stone, A.J. (1981), Chem. Phys. Lett. **83**, 233.

[55] Price, S.L. and Stone, A.J. (1983), Chem. Phys. Lett. **98**, 419.

[56] Sokalski, W.A., and Poirier, R.A. (1983), Chem. Phys. Lett. **98**, 86.

[57] Vigné-Maeder, F. and Claverie, P. (1985), "The exact multipolar part of a molecular charge distribution and its simplified representations", submitted for publication to J. Chem. Phys..

[58] Vigné-Maeder, F. (1985), "Multicenter Multipole Expansion for Conjugated Molecules", submitted for publication to J. Chem. Phys..

[59] Longuet-Higgins, H.C. (1956), Proc Roy. Soc. (London) A **235**, 537.

[60] Messiah, A. (1959-1960), Mécanique Quantique, Dunod, Paris (part I: 1959; part II:1960) (English translation: Quantum Mechanics, North-Holland, Amsterdam, 1961). See part I, appendix B, section IV.10 and part II, appendix D, section I.3.

[61] Edmonds, A.R. (1960), Angular Momentum in Quantum Mechanics, Princeton University Press, Princeton, New Jersey. See section 4.6, p. 62-63.

[62] Shavitt, I. (1963), p.1 in Methods in Computational Physics, Academic Press, New York.

[63] Gresh, N., Claverie, P. and Pullman, A. (1984), Theoret. Chim. Acta (Berlin) **66**, 1.

[64] Gresh, N., Pullman, A. and Claverie, P. (1985), Theoret. Chim. Acta (Berlin) **67**, 11.

[65] Dalgarno, A. and Lewis, J.T. ( ) Proc. Roy. Soc. (London) A **233**, 70.

[66] Dalgarno, A. and Lewis, J.T (1956), Proc. Phys. Soc. (London) A **69**, 57.

[67] Dalgarno, A. and Lynn, N. (1957), Proc. Phys. Soc. (London) A **70**, 223.

[68] Erdelyi, A. (1956), Asymptotic Expansions, Dover, New York.

[69] Bender, C.M. and Orszag, S.A. (1978), Advanced Mathematical Methods for Scientists and Engineers, Mc Graw-Hill, New York (section 3.8).

[70] Kreek, H. and Meath, W.J. (1969), J. Chem. Phys. **50**, 2289.

[71] Singh, T.R., Kreek, H. and Meath, W.J. (1970), J. Chem. Phys. **53**, 4121.

[72] Chalasinski, G. and Jeziorski, B. (1974), Mol. Phys. **27**, 649.

[73] Linderberg, J. (1964), Arkiv Fysik **26**, 323.

[74] Linderberg, J. (1967), Intern. J. Quantum Chem. **1S**, 719.

[75] Linderberg, J. and Ohrn, Y. (1973), Propagators in Quantum Chemistry, Academic Press, New York (Chaps. 5 and 11).

[76] Mahan, G.D. (1965), J. Chem. Phys. **43**, 1569.

[77] Linder, B. (1967), Advances in Chemical Physics **12**, 203.

[78] Linder, B. and Rabenold, D. (1972), Advances in Quantum Chemistry **6**, 203.

[79] Boehm, R. and Yaris, R. (1971), J. Chem. Phys. **55**, 2620.

[80] Hirschfelder, J.O., Byers-Brown, W. and Epstein, S.T. (1964), Advances in Quantum Chemistry **1**, 255 (Academic Press, New York).

[81] Hameka, H.F. (1967), J. Chem. Phys. **47**, 2728.

[82] Casimir, H.B.G. and Polder, D. (1984), Phys. Rev. **73**, 360.

[83] Mavroyannis, C. and Stephen, M.J. (1962), Mol.Phys. **5**, 629.

[84] McLachlan, A.D. (1963), Proc. Roy. Soc. (London) **A 271**, 387.

[85] Mc Lachlan, A.D., Gregory, R.D. and Ball, M.A. (1963), Mol. Phys. **7**, 119.

[86] Dalgarno, A. and Davison, W.D. (1966), Advances in Atomic and Molecular Physics **2**, 1 (eds. D.R. Bates and I. Estermann, Academic Press, New York).

[87] Callen, H.B. and Welton, T.A. (1951), Phys. Rev. **83**, 34.

[88] Zubarev, D.N. (1974), Nonequilibrium Statistical Thermodynamics, Consultants Bureau, New York (Chap. 3, section 17).

[89] Forster, D. (1975), Hydrodynamic Fluctuations, Broken Symmetry and Correlation Functions, Benjamin/Cummings, Advanced Book Program, Reading, Massachusetts (Chap. 3, section 3.5).

[90] London, F. (1936), Trans Faraday Soc. **33**, 8.

[91] Jehle, H. (1965), Advances in Quantum Chemistry **2**, 195 (ed. P.O. Lowdin, Academic Press, New York).

[92] Le Fevre, R.J.W. (1965), Advances in Physical Organic Chemistry **3**, 1 (ed. V. Gold, Academic Press, New York).

SYMMETRY ANALYSIS AND CONFORMATIONAL DEPENDENCE OF THE PROPERTIES OF RIGID MOLECULES EMBEDDED IN CRYSTAL SITES

A. Toro-Labbé[+] and J. Maruani[x]
Centre de Mécanique Ondulatoire Appliquée, 23, rue du Maroc
75019 Paris, France

ABSTRACT. We present a group theory for rigid molecules of arbitrary symmetry embedded in crystal sites of any symmetry. We assume that the potential barrier to the rotation of the molecule relative to the crystal remains lower than the potential barriers to internal rearrangements in both the molecule and the crystal. We use Altmann's formalism to investigate the operator structure of the isodynamic group of the composite system in terms of the symmetry operations of the point groups of the molecule and the crystal. The effects of the isodynamic operations on the Wigner D-functions help us derive selection rules which can be used in reducing the harmonic expansions of the various properties of the composite system in terms of the relative-orientation Euler angles. Applications to the interpretation of scattering and spectroscopic experimental data are briefly outlined.

1. INTRODUCTION

A variety of crystals have lattice nodes occupied by neutral or ionic atoms or quasi-spherical molecules (e.g. : Ar, Cl^-, CH_4, NH_4^+, adamantane, etc. ; hereafter referred to as host molecules), making up unit cells of some particular symmetry housing at their center rigid molecules of the same or another symmetry (hereafter called guest molecules). In such cases the symmetry of a property which is dependent on the orientation of the rigid molecule with respect to the unit cell will be completely determined by the symmetries of both the molecule and the cell. Such properties are usually expanded in terms of a complete set of orthonormal functions of the relative-orientation variables $\Omega = \{\alpha,\beta,\gamma\}$ (Figure 1) /1-14/. For rigid molecules embedded in regular crystals these variables can be separated from the internal molecular variables as well as from the phonon crystal variables, due to the observed large differences between the corresponding motion frequencies. The properties of interest include the orientational distribution and correlation functions, the molecule-site energy and torque, and whichever theoretical property which may be of use in the interpretation of experimental data.

R. Daudel et al. (eds.), Structure and Dynamics of Molecular Systems – II, 35–56.

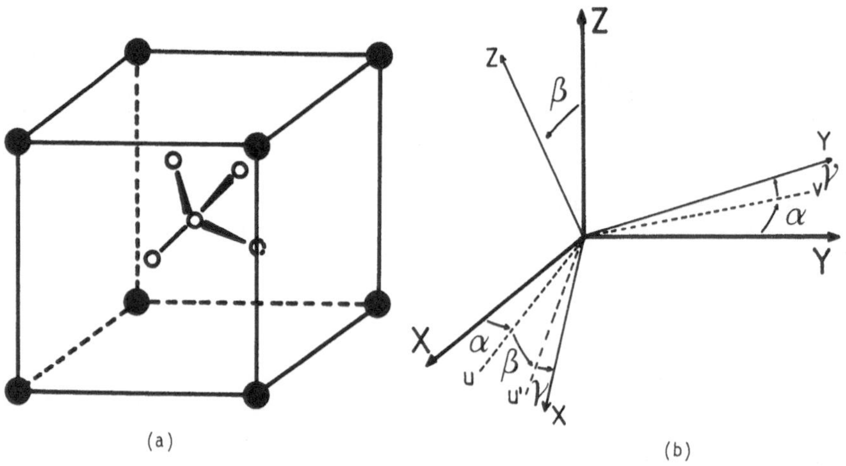

Figure 1. (a) An example of a rigid molecule embedded in a crystal site. (b) The Euler angles α,β,γ defining the relative orientation of the molecule reference axes x,y,z with respect to the site reference axes X,Y,Z.

Hindered molecular rotations in crystal fields have been the subject of a number of theoretical studies aiming at interpreting high-resolution vibration-rotation or pure rotation spectra for molecules trapped in crystals /2,5,15/. Librational states were generally classified under a symmetry group G_{SM} which was the direct product of the rotational subgroups of the point groups of the molecule and the site /2,5/. Using the group theoretical approach initially proposed for non-rigid molecules by Longuet-Higgins /16/, Miller and Decius /15/ have determined the complete groups for molecules of arbitrary symmetry embedded in sites of any symmetry. Similar approaches have recently been developed in the investigation of the orientational distributions and motions of molecular crystals as studied by neutron or x-ray scattering techniques /8-12/. Orientational distribution and correlation functions were generally considered as transforming as the totally symmetric irreducible representation of a symmetry group G_{SM} which was either reduced to the molecular group /8,12/ or taken as the direct product of the molecule and the site point groups /10,11/. Pick and Yvinec /9/ have introduced the condition that, for an improper rotation to be effective, it is necessary that both the molecule and the site possess compatible improper rotations.

In this paper we give an overall description of how the symmetry properties of a composite system (SM) made up of a rigid molecule (M) embedded in a crystal site (S) can yield significant (and sometimes drastic) reductions in the generalized spherical harmonic expansions of its various properties (A). Using the group theoretical approach initially proposed for non-rigid molecules by Altmann /17/, we have determined the isodynamic groups ($\mathcal{I}_{SM}$) relative to molecules of arbitrary symmetry (G_M) centered at sites of any symmetry (G_S). From the action of all possible isodynamic operations on the Wigner D-functions making up the relevant expansions one can derive explicit relations between the expansion coefficients. More precisely, any property of the composite system may be given the following form :

$$A(\Omega) = \sum_{l,m,m'} A_l^{mm'} D_l^{mm'}(\Omega) , \qquad (1)$$

where for real properties the coefficients satisfy the conjugation relations :

$$A_l^{mm'*} = (-1)^{m+m'} A_l^{-m-m'} .$$

A symmetry-adapted expansion for a property transforming as an irreducible representation Γ_{SM} of the isodynamic group $\mathcal{I}_{SM}$ can be derived by applying to Eq. (1) the projection operator for the composite system :

$$\mathcal{O}^{\Gamma_{SM}} = \frac{1}{h_{SM}} \sum_{i=1}^{h_{SM}} \chi_i^{\Gamma_{SM}} \mathcal{O}_i , \qquad (2)$$

where h_{SM} is the dimension of the isodynamic group $\mathcal{I}_{SM}$ and $\chi_i^{\Gamma_{SM}}$ is the character corresponding to the isodynamic operator $\mathcal{O}_i$ in the irreducible representation Γ_{SM} /17,18/.

In the next section we will investigate the operator structure of the isodynamic group $\mathcal{I}_{SM}$ in terms of the symmetry operations of the point groups G_M and G_S. In Section 3 we will consider the effects of the isodynamic operations of $\mathcal{I}_{SM}$ on the Wigner D-functions and derive the associated selection rules. In Section 4 we will show how these selection rules can be used in deriving appropriate expansions of the various properties for the composite system. In the last section we will outline the applications of symmetry-adapted harmonic expansions to scattering and spectroscopic experimental problems.

2. ISODYNAMIC GROUPS OF SITE-MOLECULE SYSTEMS

In this section we shall apply Altmann's formalism in a way similar to that previously followed for molecules with internal rotations /13,20, 21/. Among all symmetry operations Altmann /17/ distinguishes those which reduce to changes of the laboratory or system axes, leaving the electronic Hamiltonian invariant, and the internal motions of the system which leave the nuclear Hamiltonian unaltered. He calls "isodynamic" these latter operations /17a/ because they transform a conformation of the system having broken symmetry into an isoenergetic conformation having the same symmetry. The "isodynamic" group made up by these operations may be written as the semi-direct product of two further groups /17c/ :

$$\mathcal{I}^{\sim} = \mathcal{I}^{\sim}_{r} \wedge \mathcal{I}^{\sim}_{d} \qquad (3)$$

where $\mathcal{I}^{\sim}_{r}$ is the group of internal rearrangements which transfer a set of atoms from one equilibrium position onto another and $\mathcal{I}^{\sim}_{d}$ is the group of restricted distorsions which leave the equilibrium positions of all atoms unchanged.

For a rigid molecule embedded in a crystal site we shall only consider those motions of the molecule with respect to the site that leave the centers of mass coincident. Let us call OXYZ the reference frame of the site and Oxyz that of the molecule (Ox and Oy being degenerate for a linear molecule) and define the relative orientation of M with respect to S by the Euler angles α,β,γ (Figure 1) (γ is undefined when Ox and Oy are degenerate). The 34 possible point groups G_M and 7 possible point groups G_S may involve symmetry centers, axes and planes and, whenever possible, the reference axes of M and those of S will be chosen, in the usual complementary and consistent manner, in relation with symmetry elements of M and of S. If M or S has a symmetry center, this latter coincides with the center of mass and is taken as the origin of axes. If M or S has symmetry axes (involving proper or improper rotations), the z or Z reference axis is chosen along the principal axis (or anyone of several equivalent axes with lower multiplicity), and the x or X reference axis along one of the orthogonal twofold axes (if there are any). If M or S has symmetry planes, the xy or XY reference plane is chosen in the horizontal plane (if there is a single plane it is considered as horizontal), and the xz or XZ reference plane in anyone of several equivalent vertical planes (if there are only two planes they are considered as vertical).

Assume S possesses an n-fold principal axis and M an n'-fold principal axis, these two axes making an angle β (Figure 1). If $E(\alpha,\beta,\gamma)$ is the total energy of the composite system, an isodynamic rotation of M around C_n is defined by :

$$\mathfrak{C}^{q}_{n}\, E(\alpha,\beta,\gamma) = E(\alpha + q\,\frac{2\pi}{n},\ \beta,\ \gamma) \quad , \qquad (4)$$

and an isodynamic rotation of M around $C'_{n'}$, by :

$$\mathfrak{C}'^{q'}_{n'}\, E(\alpha,\beta,\gamma) = E(\alpha,\ \beta,\ \gamma + q'\,\frac{2\pi}{n'}) \quad . \qquad (5)$$

The first operation generates the isodynamic cyclic group $\mathfrak{C}^{\sim}_{n}$ and the second operation the isodynamic cyclic group $\mathfrak{C}'^{\sim}_{n'}$. If S and M also possess twofold axes orthogonal to their principal axis, one can define the additional isodynamic operations :

$$^{o}\mathfrak{C}_{2}\, E(\alpha,\beta,\gamma) = E(-\alpha,\ \pi-\beta,\ \gamma+\pi) \quad , \qquad (6)$$

$$ {}^{o}\mathfrak{C}_2' \, E(\alpha,\beta,\gamma) = E(\alpha+\pi,\ \pi-\beta,\ -\gamma) \quad . \tag{7}$$

These two sets of idempotent operations generate the respective isodynamic groups ${}^{o}\mathfrak{C}_2^{\sim}$ and ${}^{o}\mathfrak{C}_2'^{\sim}$. The T and O cubic groups involve also diagonal threefold axes, which induce the following isodynamic operations :

$$ {}^{a}\mathfrak{C}_3 \, E(\alpha,\beta,\gamma) = E(\alpha_S,\ \beta_S,\ \gamma_S) \quad /XYZ \rightarrow ZXY \quad , \tag{8}$$

$$ {}^{a}\mathfrak{C}_3' \, E(\alpha,\beta,\gamma) = E(\alpha_M,\ \beta_M,\ \gamma_M) \quad /xyz \rightarrow zxy \quad . \tag{9}$$

These operations generate the isodynamic cyclic groups ${}^{a}\mathfrak{C}_3^{\sim}$ and ${}^{a}\mathfrak{C}_3'^{\sim}$. As a consequence of Altmann's representation of symmetry point groups as semi-direct products /23/, the full isodynamic groups generated by all proper rotations in G_S and G_M can then be written as :

$$ \mathfrak{R}_S^{\sim} = \mathfrak{C}_n^{\sim} \quad (n = 1,2,3,4,6), \quad \mathfrak{D}_n^{\sim} = \mathfrak{C}_n^{\sim} \wedge {}^{o}\mathfrak{C}_2^{\sim} \ , $$
$$ \mathfrak{T}^{\sim} = \mathfrak{D}_2^{\sim} \wedge {}^{a}\mathfrak{C}_3^{\sim} \ , \quad \mathfrak{O}^{\sim} = \mathfrak{D}_2^{\sim} \wedge {}^{a}\mathfrak{D}_3^{\sim} \quad ; \tag{10}$$

$$ \mathfrak{R}_M^{\sim} = \mathfrak{C}_{n'}'^{\sim} \quad (n' = 1,2,3,4,6), \quad \mathfrak{D}_{n'}'^{\sim} = \mathfrak{C}_{n'}'^{\sim} \wedge {}^{o}\mathfrak{C}_2'^{\sim} , $$
$$ \mathfrak{T}'^{\sim} = \mathfrak{D}_2'^{\sim} \wedge {}^{a}\mathfrak{C}_3'^{\sim} , \quad \mathfrak{O}'^{\sim} = \mathfrak{D}_2'^{\sim} \wedge {}^{a}\mathfrak{D}_3'^{\sim} . \tag{11}$$

The direct product of the site and the molecule proper-rotation subgroups generates the isodynamic group of internal rearrangements :

$$ \mathfrak{I}_r^{\sim} = \mathfrak{R}_S^{\sim} \times \mathfrak{R}_M^{\sim} \ . \tag{12}$$

Whereas isodynamic internal rotations can be defined for the composite system whenever anyone of the system components possesses proper-rotation symmetries, isodynamic operations resulting from rotation-reflections, reflections and inversions can be defined only if both molecule and site posses improper-rotation symmetries. Now, according to Altmann /23/, all improper point groups can be written as a direct or semi-direct product of a proper point group with either C_i or an appropriate C_s, with the single exception of S_4 (Table 1). All isodynamic operations will thus be some product of an isodynamic operation of the first kind (Eqs 4-12) with an isodynamic operation of a second kind which we shall now consider.

TABLE I

Improper point groups as direct or semi-direct products of proper point groups with C_i or C_s, respectively. The type of plane involved in C_s, C'_s and C''_s is σ_h, σ_v and σ_d, respectively.

$S_4 = \{E, S_4^1, C_2, S_4^3\}$			
$S_{2n} = C_{ni}$	$= C_n \times C_i$	$(n = 1, 3)$	$S_2 = C_i$
C_{nh}	$= C_n \times C_i$	$(n = 2, 4, 6)$	
	$= C_n \wedge C_s$	$(n = 1, 3 ;\ C_n \perp \sigma_h)$	$C_{1h} = C_s$
D_{nh}	$= D_n \times C_i$	$(n = 2, 4, 6)$	
	$= D_n \wedge C''_s$	$(n = 3 ;\ C_3 // \sigma_d)$	
T_h	$= T \times C_i$		
O_h	$= O \times C_i$		
C_{nv}	$= C_n \wedge C'_s$	$(n = 2, 3, 4, 6 ;\ C_n // \sigma_v)$	
D_{nd}	$= D_n \wedge C''_s$	$(n = 2)$	
	$= D_n \times C_i$	$(n = 3)$	
T_d	$= T \wedge C''_s$		
$C_{\infty v}$, $D_{\infty h}$			

To define the isodynamic operations resulting from appropriate combinations of reflection symmetries in the molecule and the site, one may start from a conformation of the system in which symmetry planes coincide and vary the conformational angles so as to obtain two distorted conformations superimposable to each other through a bodily rotation or a change of axes of the whole system /17a,21a/. We shall only consider those isodynamic operations induced by a $\sigma_{v/h}$ plane in S combined with a $\sigma_{v/h}$ plane in M, the σ_d planes occurring in the point groups D_{nd} and T_d inducing isodynamic operations similar to those induced by σ_v planes after a shift of α by $2\pi/n$ and/or of γ by $2\pi/n'$.

Starting from a conformation in which the reference σ_v and σ'_v planes coincide (i.e., XZ // xz // Xx // Zz : $\alpha = \gamma = 0$, $\beta \neq 0$), it can be seen (Figure 1) that the two distorted conformations (α,β,γ) and $(-\alpha,\beta,-\gamma)$ are connected by the isodynamic operation :

$$\mathfrak{V}_v \; E(\alpha,\beta,\gamma) = E(-\alpha, \beta, -\gamma) = E(\pi-\alpha, -\beta, \pi-\gamma) \; . \qquad (13)$$

Starting from a conformation in which the σ_h and σ'_h planes coincide (i.e., XY // xy // Xx // Yy : $\beta = 0$, $\alpha + \gamma \neq 0$), it can similarly be seen (Figure 1) that the two distorted conformations (α,β,γ) and $(\alpha,-\beta,\gamma)$ are connected by the isodynamic operation :

$$\mathfrak{H}_h \; E(\alpha,\beta,\gamma) = E(\alpha+\pi, \beta, \gamma+\pi) = E(\alpha, -\beta, \gamma) \; . \qquad (14)$$

One can similarly define, with a σ_v and a σ'_h planes, the isodynamic operation :

$$\mathfrak{V}_h \; E(\alpha,\beta,\gamma) = E(\pi-\alpha, \pi-\beta, \gamma) \; , \qquad (15)$$

and, with a σ_h and a σ'_v planes, the isodynamic operation :

$$\mathfrak{H}_v \; E(\alpha,\beta,\gamma) = E(\alpha, \pi-\beta, \pi-\gamma) \; . \qquad (16)$$

These four idempotent operations generate the respective isodynamic groups $\tilde{\mathfrak{V}}_v$, $\tilde{\mathfrak{H}}_h$, $\tilde{\mathfrak{V}}_h$, $\tilde{\mathfrak{H}}_v$.

Two properties of the inversion will be of interest in building up additional operations. First, since we assumed that the centers of inversion of the molecule and the site coincide, an inversion in M and an inversion in S will have identical effects. Second, the inversion in either subsystem is an idempotent symmetry operation. A first consequence of these two properties is that a combination of inversions in M and in S will not produce any effective isodynamic operation. We are thus left with combinations of the inversion I with reflections through either σ_v or σ_h planes in either M (primed) or S (unprimed). It can be shown, through simple trigonometric considerations, that these combinations produce the following isodynamic operations :

$$\mathfrak{I}_v \; E(\alpha,\beta,\gamma) = E(\pi-\alpha, \pi-\beta, \gamma+\pi) \; , \qquad (17)$$

$$\mathfrak{I}_h \; E(\alpha,\beta,\gamma) = E(\alpha+\pi, \beta, \gamma) \; , \qquad (18)$$

$$\mathcal{I}'_{v}\ E(\alpha,\beta,\gamma) = E(\alpha+\pi,\ \pi-\beta,\ \pi-\gamma)\ , \tag{19}$$

$$\mathcal{I}'_{h}\ E(\alpha,\beta,\gamma) = E(\alpha,\ \beta,\ \gamma+\pi)\,. \tag{20}$$

These four idempotent operations generate the respective isodynamic groups $\mathcal{I}_{v}^{\sim}$, $\mathcal{I}_{h}^{\sim}$, $\mathcal{I}_{v}'^{\sim}$, $\mathcal{I}_{h}'^{\sim}$.

The case of molecules having symmetry S_4, which defines a purely improper rotation group (Table 1), desserves special attention. Combination with site inversion can be written :

$$\{E,\ S_4^1,\ C_2,\ S_4^3\}\ \mathsf{X}\ \{E,\ I\} = \{\mathfrak{E},\ \mathfrak{C}_4^3,\ \mathfrak{C}_2,\ \mathfrak{C}_4^1\} = \mathfrak{C}_4^{\sim}\ , \tag{21}$$

where the sign X means a direct product followed by selection of the "feasible" (proper rotation) operations. Similarly, combination with site reflection can be written :

$$\{E,\ S_4^1,\ C_2,\ S_4^3\}\ \mathsf{X}\ \{E,\sigma_v\} = \{\mathfrak{E},\ \mathcal{D}_q^1,\ \mathfrak{C}_2,\ \mathcal{D}_q^3\} = \mathcal{D}_q^{\sim}\ , \tag{22}$$

where $\mathcal{D}_q^n = \mathfrak{C}_4^n\,\mathcal{D}_h^n$, $\mathcal{D}_h$ being the idempotent internal-deflection operation defined in Eq. (15).

To every plane σ_h one may associate a (real or fictive) twofold axis $\underline{C}_2$ such that $I\sigma_h = \underline{C}_2$, and to every plane $\sigma_{v/d}$ one may similarly associate a twofold axis $^{\sigma}\underline{C}_2$. Using the properties of the inversion and our convention of axes one may then write the above-defined isodynamic operations in the following alternative forms :

$$\mathcal{I}_{v} = I \times \sigma_v = \underline{C}_2^{Y}\ , \tag{17'}$$

$$\mathcal{I}_{h} = I \times \sigma_h = \underline{C}_2^{Z}\ , \tag{18'}$$

$$\mathcal{I}'_{v} = I \times \sigma'_v = \underline{C}_2^{y}\ , \tag{19'}$$

$$\mathcal{I}'_{h} = I \times \sigma'_h = \underline{C}_2^{z}\ , \tag{20'}$$

$$\mathcal{D}_{v} = \sigma_v \times \sigma'_v = I\sigma_v \times I\sigma'_v = \underline{C}_2^{Y} \times \underline{C}_2^{y}\ , \tag{13'}$$

$$\mathfrak{H}_{h} = \sigma_h \times \sigma'_h = I\sigma_h \times I\sigma'_h = \underline{C}_2^{Z} \times \underline{C}_2^{z}\ , \tag{14'}$$

$$\mathcal{V}_h = \sigma_v \times \sigma'_h = I\sigma_v \times I\sigma'_h = \underline{C}_2^Y \times \underline{C}_2^z , \tag{15'}$$

$$\mathcal{H}_v = \sigma_h \times \sigma'_v = I\sigma_h \times I\sigma'_v = \underline{C}_2^Z \times \underline{C}_2^y . \tag{16'}$$

In addition to the proper-rotation symmetry groups $R_M = C'_n$, D'_n, T' and O' (from which the isodynamic groups given in Eqs (11) directly derive), a rigid molecule may belong to one of the improper-rotation symmetry groups expressed in Table 1. Altogether there are 11 point groups of type R_M (a), 11 of type $R_M \times C_i$ (b), 9 of type $R_M \wedge C_s$ (c) and 3 unclassified (S'_4, $C'_{\infty v}$ and $D'_{\infty h}$) (d). For a lattice site, only 7 out of these 34 point groups are possible : 6 of type $R_S \times C_i$ (S_2, C_{2h}, D_{nh} with n = 2, 4 or 6 and O_h), and 1 of type $R_S \wedge C_s$ (C_{3v}). While the isodynamic group of internal rearrangements introduced in Eq. (3) is given by Eq. (12), the isodynamic group of restricted distorsions $\mathcal{I}_n^{\sim}$ is generated by the above-considered factors of R_M and R_S in the molecule and site point groups. First, as we already know, an improper-rotation symmetry operation in only M or S cannot yield any "feasible" (proper rotation) isodynamic operation, and the product $C'_i \times C_i$ yields the identity $\mathcal{E}$. Second, a product of type $C'_i \times C_s$ yields $\mathcal{I}_v$ or $\mathcal{I}_h$ depending on the plane to which C_s refers and, similarly, $C'_s \times C_i$ yields $\mathcal{I}'_v$ or $\mathcal{I}'_h$, whereas a product of type $C'_s \times C_s$ yields one of the isodynamic operations defined in Eqs (13-16). Finally, the products $S'_4 \times C_{i/s}$ yield the isodynamic operations defined in Eqs (21,22). We have gathered in Table 2 the structures of the complete isodynamic groups $\mathcal{I}^{\sim}$ for all possible combinations of molecule and site point groups. The isodynamic subgroups $\mathcal{H}_h$, $\mathcal{H}_v$ and $\mathcal{I}_h$ do not appear in this table because there is no site group of type C_{1h} or C_{3h} (Table 1).

3. WIGNER-HARMONIC TRANSFORMATIONS AND SELECTION RULES

Let us first consider an isodynamic rotation of M around the symmetry axis Z of S ; its effect on a Wigner D-function is /24/ :

$$\mathcal{C}_n^q \, D_l^{mm'}(\Omega) = D_l^{mm'}(\alpha + q\,\frac{2\pi}{n}, \beta, \gamma) = (-1)^{2mq/n} \, D_l^{mm'}(\Omega) . \tag{23}$$

More generally, any rotation of M around any axis of S making angles (η,ζ) with the site-fixed axes XYZ transforms the Wigner harmonics as follows /13/ :

$$\mathcal{C}_n^q(\eta,\zeta) \, D_l^{mm'}(\Omega) = \sum_{r,r'} D_l^{mr}(\pi-\zeta, \eta, \pi + q\,\frac{2\pi}{n}) \, D_l^{rr'}(0, \eta, \zeta) \times \times D_l^{r'm'}(\Omega) . \tag{24}$$

TABLE II

Structure of the isodynamic group of a site-molecule system for a site and a molecule of any symmetry point group. In D_{nh}, n = 2, 4 or 6. The case of infinite n' is formally included in $C'_{n'v}$ and $D'_{n'h}$.

	M \ S	C_i (b)	C_{3v} (c)	C_{2h} (b)	D_{nh} (b)	O_h (b)
(a)	$C'_{n'}$	$(\mathfrak{C}'_{n'} \times \mathfrak{C}_1)^{\sim}$	$(\mathfrak{C}'_{n'} \times \mathfrak{C}_3)^{\sim}$	$(\mathfrak{C}'_{n'} \times \mathfrak{C}_2)^{\sim}$	$(\mathfrak{C}'_{n'} \times \mathfrak{D}_n)^{\sim}$	$(\mathfrak{C}'_{n'} \times \mathfrak{O})^{\sim}$
(a)	$D'_{n'}$	$(\mathfrak{D}'_{n'} \times \mathfrak{C}_1)^{\sim}$	$(\mathfrak{D}'_{n'} \times \mathfrak{C}_3)^{\sim}$	$(\mathfrak{D}'_{n'} \times \mathfrak{C}_2)^{\sim}$	$(\mathfrak{D}'_{n'} \times \mathfrak{D}_n)^{\sim}$	$(\mathfrak{D}'_{n'} \times \mathfrak{O})^{\sim}$
(a)	T'	$(\mathfrak{T}' \times \mathfrak{C}_1)^{\sim}$	$(\mathfrak{T}' \times \mathfrak{C}_3)^{\sim}$	$(\mathfrak{T}' \times \mathfrak{C}_2)^{\sim}$	$(\mathfrak{T}' \times \mathfrak{D}_n)^{\sim}$	$(\mathfrak{T}' \times \mathfrak{O})^{\sim}$
(a)	O'	$(\mathfrak{O}' \times \mathfrak{C}_1)^{\sim}$	$(\mathfrak{O}' \times \mathfrak{C}_3)^{\sim}$	$(\mathfrak{O}' \times \mathfrak{C}_2)^{\sim}$	$(\mathfrak{O}' \times \mathfrak{D}_n)^{\sim}$	$(\mathfrak{O}' \times \mathfrak{O})^{\sim}$
(d)	S'_4	$(\mathfrak{C}'_4 \times \mathfrak{C}_1)^{\sim}$	$(\mathfrak{C}'_1 \times \mathfrak{C}_3)^{\sim} \wedge \mathfrak{V}^{\sim}_q$	$(\mathfrak{C}'_4 \times \mathfrak{C}_2)^{\sim}$	$(\mathfrak{C}'_4 \times \mathfrak{D}_n)^{\sim}$	$(\mathfrak{C}'_4 \times \mathfrak{O})^{\sim}$
(b)	$S'^{\,odd}_{2n'}$	$(\mathfrak{C}'_{n'} \times \mathfrak{C}_1)^{\sim}$	$(\mathfrak{C}'_{n'} \times \mathfrak{C}_3)^{\sim} \wedge \mathfrak{I}^{\sim}_v$	$(\mathfrak{C}'_{n'} \times \mathfrak{C}_2)^{\sim}$	$(\mathfrak{C}'_{n'} \times \mathfrak{D}_n)^{\sim}$	$(\mathfrak{C}'_{n'} \times \mathfrak{O})^{\sim}$
(b)	$C'^{\,even}_{n'h}$	$(\mathfrak{C}'_{n'} \times \mathfrak{C}_1)^{\sim}$	$(\mathfrak{C}'_{n'} \times \mathfrak{C}_3)^{\sim} \wedge \mathfrak{I}^{\sim}_v$	$(\mathfrak{C}'_{n'} \times \mathfrak{C}_2)^{\sim}$	$(\mathfrak{C}'_{n'} \times \mathfrak{D}_n)^{\sim}$	$(\mathfrak{C}'_{n'} \times O)^{\sim}$
(c)	$C'^{\,odd}_{n'h}$	$(\mathfrak{C}'_{n'} \times \mathfrak{C}_1)^{\sim} \wedge \mathfrak{I}'^{\sim}_h$	$(\mathfrak{C}'_{n'} \times \mathfrak{C}_3)^{\sim} \wedge \mathfrak{V}^{\sim}_h$	$(\mathfrak{C}'_{n'} \times \mathfrak{C}_2)^{\sim} \wedge \mathfrak{I}'^{\sim}_h$	$(\mathfrak{C}'_{n'} \times \mathfrak{D}_n)^{\sim} \wedge \mathfrak{I}'^{\sim}_h$	$(\mathfrak{C}'_{n'} \times \mathfrak{O})^{\sim} \wedge \mathfrak{I}'^{\sim}_h$
(c)	$C'_{n'v}$	$(\mathfrak{C}'_{n'} \times \mathfrak{C}_1)^{\sim} \wedge \mathfrak{I}'^{\sim}_v$	$(\mathfrak{C}'_{n'} \times \mathfrak{C}_3)^{\sim} \wedge \mathfrak{V}^{\sim}_v$	$(\mathfrak{C}'_{n'} \times \mathfrak{C}_2)^{\sim} \wedge \mathfrak{I}'^{\sim}_v$	$(\mathfrak{C}'_{n'} \times \mathfrak{D}_n)^{\sim} \wedge \mathfrak{I}'^{\sim}_v$	$(\mathfrak{C}'_{n'} \times \mathfrak{O})^{\sim} \wedge \mathfrak{I}'^{\sim}_v$
(c)	D'_{2d}	$(\mathfrak{D}'_2 \times \mathfrak{C}_1)^{\sim} \wedge \mathfrak{I}'^{\sim}_v$	$(\mathfrak{D}'_2 \times \mathfrak{C}_3)^{\sim} \wedge \mathfrak{V}^{\sim}_v$	$(\mathfrak{D}'_2 \times \mathfrak{C}_2)^{\sim} \wedge \mathfrak{I}'^{\sim}_v$	$(\mathfrak{D}'_2 \times \mathfrak{D}_n)^{\sim} \wedge \mathfrak{I}'^{\sim}_v$	$(\mathfrak{D}'_2 \times \mathfrak{O})^{\sim} \wedge \mathfrak{I}'^{\sim}_v$
(b)	D'_{3d}	$(\mathfrak{D}'_3 \times \mathfrak{C}_1)^{\sim}$	$(\mathfrak{D}'_3 \times \mathfrak{C}_3)^{\sim} \wedge \mathfrak{I}^{\sim}_v$	$(\mathfrak{D}'_3 \times \mathfrak{C}_2)^{\sim}$	$(\mathfrak{D}'_3 \times \mathfrak{D}_n)^{\sim}$	$(\mathfrak{D}'_3 \times \mathfrak{O})^{\sim}$
(c)	T'_d	$(\mathfrak{T}' \times \mathfrak{C}_1)^{\sim} \wedge \mathfrak{I}'^{\sim}_v$	$(\mathfrak{T}' \times \mathfrak{C}_3)^{\sim} \wedge \mathfrak{V}^{\sim}_v$	$(\mathfrak{T}' \times \mathfrak{C}_2)^{\sim} \wedge \mathfrak{I}'^{\sim}_v$	$(\mathfrak{T}' \times \mathfrak{D}_n)^{\sim} \wedge \mathfrak{I}'^{\sim}_v$	$(\mathfrak{T}' \times \mathfrak{O})^{\sim} \wedge \mathfrak{I}'^{\sim}_v$
(b)	$D'^{\,even}_{n'h}$	$(\mathfrak{D}'_n \times \mathfrak{C}_1)^{\sim}$	$(\mathfrak{D}'_n \times \mathfrak{C}_3)^{\sim} \wedge \mathfrak{I}^{\sim}_v$	$(\mathfrak{D}'_n \times \mathfrak{C}_2)^{\sim}$	$(\mathfrak{D}'_n \times \mathfrak{D}_n)^{\sim}$	$(\mathfrak{D}'_n \times \mathfrak{O})^{\sim}$
(c)	D'_{3h}	$(\mathfrak{D}'_3 \times \mathfrak{C}_1)^{\sim} \wedge \mathfrak{I}'^{\sim}_v$	$(\mathfrak{D}'_3 \times \mathfrak{C}_3)^{\sim} \wedge \mathfrak{V}^{\sim}_v$	$(\mathfrak{D}'_3 \times \mathfrak{C}_2)^{\sim} \wedge \mathfrak{I}'^{\sim}_v$	$(\mathfrak{D}'_3 \times \mathfrak{D}_n)^{\sim} \wedge \mathfrak{I}'^{\sim}_v$	$(\mathfrak{D}'_3 \times \mathfrak{O})^{\sim} \wedge \mathfrak{I}'^{\sim}_v$
(b)	T'_h	$(\mathfrak{T}' \times \mathfrak{C}_1)^{\sim}$	$(\mathfrak{T}' \times \mathfrak{C}_3)^{\sim} \wedge \mathfrak{I}^{\sim}_v$	$(\mathfrak{T}' \times \mathfrak{C}_2)^{\sim}$	$(\mathfrak{T}' \times \mathfrak{D}_n)^{\sim}$	$(\mathfrak{T}' \times \mathfrak{O})^{\sim}$
(b)	O'_h	$(\mathfrak{O}' \times \mathfrak{C}_1)^{\sim}$	$(\mathfrak{O}' \times \mathfrak{C}_3)^{\sim} \wedge \mathfrak{I}^{\sim}_v$	$(\mathfrak{O}' \times \mathfrak{C}_2)^{\sim}$	$(\mathfrak{O}' \times \mathfrak{D}_n)^{\sim}$	$(\mathfrak{O}' \times \mathfrak{O})^{\sim}$

In particular, for a tworold rotation of M around a reference axis of S one has the following relations :

$$\underline{C}_2^X D_l^{mm'}(\Omega) = D_l^{mm'}(-\alpha, \pi-\beta, \gamma+\pi) = (-1)^l D_l^{-mm'}(\Omega) , \quad (25)$$

$$\underline{C}_2^Y D_l^{mm'}(\Omega) = D_l^{mm'}(\pi-\alpha, \pi-\beta, \gamma+\pi) = (-1)^{l+m} D_l^{-mm'}(\Omega) , \quad (26)$$

$$\underline{C}_2^Z D_l^{mm'}(\Omega) = D_l^{mm'}(\alpha+\pi, \beta, \gamma) = (-1)^m D_l^{mm'}(\Omega) . \quad (27)$$

Let us now consider an isodynamic rotation of M around its own symmetry axis z ; its effect on a Wigner D-function is :

$$\mathfrak{C}'^{q'}_{n'} D_l^{mm'}(\Omega) = D_l^{mm'}(\alpha, \beta, \gamma + q' \frac{2\pi}{n'}) = (-1)^{2m'q'/n'} D_l^{mm'}(\Omega) . \quad (28)$$

More generally, any rotation of M around any proper axis making angles (η',ζ') with the molecule-fixed axes xyz transforms the Wigner harmonics as follows :

$$\mathfrak{C}'^{q'}_{n'}(\eta',\zeta') D_l^{mm'}(\Omega) = \sum_{r,r'} D_l^{mr}(\Omega) D_l^{rr'}(-\zeta', \eta', \pi + q' \frac{2\pi}{n'}) \times$$
$$\times D_l^{r'm'}(0, \eta', \zeta'+\pi) . \quad (29)$$

In particular, for a twofold rotation of M around one of its own reference axes one has the following relations :

$$\underline{C}_2^x D_l^{mm'}(\Omega) = D_l^{mm'}(\alpha+\pi, \pi-\beta, -\gamma) = (-1)^l D_l^{m-m'}(\Omega) , \quad (30)$$

$$\underline{C}_2^y D_l^{mm'}(\Omega) = D_l^{mm'}(\alpha+\pi, \pi-\beta, \pi-\gamma) = (-1)^{l-m'} D_l^{m-m'}(\Omega) , \quad (31)$$

$$\underline{C}_2^z D_l^{mm'}(\Omega) = D_l^{mm'}(\alpha, \beta, \gamma+\pi) = (-1)^{m'} D_l^{mm'}(\Omega) . \quad (32)$$

According to Eqs (17'-20') the isodynamic operators $\mathcal{I}_v$, $\mathcal{I}_h$, $\mathcal{I}'_v$, $\mathcal{I}'_h$ transform the Wigner harmonics as in Eqs (26), (27), (31), (32), respectively. The effects of the isodynamic operators $\mathfrak{D}_v$, $\mathfrak{H}_h$, $\mathfrak{D}_h$, $\mathfrak{H}_v$ can be directly derived from Eqs (13-16) :

$$\mathfrak{D}_v D_l^{mm'}(\Omega) = D_l^{mm'}(-\alpha, \beta, -\gamma) = (-1)^{m-m'} D_l^{-m-m'}(\Omega) , \quad (33)$$

$$\mathfrak{H}_h D_l^{mm'}(\Omega) = D_l^{mm'}(\alpha+\pi,\ \beta,\ \gamma+\pi) = (-1)^{m+m'} D_l^{mm'}(\Omega) \ , \qquad (34)$$

$$\mathfrak{D}_h D_l^{mm'}(\Omega) = D_l^{mm'}(\pi-\alpha,\ \pi-\beta,\ \gamma) = (-1)^{l+m+m'} D_l^{-mm'}(\Omega) \ , \qquad (35)$$

$$\mathfrak{H}_v D_l^{mm'}(\Omega) = D_l^{mm'}(\alpha,\ \pi-\beta,\ \pi-\gamma) = (-1)^{l+m-m'} D_l^{m-m'}(\Omega) \ . \qquad (36)$$

Finally the effect of the isodynamic operator $\mathfrak{D}_q$ introduced in Eq. (22) can be written :

$$\mathfrak{D}_q D_l^{mm'}(\Omega) = D_l^{mm'}(\pi-\alpha,\ \pi-\beta,\ \gamma+\pi/2) = (-1)^{l+m+3m'/2} D_l^{-mm'}(\Omega) \ . \qquad (37)$$

The symmetry-adapted expansion of a molecule-site property will be obtained by applying to its Wigner harmonic expansion, given by Eq. (1), the corresponding projection operator, given by Eq. (2), that is :

$$A^{\Gamma}(\Omega) = \mathfrak{O}^{\Gamma} A(\Omega) \ , \qquad (38)$$

where the index SM has been dropped for conveniency. This expansion involves less independent non-zero coefficients, to a given order, than the starting expansion, due to a set of selection rules expressing the fact that some of the coefficients annihilate while some of the remaining ones are connected by various relations. These selection rules are obtained by stating for each operator $\mathfrak{O}_i$ in $\mathfrak{O}^{\Gamma}$ that :

$$\mathfrak{O}_i A(\Omega) = A(\Omega) \ , \qquad (39)$$

and then using Eqs (23-37) to compare the coefficients of the corresponding D-fuctions on both sides of Eq. (39).

The selection rules concerning the coefficients of the Wigner D-functions in the symmetry-adapted expansions are gathered in Table 3. Most of the isodynamic operations introduced in Section 2 yield relations of dependence rather than annihilation of coefficients. Other selection rules can be obtained by combining the relations given in the Table. Note that when cubic groups are involved some relations take the form of linear relations between the coefficients : it may then be more appropriate to use, in the harmonic expansions, symmetry-adapted combinations of spherical harmonics, called tetrahedric or cubic harmonics /2,14,25/.

TABLE III

Relations between the coefficients of the harmonic expansions induced by the various isodynamic operations. In the relations occurring for the cubic groups $\cos^2\eta_a = 1/3$.

$$\tilde{\mathfrak{C}}_n : \quad A_l^{mm'} = (-1)^{2m/n} A_l^{mm'} = 0 \text{ unless } m = kn$$

$$\tilde{\mathfrak{C}}'_{n'} : \quad A_l^{mm'} = (-1)^{2m'/n'} A_l^{mm'} = 0 \text{ unless } m' = k'n' \quad (m' = 0 \text{ if } n' = \infty)$$

$${}^{o}\tilde{\mathfrak{C}}_2 : \quad A_l^{mm'} = (-1)^{l} A_l^{-m\,m'}$$

$${}^{o}\tilde{\mathfrak{C}}'_2 : \quad A_l^{mm'} = (-1)^{l} A_l^{m\,-m'}$$

$${}^{a}\tilde{\mathfrak{C}}_3 : \quad A_l^{mm'} = \sum_{r,r'} (-1)^{(m+3r')/4-r/3}\, d_l^{mr}(\eta_a)\, d_l^{rr'}(\eta_a)\, A_l^{r'm'}$$

$${}^{a}\tilde{\mathfrak{C}}'_3 : \quad A_l^{mm'} = \sum_{r,r'} (-1)^{(m'+3r)/4-r'/3}\, d_l^{r'm'}(\eta_a)\, d_l^{rr'}(\eta_a)\, A_l^{mr}$$

$$\tilde{\mathfrak{D}}_v : \quad A_l^{mm'} = (-1)^{m-m'} A_l^{-m\,-m'}$$

$$\tilde{\mathfrak{H}}_h : \quad A_l^{mm'} = (-1)^{m+m'} A_l^{mm'}$$

$$\tilde{\mathfrak{D}}_h : \quad A_l^{mm'} = (-1)^{l+m+m'} A_l^{-m\,m'}$$

$$\tilde{\mathfrak{H}}_v : \quad A_l^{mm'} = (-1)^{l+m+m'} A_l^{m\,-m'}$$

$$\tilde{\mathfrak{I}}_v : \quad A_l^{mm'} = (-1)^{l+m} A_l^{-m\,m'}$$

$$\tilde{\mathfrak{I}}_h : \quad A_l^{mm'} = (-1)^{m} A_l^{mm'}$$

$$\tilde{\mathfrak{I}}'_v : \quad A_l^{mm'} = (-1)^{l-m'} A_l^{m\,-m'}$$

$$\tilde{\mathfrak{I}}'_h : \quad A_l^{mm'} = (-1)^{m'} A_l^{mm'}$$

---

4. SYMMETRY-ADAPTED EXPANSIONS OF SITE-MOLECULE PROPERTIES

Our basic interest is in the conformational dependence of the various properties of a molecule in a site. These properties can be classified in different ways. On one side, they may be either scalar (i.e., invariant over all overall rotations), or polar (i.e., transforming as a translation vector X, Y, Z), or axial (i.e., transforming as a rotation vector R_X, R_Y, R_Z), or, more generally, tensorial (e.g., representable by a symmetric second-rank tensor). On the other side, the properties may be either of aggregate type (i.e., characterizing the molecule in the site as a whole), or of mononuclear type (i.e., defining the interaction of a particular nucleus with the perturbed-molecule electronic distribution), or of binuclear type (i.e., defining the interaction between two particular nuclei through the perturbed-molecule electronic distribution). The basic example of a scalar aggregate property is the site-molecule interaction energy, V, a significant part of which may depend on the conformation of the system. Examples of vectorial aggregate properties are the site-molecule forces $\bar{F}$ and torques $\breve{G}$ and perturbed-molecule electric dipole moment $\bar{\mu}$. Examples of tensorial aggregate properties are the electric quadrupole moment $\bar{\bar{\Theta}}$, electric polarizability tensor $\bar{\bar{\alpha}}$, spectroscopic splitting tensor $\bar{\bar{g}}$ if the molecule possesses a total electronic spin $S \geqq 1/2$, and spin-spin coupling tensor $\bar{\bar{D}}$ if $S \geqq 1$. Examples of mononuclear properties are the chemical shifts $\bar{\bar{\sigma}}_i$ for nuclei with effective spin $I_i \geqq 1/2$, quadrupole couplings $\bar{\bar{Q}}_i$ for nuclei with $I_i \geqq 1$, and hyperfine couplings $\bar{\bar{c}}_i$ when both S and $I_i \geqq 1/2$. Binuclear properties are illustrated by the magnetic interactions $\bar{\bar{J}}_{ij}$ when both I_i and $I_j \geqq 1/2$.

As for non-rigid molecules /13,21/, the fact that a property is of aggregate, mononuclear or binuclear type determines the effective isodynamic group $\mathcal{I}^{\sim}$ to which the operators $\mathcal{O}_i$ in Eq. (2) belong, and the fact that a property is scalar, vectorial or tensorial determines the irreducible representation Γ to which the characters χ_i^{Γ} belong. For an aggregate property A, the effective isodynamic group $\mathcal{I}_A^{\sim}$ is the full isodynamic group of the site-molecule system as determined in Section 2. For a mononuclear property M_i or S_i (depending on the location of the nucleus, in the molecule or in the site, respectively), one can no longer consider as superimposable two conformations of the system that differ by the interchange of nucleus i with an equivalent nucleus of the same subsystem. As a result, the corresponding isodynamic group $\mathcal{I}_{M_i}^{\sim}$ or $\mathcal{I}_{S_i}^{\sim}$ is the same as that of A but for a system in which the symmetry of M or of S has been reduced by singling out nucleus i. For a binuclear property one must distinguish the cases B_{ij}, C_{ij} and D_{ij}, where i and j are both on M, shared between M and S and both on S, respectively. In all three cases the effective isodynamic group ($\mathcal{I}_{B_{ij}}^{\sim}$, $\mathcal{I}_{C_{ij}}^{\sim}$ or $\mathcal{I}_{D_{ij}}^{\sim}$) is the same as that of A but for a system in which nuclei i and j have been singled out. As an example, if a molecule contributes by T_d symmetry elements to the effective group $\mathcal{I}^{\sim}$ for a property A, it may well contribute only by C_{3v} symmetry elements for a property M_i and by C_{2v} symmetry elements for a property B_{ij}.

Once the isodynamic operators $\mathcal{O}_i$ are known, one can express the projection operator $\mathcal{O}^{\Gamma}$ given by Eq. (2) if one knows the character table of the effective group $\mathcal{I}^{\sim}$. However, the character tables of isodynamic groups are usually rather involved already for non-rigid molecules /26-29/, and only few have been derived for rigid molecules in crystal sites /15/. Nevertheless, as for non-rigid molecules /20,21/, some simple rules can be obtained to derive the characters χ_i^{Γ} for our cases of interest. For a scalar property A, the relevant irreducible representation Γ_A is the totally symmetric representation of $\mathcal{I}_A^{\sim}$, all characters of which are equal to 1. For a vectorial or tensorial property $\bar{A}$, $\breve{A}$ or $\bar{\bar{A}}$, one can introduce symmetry indices similar to those previouly defined for non-rigid molecules /13,21/. In order to do this, one first considers the components of the representative vector or tensor on reference axes of the perturbed molecule related to the isodynamic operations defined in Eqs (13-16) : y for $\mathfrak{V}_v$ and $\mathfrak{H}_v$, z for $\mathfrak{H}_h$ and $\mathfrak{V}_h$, x orthogonal to these both. For all components of all vectors and tensors, the characters χ_i^Γ in Eq. (2) remain equal to 1 for all operators $\mathcal{O}_i$ belonging to the rotational subgroup $\mathcal{I}_r^\sim$ given by Eq. (12). For those operators $\mathcal{O}_i$ belonging to a deflectional subgroup $\mathcal{I}_d^\sim$ defined with either Eq. (13-16), the characters χ_i^Γ in Eq. (2) change their sign if the axis along which the component is measured is related (for polar vectors) or unrelated (for axial vectors) to the corresponding operation. Similar rules can be established for the operators $\mathcal{O}_i$ defined in Eqs (17-22). For tensorial properties the diagonal components transform like scalars and the off-diagonal components like axial vectors.

It may be of interest to evaluate the numbers of independent non-zero coefficients remaining in Eq. (1), up to a certain order, after applying the projection operator given by Eq. (2), for a particular type of property. In Table 4 we have displayed these numbers for all combinations of site and molecule non-cubic point groups, for the harmonic expansions of scalar aggregate properties up to the order $l = 6$. Comparison of these numbers with the isodynamic groups shown in Table 2 helps assess the efficiency of the various operations in establishing the numbers of significant harmonics. In Table 5 we have summarized the rules which help select these harmonics for a few typical site and molecule point groups. As an example, the explicit reduced expansion for the particular combination $C'_{\infty v}$ / D_{6h} can be written :

$$A(\Omega) = A_o^{oo}\, D_o^{oo}(\Omega) + A_2^{oo}\, D_2^{oo}(\Omega) + A_4^{oo}\, D_4^{oo}(\Omega) + A_6^{oo}\, D_6^{oo}(\Omega) + A_6^{6o}\, D_6^{6o}(\Omega) =$$

$$= A_o^{oo} + A_2^{oo}\,(3\cos^2\beta - 1)\,/\,2 + A_4^{oo}\,(35\cos^4\beta - 30\cos^2\beta + 3)\,/\,8 + A_6^{oo}\,(231\cos^6\beta - 315\cos^4\beta + 105\cos^2\beta - 5)\,/\,16 - 231\,A_6^{6o}\cos 6\alpha\,(\cos^6\beta - 3\cos^4\beta + 3\cos^2\beta - 1)\,/\,8\,.$$

5. APPLICATIONS TO SCATTERING AND SPECTROSCOPIC PROBLEMS

The applications of the above symmetry considerations to scattering and spectroscopic problems rest on the possibility to expand the orientational distribution and correlation functions in terms of generalized spherical harmonics. For distribution functions the harmonic expansion is an obvious direct physical assumption, whereas for correlation functions it results from the harmonic expansion of a transition moment.

TABLE IV

Numbers of independent non-zero coefficients up to the order $l = 6$ for scalar aggregate properties.

M \ S	C_i	C_{3v}	C_{2h}	D_{2h}	D_{4h}	D_{6h}
C'_1	455	155	231	119	59	41
C'_2	231	64	109	63	33	23
C'_3	155	55	79	41	21	15
C'_4	115	39	59	31	17	11
C'_6	75	27	39	21	11	9
D'_2	118	40	63	35	19	14
D'_3	81	28	43	24	13	10
D'_4	61	21	33	19	11	8
D'_6	41	15	23	14	8	7
S'_2	455	155	231	119	59	41
S'_4	115	43	59	31	17	11
S'_6	155	55	79	41	21	15
C'_{2v}	231	43	63	35	19	14
C'_{3v}	155	31	43	24	13	10
C'_{4v}	115	23	33	19	11	8
C'_{6v}	75	17	23	14	8	7
$C'_{\infty v}$	49	12	16	10	6	5
D'_{2d}	118	27	63	35	18	14
D'_{3d}	81	10	43	24	13	10
C'_{1h}	455	93	119	63	33	23
C'_{2h}	231	43	119	63	33	23
C'_{3h}	155	28	39	21	11	9
C'_{4h}	115	21	59	31	17	11
C'_{6h}	75	15	39	21	11	9
D'_{2h}	118	24	63	35	17	14
D'_{3h}	81	20	23	14	8	7
D'_{4h}	61	13	33	17	11	8
D'_{6h}	41	10	23	14	8	7
$D'_{\infty h}$	28	7	16	10	6	5

TABLE V

Summary of selection rules for a few typical site and molecule point groups.

S / M	C_{3v}	D_{6h}
C'_{3v}	$m = 3k$ $m' = 3k'$ $A_1^{mm'} = (-1)^{m-m'} A_1^{-m-m'}$	$m = 6k$ $m' = 3k'$ $A_1^{mm'} = (-1)^{m-m'} A_1^{-m-m'}$ $= (-1)^{1} A_1^{-m-m'}$ $1 + m - m' = 2k''$
$C'_{\infty v}$	$m = 3k$ $m' = 0$ $A_1^{m0} = (-1)^{m} A_1^{-m0}$	$m = 6k$ $m' = 0$ $A_1^{m0} = (-1)^{m} A_1^{-m0}$ $= (-1)^{1} A_1^{-m0}$ $1 - m = 2k''$
D'_{3h}	$m = 3k$ $m' = 3k'$ $A_1^{mm'} = (-1)^{m-m'} A_1^{-m-m'}$ $= (-1)^{1} A_1^{-m-m'}$ $1 + m - m' = 2k''$	$m = 6k$ $m' = 3k'$ $A_1^{mm'} = A_1^{m-m'} = A_1^{-mm'} = A_1^{-m-m'}$ 1 and $m - m'$ even integers $1 + m - m' = 2k''$
$D'_{\infty h}$	$m = 3k$ $m' = 0$ $A_1^{m0} = (-1)^{m \text{ or } 1} A_1^{-m0}$ 1 and m even integers	$m = 6k$ $m' = 0$ $A_1^{m0} = A_1^{-m0}$ 1 and m even integers

5.1 Scattering Problems

X-ray, electron and neutron scattering can be used to investigate the orientational disorder of molecules in crystals. The diffuse scattering intensity $I(\bar{q})$, with scattering vector $\bar{q}$, due to the orientational disorder of molecule m, can be written /12/ :

$$I(\bar{q}) = N \sum_{a,b} \{< e^{i\bar{q}.(\bar{r}_{ma} - \bar{r}_{mb})}> - <e^{i\bar{q}.\bar{r}_{ma}}> <e^{i\bar{q}.\bar{r}_{mb}}>\} \times$$
$$\times f_a(\bar{q}) \; f_b(\bar{q}) \;, \qquad (40)$$

where $f_a(\bar{q})$ is the scattering factor of atom a and $\bar{r}_{ma}$, its position vector in molecule m, N being the number of such molecules in the crystal, and the brackets denoting the thermodynamic average :

$$<A> = \int A \; \rho(\Omega) \; d\Omega \;, \qquad (41)$$

where $\rho(\Omega)$ is the single-molecule orientational distribution function. Kobashi and Etters /12/ have shown that Eq. (40) can be expressed using symmetry-adapted distribution functions but considered only molecule symmetry. To consider also site symmetry it seems natural to start from an expansion of $\rho(\Omega)$ in generalized spherical harmonics :

$$\rho(\Omega) = \sum_{l,m,m'} \rho_l^{mm'} \; D_l^{mm'}(\Omega) \;, \qquad (42)$$

with, due to orthonormalization and Eq. (41) :

$$\rho_l^{mm'} = < D_l^{mm'*}(\Omega)> \;.$$

If one uses the spherical wave expansion of $e^{i\bar{q}.\bar{r}}$:

$$e^{i\bar{q}.\bar{r}} = 4\pi \sum_{l,m} i^l \; J_l(\bar{q}.\bar{r}) \; Y_l^{m*}(\omega_r) \; Y_l^{m}(\omega_q) \;, \qquad (43)$$

where $J_l(x)$ is a spherical Bessel function and ω_r is the set of angles defining the position of an atom with respect to the coordinate system attached to the molecule, then the thermal averages in Eq. (40) take the form :

$$< e^{i\bar{q}.\bar{r}}> = 4\pi \sum_{l,m,m'} i^l \; J_l(\bar{q}.\bar{r}) \; Y_l^{m*}(\omega_r) \; Y_l^{m'}(\omega_q) \times$$
$$\times \rho_l^{mm'} \;. \qquad (44)$$

Using Eq. (44) in Eq. (40) yields the expression :

$$I(\bar{q}) = 8\pi N \sum_{a,b} \{ \sum_{l,m,m'} B_l^{*} Y_l^{m*}(\omega_{r_{ab}}) Y_l^{m'}(\omega_q) \rho_l^{mm'} \} \times f_a(\bar{q}) f_b(\bar{q}) , \quad (45)$$

where : (46)

$$B_l^{*} = \sum_{\text{odd } j,k} (-1)^k i^{j+k} \begin{Bmatrix} 1 & j & k \\ o & o & o \end{Bmatrix}^2 (2j+1)(2k+1) J_j(\bar{q}.\bar{r}_a) J_k(\bar{q}.\bar{r}_b)$$

with $\begin{Bmatrix} 1 & j & k \\ o & o & o \end{Bmatrix}$ denoting a 3-j symbol. The intensity coefficients $\rho_l^{mm'}$ in Eq. (45) must satisfy the selection rules imposed by the symmetry properties of the conjugated system. They can be estimated from experimental spectra if detailed information about the intensity distribution of diffuse scattering due to orientational disorder is known.

5.2 Spectroscopic Problems

Infrared and Raman spectra of rigid molecules embedded in crystal sites can yield rotational information through band shapes of internal modes. The band shape of any system interacting with radiation is the exponential Fourier transform of a transition moment correlation function /30/ :

$$I(\omega) = \frac{1}{2\pi} \int_{-\infty}^{+\infty} e^{-i\omega t} G(t)\, dt , \quad (47)$$

where ω is the angular frequency of the incident light and G(t) is the correlation function of the band transition dipole moment :

$$G(t) = < \bar{\mu}(0).\bar{\mu}(t)> . \quad (48)$$

If one expresses $\bar{\mu}(0)$ and $\bar{\mu}(t)$ in terms of Wigner D-functions, the correlation function becomes :

$$G(t) = \sum_{m,m'} \mu_m(0)\mu_{m'}(t) < D_1^{m0}(\theta_0, \phi_0) D_1^{m'0*}(\theta_t, \phi_t)> , \quad (49)$$

where (θ_0, ϕ_0) and (θ_t, ϕ_t) are the orientations at times 0 and t of the dipole vector with respect to spaced-fixed axes. To express the correlation function for reorientations of the molecule principal axes one may use the following identities :

$$D_1^{m0}(\theta_0, \phi_0) = \sum_r D_1^{mr}(\Omega_0)\, D_1^{r0}(\theta, \phi) \ , \tag{50a}$$

$$D_1^{m'0*}(\theta_t, \phi_t) = \sum_{r'} D_1^{m'r'*}(\Omega_t)\, D_1^{r'0*}(\theta, \phi) \ , \tag{50b}$$

this leading to :

$$G(t) = \sum_{m,m'} \sum_{r,r'} \mu_m(0)\mu_{m'}(t)\, D_1^{r0}(\theta, \phi)\, D_1^{r'0*}(\theta, \phi) \times \\ \times < D_1^{mr}(\Omega_0)\, D_1^{m'r'*}(\Omega_t)> \ , \tag{51}$$

where Ω_0 and Ω_t are the orientations at times 0 and t of the molecule principal axes.

At this point it must be recalled that the correlation function depends on the molecular reorientation $d\Omega$ but not on the initial orientation Ω_0. Averaging over a random distribution of initial orientations one obtains :

$$< D_1^{mr}(\Omega_0)\, D_1^{m'r'*}(\Omega_t)> \; = \; (1/3) < D_1^{rr'*}(d\Omega)> \delta_{mm'} . \tag{52}$$

Substituting Eq. (52) into Eq. (51) one obtains :

$$G(t) = (1/3) \sum_m \mu_m(0)\mu_m(t) \sum_{r,r'} D_1^{r0}(\theta, \phi)\, D_1^{r'0*}(\theta, \phi) \times \\ \times < D_1^{rr'*}(d\Omega)> \ , \tag{53}$$

and substitution of this equation into Eq. (47) yields :

$$I(\omega) = \frac{1}{6\pi} \sum_m \mu_m(0)\mu_m(t) \sum_{r,r'} D_1^{r0}(\theta, \phi)\, D_1^{r'0*}(\theta, \phi) \times \\ \times \int_{-\infty}^{+\infty} e^{-i\omega t} < D_1^{rr'*}(d\Omega)> dt \ . \tag{54}$$

From Eqs (53) and (54) one can derive the expressions of the components of the correlation and intensity functions written as follows :

$$G(t) = \sum_{m=-1}^{+1} G_m(t) \ , \quad I(\omega) = \sum_{m=-1}^{+1} I_m(\omega) \ . \tag{55, 56}$$

It may be noted that $\mu_m(t)$ can be derived from $\mu_m(0)$ if one knows the process taking place. Our point here is that the intensity of a band can be expressed in terms of Wigner D-functions /31/, for which selection rules can be obtained using the transformation properties of the transition dipole-moment components under the isodynamic operations of the conjugated system. Essentially one measures $I(\omega)$ weighted by the factor $\omega\ (1 - \exp{-\hbar\omega/kT})$. In close analogy with the above treatment of Infrared band shapes, it is possible to obtain similar expressions for Raman band shapes, which depend on reorientations of a second-rank molecule -fixed tensor.

ACKNOWLEDGEMENTS

Professor R.M. Pick and Pr Y.G. Smeyers are gratefully acknowledged for clarifying discussions on the induction of isodynamic operations by inversion symmetries.

[+] Present address : Departamento de Química, Facultad de Ciencias Básicas y Farmacéuticas, Universitad de Chile, SANTIAGO, Chile.

[x] Present address : Laboratoire de Chimie Physique, Université Paris VI, 11, rue Pierre et Marie Curie, 75005 PARIS, France.

REFERENCES

1. W.A. Steele, J. Chem. Phys. 39 (1963) 3197.
2. H.F. King and D.F. Hornig, J. Chem. Phys. 44 (1966) 4520.
3. G.A. Stevens, Physica 44 (1969) 387.
4. L. Blum and A.J. Torruella, J. Chem. Phys. 56 (1972) 303.
5. D. Smith, J. Chem. Phys. 58 (1973) 3833 ; 65 (1976) 2568 ; 68 (1978) 3222 ; 76 (1982) 1445.
6. A.J. Stone, Molec. Phys. 36 (1978) 241.
7. J. Downs, K.E. Gubbins, S. Murad and C.G. Gray, Molec. Phys. 37 (1979) 129.
8. G. Dolling, B.M. Powell and V.F. Sears, Molec. Phys. 37 (1979) 1859.
9. (a) M. Yvinec and R.M. Pick, J. Physique 41 (1980) 1045 ; (b) R.M. Pick and M. Yvinec, J. Physique 41 (1980) 1053 ; (c) M. Yvinec and R.M. Pick, J. Physique 44 (1983) 169.
10. M. Kara and K. Kurki-Suonio, Acta Cryst. A37 (1981) 201.
11. W. Prandl, Acta Cryst. A37 (1981) 811 ; K. Vogt and W. Prandl, J. Phys. C16 (1983) 4753 ; P. Gerlash, W. Prandl and K. Vogt, Molec. Phys. 52 (1984) 383 ; Th. Brückel, W. Prandl, K. Vogt and C.M.E. Zeyen, J. Phys. C17 (1984).
12. K. Kobashi and R.D. Etters, Molec. Phys. 46 (1982) 1077.
13. J. Maruani and A. Toro-Labbé, Stud. Phys. Theor. Chem. 23 (1983) 291.
14. M.L. Klein, I.R. McDonald and Y. Ozaki, J. Chem. Phys. 79 (1983) 5579.
15. R.E. Millen and J.C. Decius, J. Chem. Phys. 59 (1973) 4871.
16. H.C. Longuet-Higgins, Molec. Phys. 6 (1963) 445.

17. S.L. Altmann, (a) Proc. Roy. Soc. (London) A298 (1967) 184 ; (b) Molec. Phys. 21 (1971) 587 ; (c) Induced Representations in Crystals and Molecules, Academic Press, New York, 1977.
18. E.P. Wigner, Group Theory and its Applications to the Quantum Mechanics of Atomic Spectra, Academic Press, New York, 1959, Chaps 12 and 19.
19. M. Hamermesh, Group Theory and its Applications to Physical Problems, Addison Wesley, Reading (Massachussets), 1962, Chap. 3.
20. (a) J. Maruani, A. Hernàndez-Laguna and Y.G. Smeyers, J. Chem. Phys. 63 (1975) 4515 ; (b) J. Maruani, Molec. Phys. 30 (1975) 1685 ; (c) J. Maruani, A. Hernàndez-Laguna and D. Bahier, J. Mol. Struct. (THEOCHEM) 92 (1983) 303.
21. (a) J. Maruani, Y.G. Smeyers and A. Hernàndez-Laguna, J. Chem. Phys. 76 (1982) 3123, erratum : J. Chem. Phys. 81 (1984) 1519 ; (b) A. Toro-Labbé and J. Maruani, Int. J. Quant. Chem. 22 (1982) 115 ; (c) A. Toro-Labbé and J. Maruani, J. Magn. Reson. 61 (1985) 254.
22. W.A. Wooster, Tensors and Group Theory for the Physical Properties of Crystals, Clarendon Press, Oxford, 1973, App. 1.
23. S.L. Altmann, Phil. Trans. Roy. Soc. A255 (1962) 216.
24. A.R. Edmonds, Angular Momentum in Quantum Mechanics, Princeton U.P., Princeton, 1960, Chap. 4.
25. F.C. von der Lage and H.A. Bethe, Phys. Rev. 71 (1947) 612.
26. A.J. Stone, J. Chem. Phys. 41 (1964) 1568.
27. J. Serre, Int. J. Quant. Chem. IIS (1968) 107.
28. C.M. Woodman, Molec. Phys. 19 (1970) 753.
29. J. Maruani and J. Serre (eds), Symmetries and Properties of Non-rigid Molecules, Elsevier S.P., Amsterdam, 1983.
30. J.I. Steinfeld, Molecules and Radiation : an Introduction to Modern Molecular Spectroscopy, MIT Press, New York, 1979.
31. W.A. Steele, Adv. Chem. Phys. 34 (1976) 1.

EXPERIMENTAL VERSUS THEORETICAL CHARGE DENSITIES IN MOLECULAR CRYSTALS

F. L. Hirshfeld
Department of Structural Chemistry
Weizmann Institute of Science
P. O. Box 26
76100 Rehovot, Israel

ABSTRACT. Formidable obstacles in the way of an accurate charge-density study by X-ray diffraction make quantitative agreement between experimental and theoretical density maps exceedingly rare. In addition to the many problems in the measurement of the X-ray intensities and their reduction to structure amplitudes, there is the difficulty of obtaining unbiased atomic coordinates and displacement amplitudes for the definition of the promolecule reference state to which the conventional deformation density is referred. Three methods for eliminating the parameter bias due to the spherical-atom model are in common use: the X-N method, requiring neutron-diffraction data to determine the atomic parameters; the X-X(H.O.) method, which uses high-order X-ray data, supposed to reflect core scattering only and to be unaffected by chemical deformation of the valence density; and the multipole-expansion method, which incorporates the charge deformation explicitly in parametric form and permits the standard structural parameters and the charge-deformation parameters to be refined simultaneously against the full set of X-ray data. This last method, using an adequately flexible model and accurate high-resolution X-ray data, allows the derivation of the *static* deformation density, through deconvolution of the vibrational smearing, thus making possible a straightforward and critical comparison of experimental and theoretical deformation densities. Illustrative experimental results are presented for iron phthalocyanine and tetrafluoroterephthalonitrile.

INTRODUCTION

The central message of my discourse this morning is that in favorable circumstances it is possible to obtain an experimental and a theoretical map of the charge density in a molecule that agree quite well in most essential details. And in fact a few such success stories may be found in the published literature. Far more often, however, X-ray diffraction leads to experimental charge-density maps that are discouragingly inaccurate. There is usually little difficulty in finding reasons, or excuses, for this situation. It may be that some

R. Daudel et al. (eds.), Structure and Dynamics of Molecular Systems – II, 57–67.

crystallographers just don't know how to do the job properly. More likely, most are not willing to invest the very great effort that is required; and the effort needed for accurate results is truly a formidable undertaking. Thirdly, in many structures, especially those containing heavy elements – and in many respects any element beyond the first row is already too heavy – quantitative accuracy is just not attainable by present methods. And finally, some chemically interesting questions seem to be answerable with the aid of charge-density maps of only qualitative reliability.

Supposing, however, that an unusually determined effort has produced a quantitatively accurate map of the charge density in some suitable compound, there remains the question whether this same molecule is amenable to a correspondingly accurate quantum-chemical calculation. This, however, is usually rather less problematic. Rapid progress in computing methods and in computer capabilities has lately extended the scope of quite sophisticated *ab initio* calculations to molecules of a size that would have been regarded, just a few years ago, as hopelessly out of reach. Yet the combined difficulties of experimental and theoretical charge-density studies suffice to make a really crucial comparison of the two approaches an unhappily rare event.

The status of charge-density research has been surveyed a number of times in recent years [1-5]. But the continued vitality of the field assures us that no published review can long remain up-to-date.

CHARGE-DENSITY AND DEFORMATION-DENSITY FUNCTIONS

Whether an X-ray diffraction study is directed at charge-density mapping or simply at a standard structure determination, we start by measuring the integrated intensities of a large set of Bragg reflections and reducing them to structure amplitudes $F_H^{obs} = |\mathbf{F}_H^{obs}|$. We then have to solve the phase problem, *i.e.* to assign appropriate phases to these quantities so as to form the structure factors $\mathbf{F}_H^{obs}$; these are generally complex quantities if the crystal is non-centrosymmetric but are real if the structure contains a center of symmetry and this has been chosen as coordinate origin. In either case we can then perform a three-dimensional Fourier transformation to obtain the electron-density function

$$\rho^{obs}(\mathbf{r}) = \frac{1}{V} \Sigma_H \mathbf{F}_H^{obs} \exp(-2\pi i \mathbf{H}\cdot\mathbf{r}) \qquad (1)$$

In this expression $\mathbf{r}$ is a real-space vector, conventionally designated by its fractional coordinates x,y,z with respect to the unit-cell axes as

$$\mathbf{r} = x\mathbf{a} + y\mathbf{b} + z\mathbf{c}.$$

Correspondingly, $\mathbf{H}$ is a reciprocal-lattice vector

$$\mathbf{H} = h\mathbf{a}^* + k\mathbf{b}^* + l\mathbf{c}^*$$

with integral components equal to the Miller indices h,k,l of the reflecting planes.

Our first concern, inevitably, is with the peaks of the function $\rho^{obs}(\mathbf{r})$. From their sizes, and other clues, we identify the several chemical elements; from their positions we derive the corresponding atomic coordinates; from their shapes we draw conclusions about the atomic vibrations. If we are interested only in the crystal structure, our task ends here. But if we are trying to study the electron distribution itself, we have scarcely begun. Our main interest is then in the less conspicuous features of the charge density, between and around the atomic maxima. To study them it is necessary to remove the atomic peaks, which so completely dominate the function ρ^{obs} as to make it virtually impossible to see anything else. So we form the difference function

$$\Delta\rho(\mathbf{r}) = \rho^{obs}(\mathbf{r}) - \Sigma_i \rho_i^{at}(\mathbf{r}) \tag{2}$$

by subtracting from ρ^{obs} the sum of the densities attributable to the individual atoms. In order for the function $\Delta\rho$ to be interpretable in terms of chemical bonding effects, it is important that the atomic densities we subtract in eq. (2) should be properly positioned in the crystal and should be smeared by the vibrational motion actually undergone by the atoms in the crystal. Only then will the difference $\Delta\rho$ truly represent the difference between the density of the bonded atoms in the crystal and that of the corresponding isolated atoms before bonding. In the case of a molecular crystal, the sum of free-atom densities in one molecule is called the promolecule

$$\rho^{pro}(\mathbf{r}) = \Sigma_i \rho_i^{at}(\mathbf{r}). \tag{3}$$

The true difference density corresponding to eq. (2), over the crystal or over a single molecule, is commonly known as the deformation density.

We see that the subtraction in eq. (2) serves two purposes. One is to eliminate the large peaks so the resulting smoother function $\Delta\rho$ may be plotted on a greatly expanded density scale to reveal features that were completely hidden in the original function ρ^{obs}. The second, and more crucial, purpose is to reveal explicitly the changes in the density of the isolated atoms that mediate the bonding of these atoms into molecular or other chemical structures.

But all this means that the chemically interesting function $\Delta\rho$ is obtained by subtraction of two large and nearly equal functions. Thus these functions themselves, *i.e.* ρ^{obs}, given by eq. (1), and ρ^{pro}, given by (3) — or by a sum of such expressions over the several molecules in the unit cell — must each be determined to the highest possible accuracy if their difference is to be at all meaningful.

Achieving the required accuracy in these two terms imposes two very different kinds of conditions. For the total observed density ρ^{obs}, we need to measure a large number, typically several thousand, of structure amplitudes F^{obs} to acceptably high accuracy. This demands adequate counting statistics, to minimize the random experimental errors in the measured intensities, plus a multitude of painstaking measures to deal with all sorts of systematic errors. These encompass careful choice of crystal specimen, meticulous alignment of the diffractometer, constant attention to the stability of the X-ray beam, detector system, and crystal quality and orientation, suitable choices of scan mode and limits, detector aperture, and the like, proper background subtraction and valid corrections for effects like sample absorption, extinction, multiple reflection, and thermal diffuse scattering, and statistically valid data reduction procedures, which include scaling of the measured intensities, averaging of symmetrically equivalent reflections, and a host of similar concerns. The list of possible errors is virtually endless and the art of the careful experimentalist consists largely in discovering which are the troublesome problems and how to deal with them effectively and reliably.

The promolecule density ρ^{pro} poses a very different kind of challenge. This requires summing the densities of the free atoms, which are known from accurate theoretical calculations, after assigning them their appropriate positions and displacement amplitudes. It is these atomic parameters that must be properly determined so as to ensure that the promolecule density does not suffer an intolerable systematic error due to biased parameters. Thus the obvious way of determining these parameters, by fitting the theoretical atomic densities to the observed total density ρ^{obs}, is totally unacceptable. Such a fitting procedure amounts to finding those parameters that make the difference density $\Delta\rho$ as small as possible, by some least-squares criterion. So we are assuming in advance that the function we hope to determine in fact vanishes, but for random experimental errors. The result can only be a set of biased structural parameters that systematically depresses the chemically significant deformation-density features we are looking for [6-10].

SEARCH FOR UNBIASED STRUCTURAL PARAMETERS

THE X-N METHOD

So how can we derive unbiased structural parameters for the evaluation of the promolecule density $\rho^{pro}(\mathbf{r})$? The methods commonly employed for this purpose are of three general types. First, we can perform a neutron-diffraction experiment on the same material. Neutrons are scattered by the atomic nuclei and are, in diamagnetic materials, entirely unaffected by the orbiting electrons. Thus they provide direct information about the positions and mean square displacements of the nuclei, just what we need for an unbiased determination of ρ^{pro}. This is the basis of the X-N method [11,12], so called because the first term on the right of eq. (2) is determined by the X-ray measurements while the second is derived from the neutron experiment. The method is beautifully simple in principle and has been used with great success for a wide variety of chemical systems, organic and inorganic. Yet it is often troublesome in practice. One reason is that neutron scattering cross sections are low, causing low counting rates and poor precision for a given measurement time. Since high-flux neutron beams are available only at a few major installations, competition for beam time severely limits the duration of a particular experiment and, hence, the precision of the resulting structural information. A second problem concerns systematic errors of various kinds, notably extinction and thermal diffuse scattering, that are of unequal importance in X-ray and neutron diffraction. One apparent consequence is that atomic displacement parameters, which are especially sensitive to such systematic errors, often disagree between the two experiments [13-16], leading to some uncertainty about how to combine the two sets of data. A third problem, primarily for non-centrosymmetric structures, is the determination of phases for the evaluation of ρ^{obs}. It is certainly wrong to assume that each structure factor $\mathbf{F}_H^{obs}$ appearing in eq. (1) has the same phase as the known Fourier coefficient $\mathbf{F}_H^{pro}$ in the corresponding sum for the promolecule density ρ^{pro} [17,18]. For this reason the X-N method has been largely limited to centrosymmetric structures, where the assignment of phases is far less troublesome. But even for these, the difficulties noted above have motivated a search for alternative methods that do not depend on neutron diffraction for the atomic parameters.

HIGH-ORDER REFINEMENT

One such approach is to derive the structural parameters from the X-ray structure amplitudes corresponding to large magnitudes of the reciprocal-lattice vector **H** only [19,7,12]. The argument behind this use of the high-order structure amplitudes — the resulting maps being designated X-X(H.O.) maps — is that the high-order Fourier coefficients $\mathbf{F}_H$ are determined essentially by sharp electron-density features due to the atomic core electrons and these are supposedly unaffected by chemical bonding and retain the spherical shapes of the free atoms. So we expect that by fitting our atomic parameters to the high-order data only, comprising structure amplitudes F_H with $H = |\mathbf{H}|$ larger than some appropriately chosen cut-off value H_{cut}, we will obtain atomic position and displacement parameters corresponding to the undeformed atomic cores and thus free of any bias arising from chemical deformation of the valence electron distribution. A central uncertainty about this method is how to choose the cut-off radius H_{cut}. Values ranging from 1.0 to 2.0 $Å^{-1}$ have been used [19,7,16], with a progressive tendency towards larger thresholds as theoretical and experimental evidence has revealed sharper and sharper deformation features, especially near terminal oxygen and fluorine atoms [20,21]. Hydrogen atoms pose a special difficulty since they completely lack core electrons. Thus, the method seems to work best for atoms of carbon and a few adjacent elements but is far from providing a totally satisfactory solution even for all light-atom structures.

MULTIPOLE REFINEMENT

The third method dispenses completely with the spherical-atom model, invoking a flexible multi-centered multipole expansion of the actual electron density $\rho(\mathbf{r})$ [22-28]. The parameters of this expansion are refined along with the standard structural parameters against the full set of X-ray data. In this way it is hoped that the expansion parameters will fully absorb the effects of chemical bonding on the electron density and so allow the structural parameters to attain their true unbiased values. The major difficulty is to choose a model that is sufficiently flexible to describe the observable details of the charge deformation while keeping the number of adjustable parameters small enough to assure convergence of the refinement. With X-ray data of sufficiently high resolution and accuracy, this challenge can usually be adequately met and some detailed charge deformation maps have been obtained in this way. The method is not dependent on neutron diffraction but neutron data, if available, can profitably be combined with the X-ray data in a joint refinement of all parameters.

Evidently, a multipole refinement is capable of providing much more than a set of unbiased structural parameters for the evaluation of ρ^{pro}. The model is usually defined so that its expansion coefficients explicitly determine the deformation density

$$\delta\rho(\mathbf{r}) = \rho^{mol} - \rho^{pro}$$

which is our principle objective. So we obtain this function directly from the refined expansion coefficients without the need of evaluating the Fourier series (1) for ρ^{obs}. This has two important advantages. First, we eliminate the the random errors in ρ^{obs} associated with the experimental noise in the measured Fourier coefficients $\mathbf{F}_{\mathbf{H}}^{obs}$. This is largely smoothed out by the least-squares process of fitting a relatively small number of adjustable parameters to a large number of measured structure amplitudes. Especially important for this smoothing process is the proper weighting of the observations in accordance with their estimated relative precision; this prevents the density $\delta\rho(\mathbf{r})$ from being excessively contaminated by large errors in those reflections that are known to have been imprecisely measured. Secondly, the expansion can be, and usually is, defined in terms of the *static* deformation density, *i.e.* the deformation density of a non-vibrating molecule relative to non-vibrating free atoms [25-28]. This static deformation map is invariably far more detailed than the corresponding dynamically smeared density and it can be more directly and critically compared with theoretically calculated deformation densities for the same molecule or, quite often, for chemically related molecules.

These considerable advantages are won at a non-negligible price. The results depend quite crucially on the chosen multipole model. So we must take care to check whether the deformation map we obtain has not been unduly distorted by the constraints built into our model. Some of the necessary checks are standard features of the least-squares process itself, such as the verification that the discrepancies

$$\Delta F_{\mathrm{H}} = F_{\mathrm{H}}^{obs} - F_{\mathrm{H}}^{calc}$$

have the expected statistical distribution consistent with the assumed accuracy of measurement. A more direct check is provided by examination of the residual density $\Delta\rho(\mathbf{r})$, computed with the Fourier coefficients $\Delta\mathbf{F}_{\mathbf{H}}$ (for which the assignment of phases is not crucial).

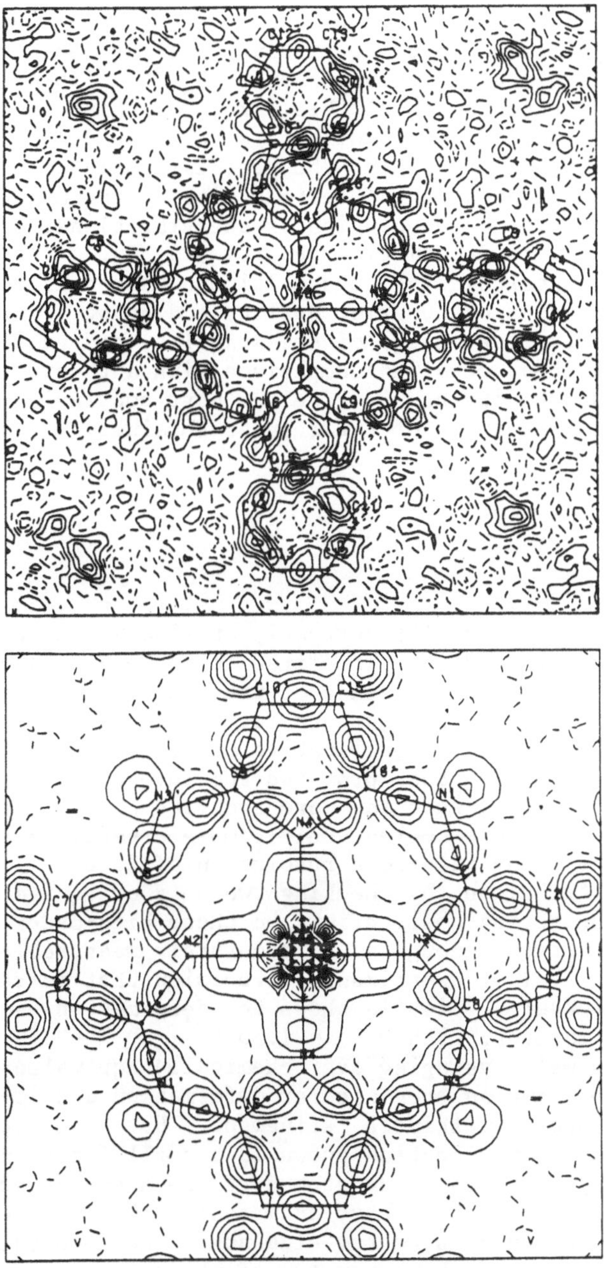

a

b

Fig. 1 . Dynamic deformation density in mean plane of iron (II) phthalocyanine, from Coppens and Li [31]. Contour interval 0.1 e$Å^{-3}$
a. X-X(H.O.) map, before symmetric averaging
b. (dynamic) model map from multipole refinement

This might show, superimposed on the random noise from the experimental errors in F_H^{obs}, chemically significant features that were suppressed by the model constraints and therefore missing from the density represented by the calculated structure factors F_H^{calc}. This is an important test but it cannot be infallible. For one thing, we lack a fully objective criterion for distinguishing systematic density features from random noise. Secondly, the test is applicable only to the dynamic density and cannot reveal deficiencies in the model that pertain specifically to the deconvolution of the vibrational smearing necessary to map the static deformation density [29,30].

In view of these uncertainties, the multipole method is best relied upon when X-ray data of superior quality and quantity are available to permit extensive checks of the stability and validity of the refinement. Even then, detailed comparison with a theoretically calculated deformation density can provide the most convincing demonstration of the correctness of the results obtained.

ILLUSTRATIVE DEFORMATION MAPS

IRON PHTHALOCYANINE

The quality of experimental results currently attainable can be illustrated by two recent studies. The first is by Coppens and Li [31] on iron(II)phthalocyanine. The presence of the heavy iron atom would seem to make this an unpromising structure for quantitative charge-density studies. And, in fact, a qualitative result was really enough to answer the central question that prompted the investigation. Yet the results actually achieved were far more accurate than might have been anticipated. The authors cooled their sample crystal to about 110 K and measured the X-ray data to a resolution of 2.36 $Å^{-1}$. By careful experimental technique, by checking two sets of equivalent reflections, by concentrating their effort on those high-order reflections that were strong enough to be accurately measurable, they managed to achieve a level of experimental precision that is rare for such a heavy-atom structure. Fig. 1a shows the deformation density from a high-order refinement, with H_{cut} = 1.50 $Å^{-1}$. A much smoother map was obtained by simply averaging over chemically equivalent regions in this X-X(H.O.) map. But an even clearer deformation density, shown in Fig. 1b, was obtained by multipole refinement. Here the estimated random error over most of the section shown is a small fraction of the contour interval. Only near the iron atom does the accuracy suffer greatly from the dominance of the heavy-atom scattering and the proximity of several molecular symmetry elements. Yet here the interesting question concerns the asymmetry of the valence density, not the quantitative measure of its features. And the positive peaks along the diagonal directions are enough to establish that the dominant electronic configuration in this compound can only be the ^3E_gA state. No theoretical calculation seems to be available for this molecule. So we have the somewhat unusual situation of a remarkably accurate X-ray determination of the deformation density in a system for which a

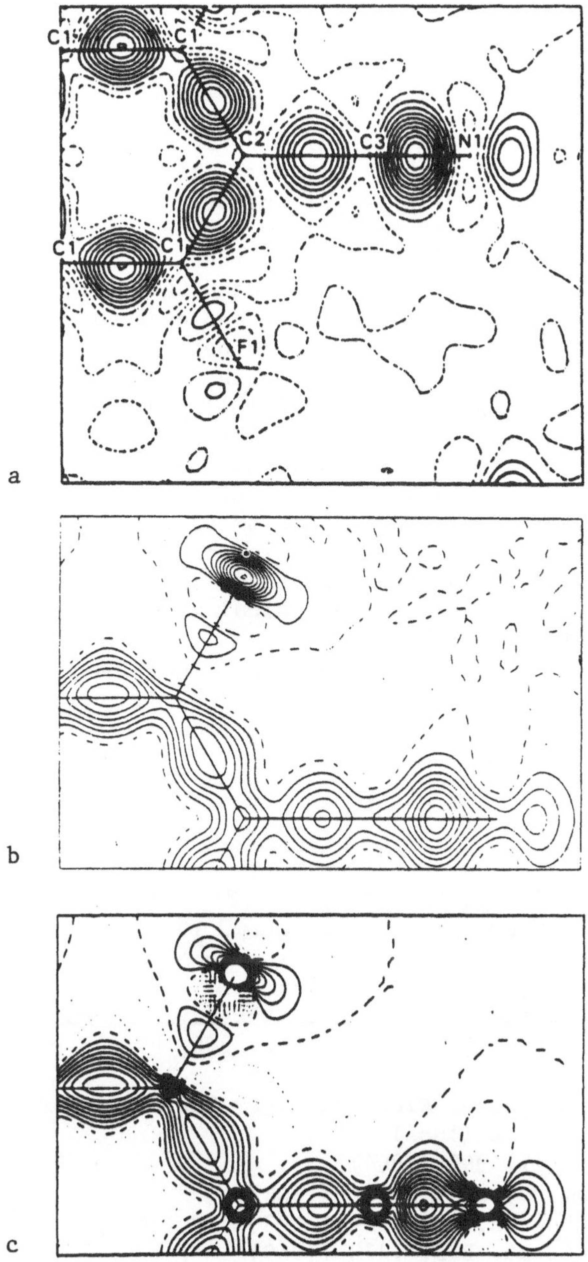

Fig. 2. Deformation density in mean plane of TFT (one quarter molecule per asymmetric unit)
a. X-X(H.O.) map, contour interval 0.075 eÅ^{-3}, from Seiler et al. [33]
b. dynamic density from multipole refinement, contour interval 0.1 eÅ^{-3}
c. static density from multipole refinement, contour interval 0.1 eÅ^{-3}
— b and c from Hirshfeld [21]

comparably accurate theoretical calculation is evidently still inaccessible.

TETRAFLUOROTEREPHTHALONITRILE

A more relevant example for the comparison of experiment and theory comes from the study, by Jack Dunitz and collaborators at the E.T.H. [32,33], of the deformation density in tetrafluoroterephthalonitrile (TFT). X-ray data of unusually superior quality were collected at 98 K to a resolution limit of 2.3 Å^{-1}. After a high-order refinement using data beyond H_{cut} = 1.7 Å^{-1}, the E.T.H. team produced the X-X(H.O.) map shown in Fig. 2a. This shows normal features in and near the C-C and C-N bonds, including a respectable lone-pair peak behind the nitrogen atom, but any deformation peak in the C-F region is virtually lost in the experimental noise. A multipole refinement [21] produced the dynamic map of Fig. 2b, which shows much more prominent features near the F atom, but also led to a map of the static deformation density (Fig. 2c). This clearly shows the main features expected in a covalently bonded fluorine atom — a redistribution of valence density from the σ to the π orbitals, strong hybridization of the singly occupied σ orbital, and a small bond peak on the C-F axis. All these features have been reproduced in theoretical calculations on formyl fluoride [34] and on TFT itself [35].

The general conclusion is somewhat mixed. Accurate charge-density maps can be obtained by X-ray diffraction but only through very hard work. Most experimental maps will likely continue to be far less accurate than the best we know how to produce.

REFERENCES

1. P. Coppens, Angew. Chemie **16**, 32 (1977)
2. P. Coppens and E.D. Stevens, Adv. Quant. Chem. **10**, 1 (1977)
3. F.L. Hirshfeld, ed. *Electron Density Mapping in Molecules and Crystals*, Isr. J. Chem. **16**, no. 2-3 (1977)
4. P. Becker, ed. *Electron and Magnetization Densities in Molecules and Crystals*, N.A.T.O. Adv. Stud. Inst. Ser. B48, Plenum Press (1980)
5. P. Coppens and M.B. Hall, eds. *Electron Distributions and the Chemical Bond*, Plenum Press (1982)
6. B. Dawson, Acta Cryst. **17**, 990 (1964)
7. R.F. Stewart, J. Chem. Phys. **48**, 4882 (1968)
8. P. Coppens and C.A. Coulson, Acta Cryst. **23**, 718 (1967)
9. A.M. O'Connell, Acta Cryst. **B25**, 1273 (1969)
10. A.F.J. Ruysink and A. Vos, Acta Cryst. **A30**, 503 (1974)
11. P. Coppens, Science **158**, 1577 (1967)
12. P. Becker, P. Coppens, and F.K. Ross, J.A.C.S. **95**, 7604 (1973)
13. J.W. Batts, P. Coppens, and T.F. Koetzle, Acta Cryst. **B33**, 37 (1977)
14. C. Scheringer, A. Kutoglu, and D. Mullen, Acta Cryst. **A34**, 481 (1978)

15. J. Epstein, J.R. Ruble, and B.M. Craven, Acta Cryst. B38, 140 (1982)
16. P. Coppens *et al.*, Acta Cryst. A40, 184 (1984)
17. E.N. Maslen, Acta Cryst. B24, 1172 (1968)
18. P. Coppens, Acta Cryst. B30, 255 (1974)
19. G.A. Jeffrey and D.W.J. Cruickshank, Quart. Rev. Lond. 7, 335 (1953)
20. B.M. Craven and R.K. McMullan, Acta Cryst. B35, 934 (1979)
21. F.L. Hirshfeld, Acta Cryst. B40, 484 (1984)
22. B. Dawson, Austr. J. Chem. 18, 595 (1965)
23. B. Dawson, Proc. Roy. Soc. A298, 255, etc. (1967)
24. K. Kurki-Suonio, Acta Cryst. A24, 379 (1968)
25. R.F. Stewart, J. Chem. Phys. 51, 4569 (1969)
26. R.F. Stewart, Acta Cryst. A32, 565 (1976)
27. F.L. Hirshfeld, Acta Cryst. B27, 769 (1971)
28. N.K. Hansen and P. Coppens, Acta Cryst. A34, 909 (1978)
29. C.A. Coulson and M.W. Thomas, Acta Cryst. B27, 1354 (1971)
30. J. Epstein and R.F. Stewart, Acta Cryst. A35, 476 (1979)
31. P. Coppens and L. Li, J. Chem. Phys. 81, 1983 (1984)
32. J.D. Dunitz, W.B. Schweizer, and P. Seiler, Helv. Chim. Acta 66, 123 (1982)
33. P. Seiler, W.B. Schweizer, and J.D. Dunitz, Acta Cryst. B40, 319 (1984)
34. M. Eisenstein and F.L. Hirshfeld, J. Comp. Chem. 4, 15 (1983)
35. B. Delley, to be published

STUDY OF MAGNETIC STRUCTURES BY X-RAY DIFFRACTION

F. de Bergevin and M. Brunel
Laboratoire de Cristallographie, associé à l'U.S.M.G.
Centre National de la Recherche Scientifique
166 X, 38042 Grenoble CEDEX
France

ABSTRACT. The magnetic moment of the electron, as well as its charge, interacts with electromagnetic radiation ; in quantum theory, the spin-photon interaction is a relativistic effect.Though very weak, this effect allows for X-ray diffraction by magnetic structures. The expression of the spin dependent amplitude is given and its properties are discussed. Experiments made by several authors are reviewed. In those made on antiferromagnets, weak superlattice peaks are observed. Measurements on Fe_2O_3 with variable spin orientations and variable polarizations give a check of theoretical formulas. On helimagnetic Holmium, a new structural model is based on results obtained with the help of the high **k** resolution obtained at a synchrotron source, and of the informations brought by polarization analysis. In experiments done on ferro(ferri)magnets, a small fraction (1/1000) of the intensity of Bragg peaks may be sensitive to the magnetization. This is observed either when the Thomson amplitude is complex (due to anomalous scattering), and such experiments made on iron, magnetite and Gd-Y multilayers are presented, or when the polarization is circular, and corresponding experiments are made with synchrotron radiation on magnetite.

1. INTRODUCTION

X-ray diffraction is usually interpreted through the Thomson scattering mechanism, i. e. the interaction between the electromagnetic radiation and the charge of the electrons. X–rays therefore seem to give information about the charge density only and not about the spin density. If the phenomenon is examined more thoroughly, the electronic spin appears as playing also some role ; its magnetic moment does interact with the magnetic field of the radiation (Fig. 1).

In quantum theory, the effect of the spin arises from relativity, where space and spin wave functions can no longer be separated. During the collision with a photon of momentum **k** (**k**' after the collision), the electron undergoes an acceleration, whose relativistic character is measured by $|\mathbf{k}'-\mathbf{k}|/mc$ (Fig. 2a). The probability of collision therefore depends on a term roughly proportional to this factor, and containing the spin S; this term is shown to be

$$\mathbf{S}.(\mathbf{k} \times \mathbf{k}') / |\mathbf{k}| mc \qquad (1)$$

Similar effects are present in other instances, such as Mott [32] or Schwinger [27,9] scattering ; electrons in the former, and neutrons in the latter, are scattered by the electric field of a nucleus, and the spin of the particle makes the scattering asymmetric (Fig. 2b, 2c). The photon polarization (or photon spin) is ignored till now; it certainly couples to

R. Daudel et al. (eds.), Structure and Dynamics of Molecular Systems – II, 69–86.

the electron spin, and the exact amplitude, as given in the next paragraph, contains a polarization dependent term, in addition to (1).

The spin scattering amplitude (1) is roughly proportional to the momentum transferred from the photon to the electron ; therefore it can reach higher values in the inelastic, or Compton, scattering than in the elastic case, where momentum transfer is limited to h times the inverse Bohr radius. In the former case, with 500 keV γ-rays, this amplitude may be one hundred times larger than in the latter. For this reason, magnetic inelastic scattering became since 1950 a standard tool for the measurement of γ-ray polarizations, essentially in nuclear and particle physics [30,32], but was considered

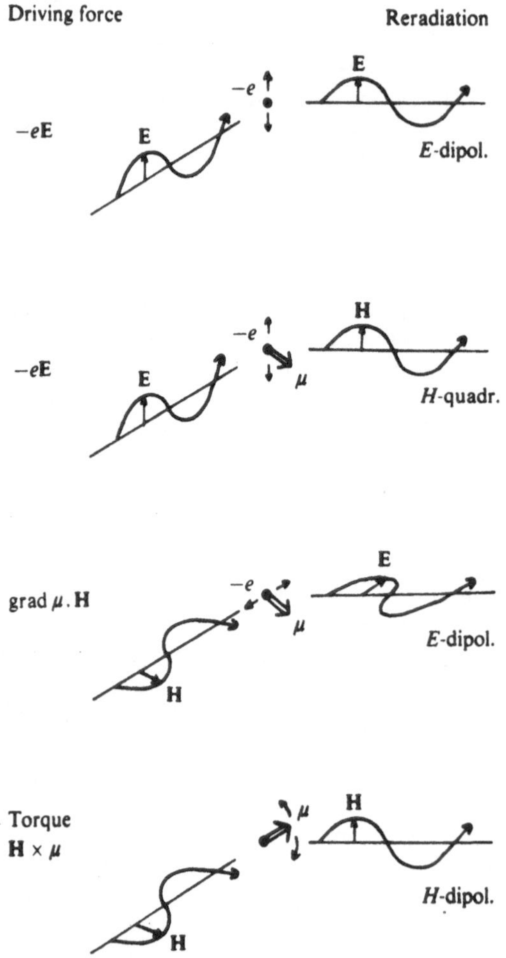

Fig. 1. The four mechanisms of scattering, in the classical description. The first is the well known Thomson scattering. In each case, the electron (charge -*e*) ot its magnetic moment is moved by the incident electromagnetic field. The force or torque is indicated in the left column. The type of reradiation (electric or magnetic, dipolar or quadrupolar) is given at the right. (From [2], © IUCr.).

much more recently in the field of diffraction. The measurement of the momentum distribution of spin up and down electrons in iron, has been made bv Sakai and Ono [26], using Compton scattering of circularly polarized γ-rays.

The precise calculation of the effects to be expected in X-ray diffraction by magnetic substances is due to Platzman and Tsoar [23], who first suggested that these effects could be observed. Subsequent experiments made by Brunel *et al* [1,2,5,6], and by Suzuki *et al* [31] on samples with known magnetic structure, demonstrated the validity of the theoretical predictions. More recent experiments bringing informations on the sample structure, hardly accessible by other methods such as neutron diffraction, were made with the use of synchrotron radiation on Holmium by Gibbs *et al* [11,12], and on Gd-Y superlattices by Vettier *et al* [33]. Further calculations were made by de Bergevin and Brunel [2] who treated the orbital momentum scattering and gave cross section formulas allowing for any polarization of the beam; Zhizhimov and Khriplovitch [34] and Blume [3] gave two different formulation suited to the representation of electron binding effects, such as orbital momentum scattering, resonant scattering [3], or modification of Thomson and spin scattering at some distance from a resonance [34]. A recent account on magnetic inelastic scattering applied to solids, has been given by Platzman and Tsoar [24].

After a discussion of the basic formulas, the experiments made on antiferromagnets (also helimagnets) and those made on ferro(ferri)magnets are described separately; for the techniques used in each of these two cases are different.

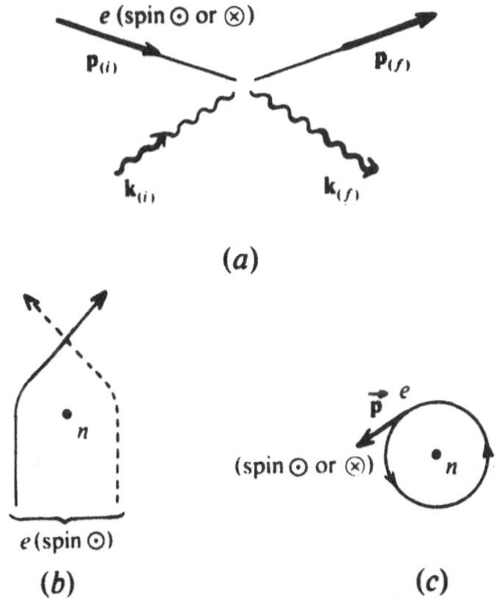

Fig. 2. All these effects rely on the relativistic correlation between spin and space wave functions. The spin is always supposed perpendicular to the plane of the figure. (a) Collision between a photon and an electron; the amplitude depends on the orientation of the spin and on **p-p'=k'-k**. (b) Scattering of an electron by the field of a nucleus (Mott scattering); the amplitude depends on the direction of scattering relative to the orientation of the spin. In Schwinger scattering, the incident particle is a neutron. (c) Spin-orbit coupling of an electron in an atom; the orientation dependent quantity is the energy of the state. (From [2], © IUCr.).

2. BASIC FORMULAS

Several calculations of the scattering amplitude of photons by the electron spin have been published ; they rely on various methods, and for some of them, on various approximations. A short review of these calculations is given in an Appendix.

The scattering amplitude as given in the following, is an approximation valid at low transferred momentum **k-k'** (unprimed and primed quantities are respectively incident and diffracted ones) ; it may not be correct at very high incident energies, >>1Mev, but this is far beyond normal experimental conditions. It is also assumed that the electron spin state is unchanged during scattering ; due to the weakness of the magnetic amplitude, the observation of spin wave scattering is unexpected in near future.

The scattering amplitude M is the sum of two terms, the Thomson and magnetic ones, which both depend linearly on the initial and final polarization vectors $\boldsymbol{\varepsilon}, \boldsymbol{\varepsilon}'$; these may be complex so as to represent circular polarizations. Furthermore, the magnetic amplitude depends linearly on the electron spin **S**.

$$M = -\frac{e^2}{m}\left[(\boldsymbol{\varepsilon}'|A|\boldsymbol{\varepsilon})F + i\frac{\lambda_c}{\lambda}(\boldsymbol{\varepsilon}'|\mathbf{B}|\boldsymbol{\varepsilon}).\mathbf{S}\right] \tag{2}$$

F and **S** are respectively the charge and spin structure factors, e and m the charge and mass of the electron, and λ_c its Compton length. The two bilinear expressions $(\boldsymbol{\varepsilon}'|A|\boldsymbol{\varepsilon})$ and $(\boldsymbol{\varepsilon}'|\mathbf{B}|\boldsymbol{\varepsilon})$ may be written as matrices, once choosen a basis for the representation of the vectors $\boldsymbol{\varepsilon}$ and $\boldsymbol{\varepsilon}'$; they may also be written as simple vector expressions, such as

$$(\boldsymbol{\varepsilon}'|A|\boldsymbol{\varepsilon}) = \boldsymbol{\varepsilon}.\boldsymbol{\varepsilon}' \tag{3}$$

Unfortunately, the magnetic term expressed in this way is not as simple as the Thomson one, and though several equivalent forms have been published [10,19,23,2,34,3], none of them is easily readable. An example is ($\hat{\mathbf{k}}$ is a unit vector, $\mathbf{k}/|\mathbf{k}|$)

$$(\boldsymbol{\varepsilon}'|\mathbf{B}|\boldsymbol{\varepsilon}) = -[\boldsymbol{\varepsilon}'\times\boldsymbol{\varepsilon} + (\hat{\mathbf{k}}'\times\boldsymbol{\varepsilon}')\hat{\mathbf{k}}'.\boldsymbol{\varepsilon} - (\hat{\mathbf{k}}\times\boldsymbol{\varepsilon})\hat{\mathbf{k}}.\boldsymbol{\varepsilon}' - (\hat{\mathbf{k}}'\times\boldsymbol{\varepsilon}')\times(\hat{\mathbf{k}}\times\boldsymbol{\varepsilon})] \tag{4}$$

On the opposite, the matrix form of **B** is much more pleasant, if written in the basis of the linear polarizations perpendicular and parallel to the diffraction plane **k, k'** ; the signs are such that $\boldsymbol{\varepsilon}'_{//} = \mathbf{k}\times\boldsymbol{\varepsilon}'_{\perp}$; i and f mean initial and final.

$$A = \begin{array}{c|cc} \mathrm{f}\backslash\mathrm{i} & \perp & // \\ \hline \perp & 1 & 0 \\ // & 0 & \cos 2\theta \end{array} \qquad \mathbf{B} = \begin{array}{c|cc} \mathrm{f}\backslash\mathrm{i} & \perp & // \\ \hline \perp & -\hat{\mathbf{k}}\times\hat{\mathbf{k}}' & 2\hat{\mathbf{k}}'\sin^2\theta \\ // & -2\hat{\mathbf{k}}\sin^2\theta & -\hat{\mathbf{k}}\times\hat{\mathbf{k}}' \end{array} \tag{5}$$

A third simple form is obtained when $\boldsymbol{\varepsilon}$ and $\boldsymbol{\varepsilon}'$ are given complex values corresponding to circular polarizations; in that case let us define **P** and **P'** as the photon spins (unit vectors along or reverse to the wave vectors), and χ equal to +1 if the photon helicity is unchanged, and -1 if it is reversed during scattering ; i **B** is now

$$i\,(\boldsymbol{\varepsilon}'\,|\mathbf{B}\,|\boldsymbol{\varepsilon}) = -[i\,\hat{\mathbf{k}}\times\hat{\mathbf{k}}' + \chi\,(\mathbf{P}+\mathbf{P}')\sin^2\theta] \tag{6}$$

The amplitude is seen to split in two parts, respectively independent and dependent of the polarizations. The first part arises from the electron acceleration during the collision ; the Schwinger scattering [27,9] is similar, if considered in the limiting case where the scatterer is a neutral atom much smaller than the wave length (the Compton length is then replaced by the atomic radius).

From the above formulas and especially from (2)(5), some practical considerations may be drawn. The ratio between magnetic and Thomson amplitudes contains essentially the factor $\lambda_c \boldsymbol{\kappa} / 2\pi$ ($\boldsymbol{\kappa}$ the scattering vector). In Compton scattering, this factor is unlimited (below 500 kev) and magnetic scattering may be high ; but these formulas are approximations which fail for $\lambda_c \boldsymbol{\kappa} \sim 1$. In elastic scattering, $\boldsymbol{\kappa}$ must be of the order of $1/r_B$ (the Bohr radius), so that the magnetic amplitude is of the order of 1/137 times the Thomson one. Since in a typical magnetic material, one electron over ten may be unpaired, the ratio between total amplitudes is roughly 1/1000. If pure magnetic Bragg peaks are to be observed, as is the case in antiferro or helimagnets, their intensity is 10^3 times smaller than Thomson peaks ; if on the opposite, magnetic and Thomson amplitudes can be coherently superposed, as is the case in ferro and ferrimagnets, the intensity contains a cross term, which can be tuned by changing the magnetization ; relative variations of the order of 1/1000 of the intensity can thus be observed.

The treatment of the polarizations, before and/or after diffraction is more complicated than in the case of Thomson scattering, because the matrix **B**, unlike A , contains non diagonal elements. In the general case, where both polarizations may be elliptical with any axis, and partial, the safe procedure is to represent them by density matrices, or equivalently by Stokes vectors. The intensity and polarization of a beam is represented by four real numbers (the components of the Stokes vector), N_0 , N_1, N_2, N_3 ; N_0 is the intensity, and N_i/N_0 three polarizations, which are respectively : 1 perpendicular to a basal plane, 2 at 45°, 3 circular. The sum of the three squared polarizations is equal to 1 when the beam is totally polarized, otherwise it is smaller. The intensity and polarization of the diffracted beam are obtained from the incident Stokes vector by a linear transformation, and the matrix which operates it (the Mueller operator) can be computed from the sample charge and spin structure factors and from the beam directions. Cascades of diffractions (as when a monochromator and/or an analyzer is interposed on the beam path) are taken into account by multiplying the successive matrices. The expression of the Mueller operator for Thomson and spin scattering (but without orbital momentum) is given in Appendix 2 of [2], along with a more detailed discussion (this discussion contains a minor error, as mentioned in [6]).

The above exposed procedure is straightfoward though the Stokes vector formalism may be considered as unusual. Some people would prefer the more direct method of summing the intensities over the final linear polarizations and averaging over the initial ones; it is emphazised that this is allowed only in some simple cases (which in fact occur the most frequently). This simple procedure can be used when only one type of polarization is present in the incident beam and eventually selected by an analyzer; "one type of polarization" means any of two perpendicular polarizations, excluding oblique and circular ones. Suppose for example that the source is unpolarized, and that a monochromator diffracting at Bragg angle α in the same plane as the sample, is put in the incident beam. The cross section is

$$d\sigma/d\Omega = r_e^{\,2} \,[|(\perp M \perp)|^2 + |(//M \perp)|^2 + (|(\perp M //)|^2 + |(//M //)|^2)\cos^2 2\alpha\,] / (1+\cos^2 2\alpha) \quad (7)$$

where symbols such as ($\perp M$ //) represents matrix elements of the amplitude (Thomson plus magnetic). When the monochromator is put after, this formula becomes

$$d\sigma/d\Omega = (r_e^{\,2}/2)\,[|(\perp M \perp)|^2 + |(\perp M //)|^2 + (|(//M \perp)|^2 + |(//M //)|^2)\cos^2 2\alpha\,] \quad (8)$$

Another remark concerns the sign of the imaginaries in the formula (2) ; it is fixed by a mere convention, but its definition must be kept consistent in the whole calculation. As discussed by some authors [9,28], two conventions are commonly used : physicists and neutron diffractionists put the minus sign in front of the frequency × time product in the expression of the waves, and they take **k**-**k**' as the scattering vector, while X-ray cristallographers usually put the opposite sign in front of the frequency × time product and use an opposite scattering vector. The structure factor has the same expression in both conventions

$$F(\boldsymbol{\kappa}) = \int d\mathbf{x}^3 \rho(\mathbf{x}) \exp i\, \boldsymbol{\kappa}.\mathbf{x} \quad (9)$$

but when anomalous scattering is present, the imaginary atomic scattering factor is positive in the X-ray cristallographers convention and negative in the physicists one. Since all papers on spin scattering use the latter, we follow it; when a dispersion correction is necessary, its imaginary part is written as $-if''$.

3. DIFFRACTION BY ANTIFERROMAGNETS AND HELIMAGNETS

Antiferro and helimagnets usually have some magnetic structure factors at new points of the reciprocal space, off the normal reciprocal lattice (there are a few exceptions, such as Cr_2O_3). At these points, pure magnetic diffracted intensities can be observed. The main difficulty is their smallness, since they are roughly 10^{-6} the normal intensities. This can be overcome if single crystals are used ; in favourable cases experiments are feasible even with a sealed laboratory X-ray tube [1,5], at the expense of some time. The conditions are of course better with a rotating anode or synchrotron radiation and the latter has been proven to give an excellent resolution in reasonable times [11,12]. Experiments have been done on antiferromagnets (NiO, Fe_2O_3) by M. Brunel and F. de Bergevin [1,5], on NiO by Suzuki, Tanokura and Onoue [31], and on helimagnetic Holmium by Doon Gibbs, Moncton, D'Amico, Bohr and Grier [11,12]. In the following are described some of the measurements made on Fe_2O_3 and Ho, which demonstrate the capabilities of the method.

3.1. Antiferromagnetic Fe_2O_3

In hematite Fe_2O_3, whose structure is trigonal, the magnetic moments are antiferromagnetically ordered below $T_N = 948$ K ; they are parallel within any (001) plane, and antiparallel to those in the adjacent planes. For temperatures below T_M ($T_M \sim 235$ K is the so called Morin transition), the moments are along the [001] axis

[29] ; for $T_M < T < T_N$ they are lying in the (001) plane, but not exactly antiparallel, so that a weak ferromagnetic moment arises in the same plane [7]. The net magnetization, though weak enough to be ignored in the calculation of the magnetic structure factor, is sufficient to allow a control of the moment direction through an external applied field.

The magnetic structure of Fe_2O_3 being well understood [29,7,21], the purpose of the X-ray measurements [5] described in the following was to check the formulas which describe the dependence of the scattering on the spin direction. Hematite was suited for this, due to the uniquely defined spin direction in the low temperature phase and the change at the Morin transition.

Let us define three axes, x along **k'-k**, y along **k'+k**, z along $\mathbf{k}' \times \mathbf{k}$ and call S_x, S_y, S_z (Fig. 3) the components of the spin density

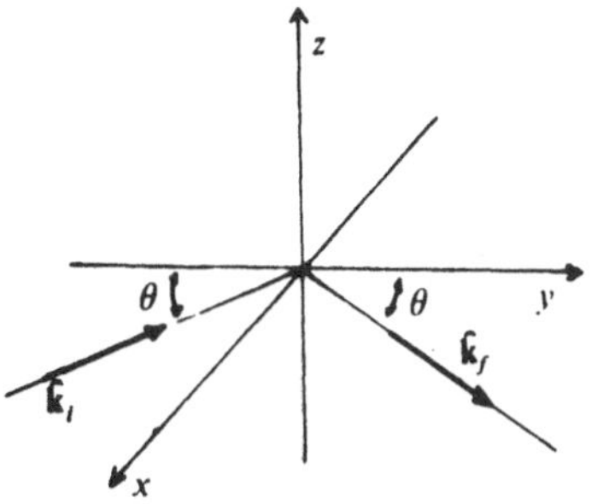

Fig. 3. Definition of the axes. x is along the scattering vector, and z perpendicular to the diffraction plane. The angular position ψ of the crystal is adjusted around the axis x . (From [5], © IUCr.)

structure factor at the Bragg angle θ. In these experiments, a monochromator with Bragg angle α is placed in the diffracted beam. The intensity of the magnetic superlattice peak is

$$I \sim (\lambda_c \sin\theta/\lambda)^2 P(S_x, S_y, S_z)$$

$$P(S_x, S_y, S_z) = S_z^2\cos^2\theta + \sin^2\theta\{S_y^2\cos^2\theta + S_x^2\sin^2\theta + [\sin^2 2\alpha/(1+\cos^2 2\alpha)]S_x S_y \sin 2\theta\} \quad (10)$$

This expression is more complicated than in the case of neutron scattering, where only two spin components, S_y and S_y, are involved ; therefore X-rays may give more complete informations on the spin direction, than neutrons. In the present experiments P can be varied either by changing the crystal temperature across the Morin transition, or by rotating it around the scattering vector or axis x ; its angular position is referred as ψ. The latter operation does not change the intensity through any other mean than the change in P (in fact intensity can also vary through the Renninger effect, i.e. variable multiple scattering; the experiment is made so as to avoid it). The points in Fig. 4 show the experimental results, which agree with the calculated intensities, in the limit of errors. In the high temperature phase, the crystal is made of several magnetic domains, containing all the symmetric magnetic directions in the hexagonal (001) plane; it is assumed in the calculation that the same amount of matter is present in every domain. The peak intensity is typically 100-300 c/min and the background 50 c/min.

The expression of P contains a cross term $S_x S_y$ which is related to the polarization brought in by the monochromator. With usual settings, this polarization is small and $S_x S_y$ is actually neglected in Fig. 4. But a higher monochromator Bragg angle makes the polarization large enough to give a quite visible asymmetry in the intensities; with $\alpha = 32°$ (the polarization rate is 0.69), the intensities measured for both spin directions represented in Fig. 5 may differ by a factor two. Such a feature cannot be observed in neutron diffraction.

A magnetic field applied to the high temperature phase modifies the volume of the various domains, whose spin directions lie in the (001) plane, and so changes the diffracted intensity. The experiment is made with this plane nearly parallel to xz (Fig. 3), and the magnetic field applied either along x (the spins then tend to be aligned along z,

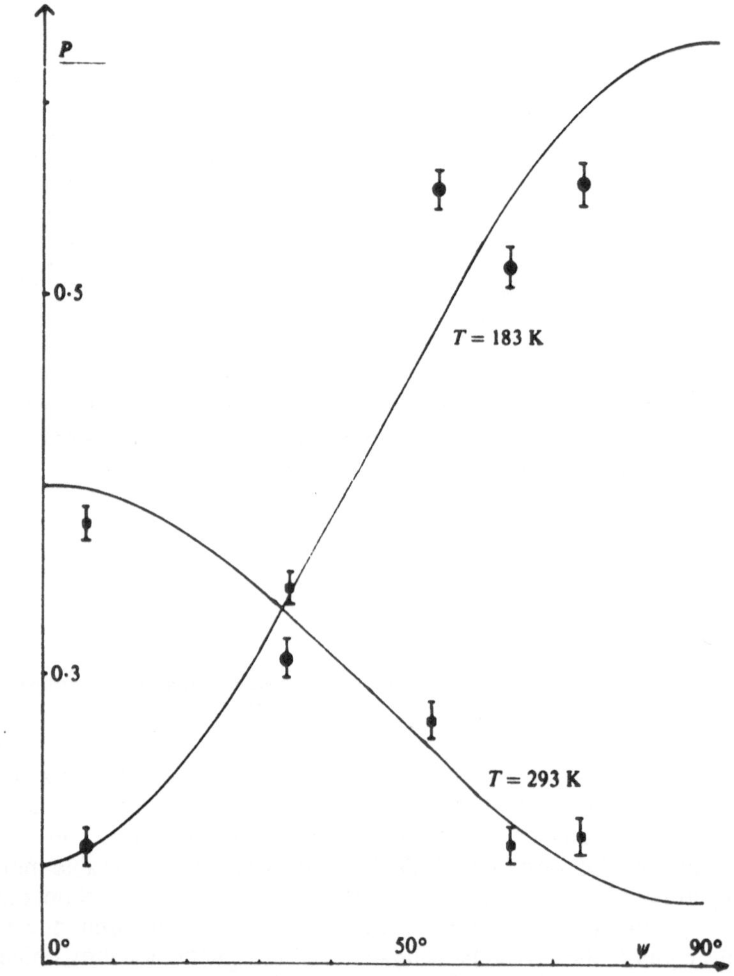

Fig. 4. Experimental (points) and theoretical (lines) values of P defined in (10). The spin directions in the crystal are different at the two temperatures. The angle ψ is the crystal position around the axis x (Fig. 3). The measurements are made on the reflection 303 at the radiation CuKα. (From [5], © IUCr.).

perpendicular to the beams), or along z (the spins then tend to make a small angle with the scattering vector). The result is shown in Fig. 6. From theory, the lower curve should saturate at a much lower value ; the reason of this discrepancy is unclear ; the magnetic anisotropy may eventually be anomalous near the surface of the crystal, in the depth viewed by X-rays.

3.2. Helimagnetic Holmium

The work made by [11,12] on metallic Holmium is the first one in which X-ray magnetic scattering brings new informations and help to improve the understanding of a magnetic structure.

Holmium is hexagonal with two layers per unit cell. Below T_N ~ 131 K, magnetic moments order in a helical magnetic structure, in which they are ferromagnetically aligned within the basal plane but rotate from plane to plane [15,16,8,25]. The period of the helix, measured in neutron diffraction by the separation of satellites from main peaks, varies with temperature, from $6c$ near 20 K to $3.6c$ near 130 K. Below 50 K the helix become distorded, as indicated by the presence of higher order satellites. This distortion is supposed to result from bunching of the moments in pairs along the six easy magnetization directions of the (001) hexagonal plane. The improvements brought by [11,12] to this model rely on experiments made by neutron and X-ray scattering ; the latter is especially useful due to (a) the high resolution in reciprocal space, 0.001 $Å^{-1}$ with synchrotron radiation as against 0.005 $Å^{-1}$ with neutrons, (b) the ability to identify by polarization analysis, the charge or spin origin of the diffracted peaks.

The high resolution allows for the detection of the helix wave vector locking in at some rational values ($2/11c$, $5/27c$, in addition to $1/6c$). The authors explain this fact by a so-called "spin-slip" model, in which one spin is missing in some periodically spaced bunches (or pairs) (Fig. 7). A consequence of this model is the presence of crystallographic distorsions, whose periodicity differs from the magnetic periodicity. They actually observe, simultaneously to the 5/27 satellite, another one at $2/9c$, as required by the model. A further confirmation is given by the polarization analysis, whose principle can be understood by looking at formula (5) ; the matrices for Thomson and spin scattering differ by the presence of non diagonal elements in the latter, which are

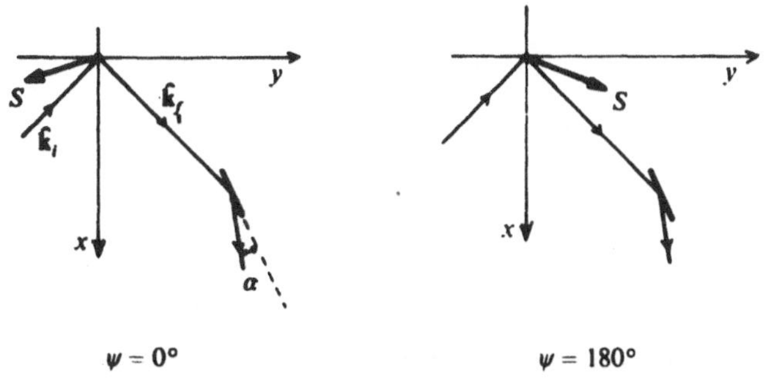

Fig. 5. In the low-temperature phase of Fe_2O_3, the spin direction is represented with respect to the axes x and y, at $\psi = 0°$ and $\psi = 180°$. The diffracted intensity changes between these two positions, due to the polarization introduced by the monochromator (at Bragg angle α) placed on the diffracted beam. (From [5], © IUCr.).

absent in the former. If the incident beam has a perpendicular polarization (as referred to the plane containing the beams), and the diffracted one is filtered by an analyzer parallel to the same reference plane, the peak corresponding to crystallographic scattering disappears, while the magnetic peak may still be observed (Fig. 8).

4. DIFFRACTION BY FERRO AND FERRIMAGNETS

In these materials, chemical and magnetic cells are equal, and the normal Bragg peaks contain both Thomson and magnetic amplitudes. These add coherently, and therefore the

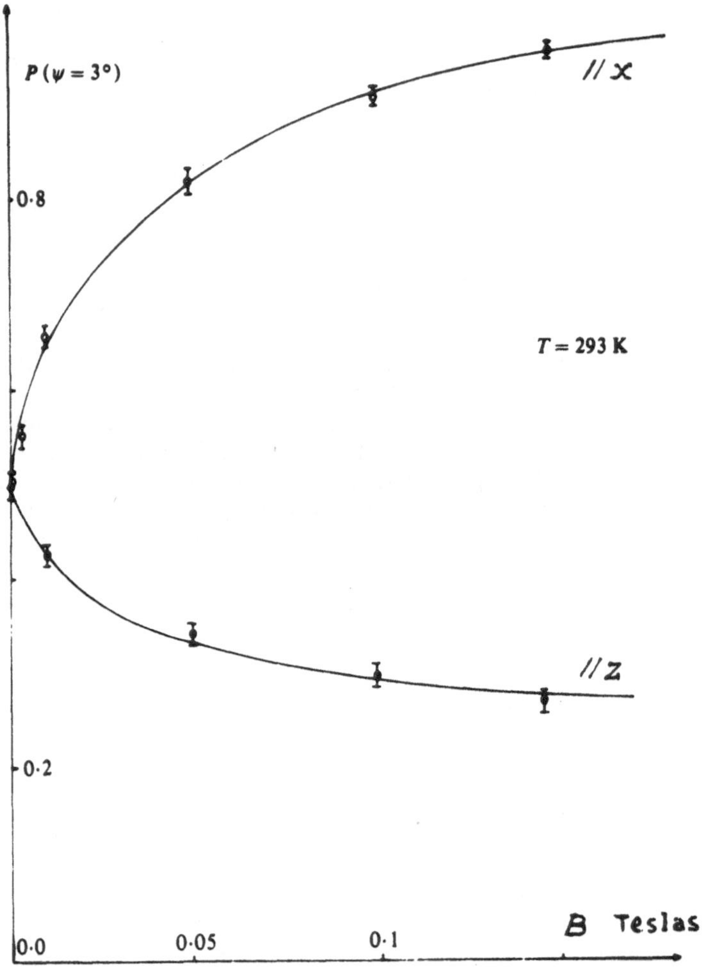

Fig. 6. Measured on the high temperature phase of Fe_2O_3, variations of the intensity of the 101 magnetic reflection, under an applied magnetic induction B along x (higher curve) or z (lower). (From [5], © IUCr.).

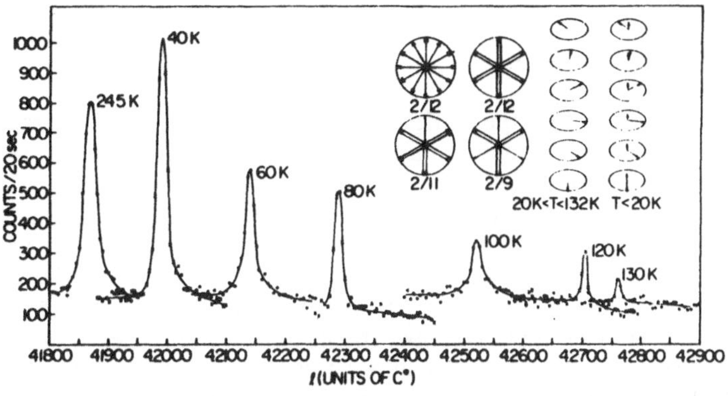

Fig. 7. Temperature dependence of the Ho(004)+ magnetic satellite, measured with the synchrotron radiation. Inset right, schematic representation of the magnetic structure (after [15]). Left, schematic projection of the magnetic unit cell for different spin slip structures ; spins in the doublet are in fact separated by a small angle. (From [12], © A.P.S.).

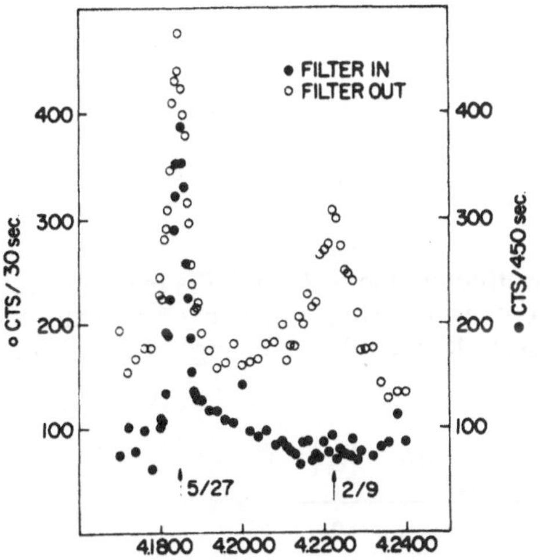

Fig. 8. Open circles: the two Ho(004) satellites, at 5/27c and 2/9c observed at 17 K. Filled circles : with polarization analysis, the second peak, coming from charge scattering, disappears, while the first magnetic one remains present. (From [12], © A. P. S.).

magnetic amplitude contributes to the intensity in the first order by a cross term, which is

$$AF\,(i\,\mathbf{B}.\mathbf{S})^* + (AF\,)^*\,(i\,\mathbf{B}.\mathbf{S}) \qquad (11)$$

It may be possible to reverse the sign of **B**, by reversing either the magnetization or a polarization, and to measure a "flipping ratio", as is done in polarized neutron scattering. The relative change in intensity is of the order of 1/1000.

A serious restriction comes from the requirement that AF and $i\,\mathbf{B}.\mathbf{S}$ must be simultaneously real or imaginary for (11) to be non zero (more exactly they must not be out of phase by $\pi/2$). The formulas (2)(5)(6) show that the magnetic amplitude is imaginary when the beam polarizations are linear, whereas it contains a real term when the polarizations are circular. Two types of experiments are thus feasible. In the first one, the beams are linearly polarized (or eventually unpolarized), and it is necessary to have an imaginary term in the Thomson contribution ; a way to obtain the occurence of this term, is to choose the wave length so that an element in the sample gives a high anomalous scattering. In another type of experiment, at least one of the beams is circularly polarized while the Thomson contribution may be real ; a circularly polarized source can only be in practice a synchrotron source.

Experiments using anomalous scattering have been made on ferromagnetic iron and ferrimagnetic magnetite by M. Brunel and F. de Bergevin [2], and on rare earth multilayers by C. Vettier, D. Mc Whan, E. Gyorgy, J. Kwo, B. Buntschuh, B. W. Batterman [33]. Experiments using circular polarization have been made on magnetite by M. Brunel, G. Patrat, F. de Bergevin, F. Rousseaux and M. Lemonnier [6].

4.1. Measurements using anomalous scattering

Formula (5) shows that when Thomson and magnetic scattering are added (with linear polarization) the cross term contains

$$F\,''\mathbf{S}.(\mathbf{k} \times \mathbf{k}') \qquad (12)$$

$F\,''$ is the imaginary part of the charge structure factor F and **S** is the spin structure factor. When the magnetization of a ferro or collinear ferrimagnet is reversed while being kept perpendicular to the plane of diffraction **k**, **k**', the relative change of the diffracted intensity is

$$\frac{\Delta I}{I} = \frac{\lambda_c}{\lambda}\,\frac{1+\cos^2 2\alpha\cos 2\theta}{1+\cos^2 2\alpha\cos^2 2\theta}\,\sin 2\theta\,\frac{4F\,''\,|\mathbf{S}|}{|F\,|^2} \qquad (13)$$

The sign of ΔI is related to the sign of $\sin 2\theta$, a feature quite similar to Mott scattering asymmetry [32] (Fig. 9).

The first experiment is made by [2] on two samples, metallic iron and zinc substituted magnetite. The spinel structure of magnetite is favourable because for some Bragg reflections, the contributions of both sublattices add in **S** and substract in F ; the composition $(Zn^{++}_{0.53}Fe^{+++}_{0.47})(Fe^{+++}_{1.53}Fe^{++}_{0.47})O_4$ further enhances $\Delta I\,/I$. Since both samples contain Fe, the CuKα radiation, which gives for this atom $f\,'' > 3$ electrons, is suitable ; it is recalled that $f\,''$ must be negative to fit the conventions used in the present account.

The measuring system is comprised of a powder diffractometer (the use of single crystals is unecessary) and a magnetic coil giving a field of a few kilooersted along its rotation axis. A monochromator put on the diffracted beam eliminates the fluorescence associated with the high f''. The magnetic field is periodically reversed, approximately every minute of time, and the diffracted intensity received during each half a cycle is stored (a cycle includes two field reversals). ΔI and I are the difference and mean of the

Table I. Values of $\Delta I/I$ for zinc ferrite

$h\,k\,l$	$\theta(°)$	$(\Delta I/I) \times 1000$ experiment	$\sigma \times 1000$	$(\Delta I/I) \times 1000$ theory
2 2 2	+18.5	+2.25	0.12	2.37
	-18.5	-2.33		
4 4 0	+31.23	+0.06	0.10	0.15
	-31.23	-0.12	0.06	
5 3 1	+32.84	+4.10	0.30	4.61
	-32.84	-4.70		

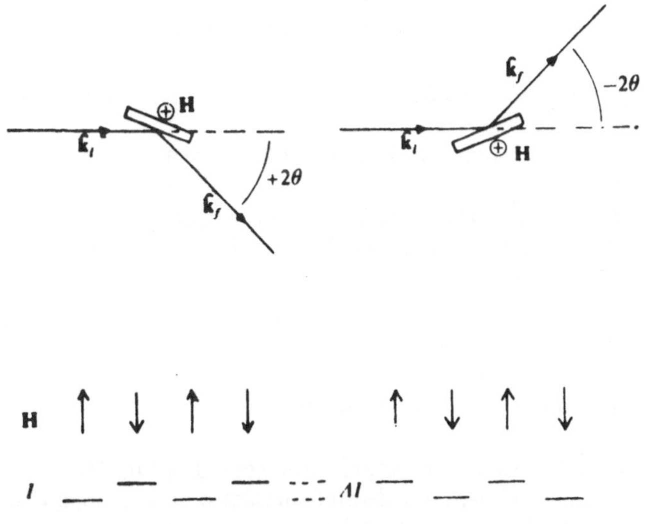

Fig. 9. The sample, diffracting in the horizontal plane, is magnetized by a vertical field H. When the field is reversed, the intensity changes. The sign of this change is correlated to the sign of θ. (From [2], © IUCr.).

integrals of odd and even results. Because the required accuracy is unusually high (the error on ΔI must be 10^{-4} to 10^{-5} of I), several checks are made. Measurements made on a non magnetic sample give a null result and a statistical analyzis of the whole set of intensities shows that no fluctuations other than normal statistical ones do affect the final result (short term fluctuations are virtually absent and long term ones have a negligible effect in the range of one minute). The result for magnetite is given in table I ; similar but lower values are obtained with iron.

Vettier *et al* [33] (see also [17]) are interested in the magnetic moment density in Gd-Y multilayers. The small volume of the samples (40 periods of 21+21 atomic layers each) and the absorbtion cross section of Gd, makes the use of neutrons difficult. Due to the superlattice structure, satellites are present around each of the $00l$ peaks. $\Delta I / I$ is measured on several of these satellites (about twenty, up to the eighth order, taken around three peaks) in an experiment similar to the preceeding one. The use of synchrotron radiation (at CHESS, Cornell University) allows for a scanning of the energy across the L absorbtion edge of Gd, and f'' can reach half the number of electrons of this atom. The results support the hypothesis of a slow decrease of the magnetization at the interfaces, associated with atomic mixing.

4.2. Measurements using circular polarization

At present time the only efficient source of circularly polarized X-rays is a bending magnet of a storage ring, provided that the observation is made above or below the electron trajectory plane. Other systems are to be developped in near future, such as perfect crystal polarizers or special insertion devices in synchrotron radiation machines [13].

The formula (6) shows that the real part of the magnetic amplitude (the one which gives a Thomson × magnetic cross term in the intensity) arises from the magnetization component which lies in the diffraction plane **k, k'**. The intensity is

$$I \sim F^2[(1-P//) + (1+P//)\cos^2 2\theta]/2 - \tau(\lambda_c/\lambda) F\, \mathbf{S}.(\hat{\mathbf{k}}\cos 2\theta + \hat{\mathbf{k}}')(1-\cos 2\theta) \quad (14)$$

where the charge density structure factor F is supposed to be real, $P//$ and τ are the incident beam linear and circular polarization rates. The sample is again a compressed powder plate. **S** is controlled by an external field which is directed along the plate, in order to get rid of the demagnetizing field ; therefore **S** is in the direction **k+k'**. When the field is reversed the intensity changes by

$$\frac{\Delta I}{I} = 4\tau \frac{\lambda_c}{\lambda} \frac{|\mathbf{S}|}{F} \frac{\cos\theta \sin^2 2\theta}{(1-P//) + (1+P//)\cos^2 2\theta} \quad (15)$$

The apparatus used at DCI (LURE, Orsay) is represented in Fig. 10. In order to select in the synchrotron radiation beam, the desired circular polarization, an horizontal slit (typically one millimeter wide) is placed on the goniometer. The whole system can be moved vertically ; by a displacement of ± 2.5 mm from the equatorial plane, τ can be made equal to ± 0.5, while the intensity is decreased by a factor lower than two.

The other features of the experiment are similar to those of the previous one. The sample is again zinc ferrite, and the frequency of the field reversal is of the same order

TABLE II. Values of $\Delta I/I$, multiplied by 1000, measured at three heights h (sample, zinc substituted magnetite)

hkl →	2 2 2	4 4 0	6 2 2
θ(°) →	21.80	37.19	45.15
h = -2.5 mm	+1.82 ± 0.3	+0.8 ± 0.2	+4.0 ± 0.5
h = 0 mm	+0.26 ± 0.3	+0.13 ± 0.1	+1.8 ± 0.6
h = +2.5 mm	-0.84 ± 0.3	-0.64 ± 0.25	-4.0 ± 0.5
Calculated τ = 0.53	1.44	0.40	3.30

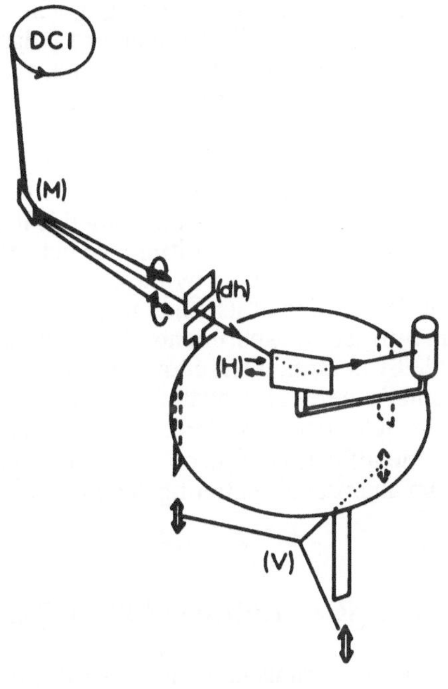

Fig. 10. The storage ring (*DCI*, far away), the monochromator (*M*), and the powder diffractometer. The sign of circular polarization, when seen from above or below the plane of the ring, is determined by the electron direction of rotation. A horizontal slit (d*h*) selects a beam slice of desired polarization rate; this one is adjusted by moving the diffractometer upwards or downwards, through the device (*V*). The sample is magnetized by a field (*H*), which is periodically reversed. (From [6], © IUCr.).

(~ 1 min). A precise correction of the decrease of the beam intensity (at a rate of 1/15 per hour) is necessary because its effects is comparable to ΔI . The result is given in the table II.

5. FURTHER DEVELOPMENTS AND CONCLUSIONS

Mention must be made of two aspects, whose corresponding experiments are quite recent, or not yet done. These are orbital momentum scattering and resonant magnetic scattering.

The effects discussed in the present conference rest on the interaction between the photon and the electronic spin ; an interaction of comparable magnitude does exist between the photon and the orbital momentum [2,3,34], and gives rise to a scattering amplitude proportional to

$$(\boldsymbol{\kappa} \times \mathbf{p}).(\boldsymbol{\varepsilon} \times \boldsymbol{\varepsilon}') \exp i\, \boldsymbol{\kappa}.\mathbf{x} \qquad (16)$$

(**p** is the electron momentum on its orbit). A similar expression is encountered in neutron scattering, where its mean value is shown to depend on the Fourier transform of the orbital magnetic moment density [29a]. In neutron scattering, spin and orbital momenta enter in the amplitude by the total magnetic density, whatever is its origin, while in X-ray scattering, they give different dependences on momentum direction and on polarizations. The use of both techniques would allow for the separation of both momenta, a separation which cannot be done from neutron results alone. At present time, no experiment has yet been performed.

Blume [3] suggested that when a core level is split by exchange with a magnetic outer shell, the core scattering becomes magnetization dependent near resonance, where the dispersion corrections become large. An experiment has been done along this line by Namikawa *et al* [20]; these authors interpret their result by the difference between spin up and down densities of states above the Fermi level, rather than by the core level splitting.

We conclude that magnetic X-ray scattering, despite of its small magnitude, gives rise to several effects, which can be observed in diffraction experiments. These effects are able to bring some informations unattainable by neutron diffraction ; it is so when very high **k** resolution is needed, or when very small samples are available ; this has been shown on 0.5 µm thick layers, but smaller samples could be used. The dependence of X-ray scattering on polarizations allows for identification of its origin (either charge or spin scattering) ; discrimination between spin and orbital scattering should also be possible.

APPENDIX. THE CALCULATIONS OF THE SCATTERING AMPLITUDE

Two types of calculation can be made. (a) the most usual method of electrodynamics consists in a perturbation expansion of the amplitude in powers of e ; only the two terms in e^2 are retained ; it seems reasonable up to a photon energy at least of the order of 500 kev (see below). (b) the second one is exact up to any order in e , but is valid only for a photon energy small as compared to the rest mass of the scattering particle (it applies to any charged particle of spin 1/2 ; for the electron, the rest mass energy is 500 kev).

The work of Klein and Nishina [14, 22], though famous by its final result, in which electron spin states are summed out, does contain also the spin dependent amplitude (formulas [14] (44) and [22] (1)) ; this source is generally not explicitly reported in subsequent litterature, including ours. It relies on a first order perturbation, of type (a), and does not include any low energy approximation; it is therefore valid up to relativistic energies. Several other works [18,23,2,3,34] follow essentially the same line, with varying techniques and presentations of the results. [18] aims to present the one electron cross section as a function of initial and final photon and electron polarizations, and does not display the amplitude. [23] and [3] present a low energy approximation; it is explicitly shown in [2] that a low momentum transfer (as compared to *mc*), even with a high photon energy, is sufficient for this approximation to be valid. Since in diffraction, momentum transfer is limited, this approximation is used in (2)(5).

The calculation (b) has been drawn by Low [19] and Gell-Mann and Goldberger [10], by two different methods. Their result is equivalent to the previous low momentum transfer approximation, except for its taking into account the anomalous magnetic moment ; they consider any spin 1/2 particle, whatever its magnetic moment; normal and anomalous parts of it enter differently in the formula.

In a discussion of the validity of the approximations, the result (b) makes sure that the formulas (2)(5) are correct at non relativistic energies, since the anomalous moment of the electron is about 1/1000 of the normal one. At relativistic energies, either the momentum transfer is still small (forward scattering), and the same formulas are used, or it is high (large angle scattering) and one should turn to those given by [14,22], or [18] ; the latter are best suited to Compton scattering experiments. But the validity of both formulations is to be discusssed, because in this regime they come from a perturbation expansion. The radiative corrections have been examined by Brown and Feynman [4], but they summed out the electron spin states. These corrections modify the amplitude, or the cross section, by a relative change increasing with the energy, and of the order of 1/137 (the electromagnetic coupling constant) at 500 kev ; it is expected that the relative change of spin dependent quantities is not higher.

Another problem is the perturbation caused by the electron binding in the atom ; it is considered as negligible because the binding energy of outer electrons is much smaller than the photon energy. An approximation valid when the latter is still higher than the binding energy, but lower than 137 times its value (that is a photon of about 300 ev) has been given by [34].

The classical calculation is finally to be mentioned. It was believed for a time to yield a wrong result, but a complete calculation by Gell-Mann and Goldberger [10] showed that it is equivalent to the low energy quantum approximation.

ACKNOWLEDGMENTS

We thank C. Vettier, Doon Gibbs, I. B. Khriplovitch, N. Namikawa for communication of their results prior to publication, and the first two of them, as well as D. E. Moncton, for discussions. We are grateful to Doon Gibbs for kindly authorizing us to reproduce several figures of paper [12]. All figures are reproduced from Acta Crystallographica, © International Union of Crystallography, and from Physical Review Letters, © American Physical Society.

REFERENCES

1. F. de Bergevin & M. Brunel, *Phys. Lett.* **A39**, 141-142 (1972)
2. F. de Bergevin & M. Brunel, *Acta Cryst.* **A37**, 314-324 (1981)
3. M. Blume, *J. Appl. Phys.* **57**, 3615-3619 (1985)

4. L. M. Brown & R. P. Feynman, *Phys. Rev.* **85**, 231-244 (1952)
5. M. Brunel & F. de Bergevin, *Acta Cryst.* **A37**, 324-331 (1981)
6. M. Brunel, G. Patrat, F. de Bergevin, F. Rousseaux & M. Lemmonier, *Acta Cryst.* **A39**, 84-88 (1983)
7. I. Dzyaloshinsky, *J. Phys. Chem. Solids* **4**, 241-255 (1958)
8. G. P. Felcher, G. H. Lander, T. Arai, S. K. Sinha & F. H. Spedding, *Phys. Rev. B* **13**, 3034-3045 (1976)
9. G. P. Felcher & S. W. Peterson, *Acta Cryst.* **A31**, 76-79 (1975)
10. M. Gell-Mann & M. L. Goldberger, M. L., *Phys. Rev.* **96**, 1433-1438 (1954)
11. D. Gibbs, D. E. Moncton & K. L. D'Amico, *J. Appl. Phys.* **57**, 3619-3622 (1985)
12. D. Gibbs, D. E. Moncton, K. L. D'Amico, J. Bohr & B. H. Grier, *Phys. Rev. Lett.* **55**, 234-237 (1985)
13. K. J. Kim, *Nucl. Instrum. Methods* **219**, 425-429 (1984)
14. O. Klein & Y. Nishina, *Z. Phys.* **52**, 853-868 (1929)
15. W. C. Koehler, J. W. Cable, H. R. Child, M. K. Wilkinson & E. O. Wollan, *Phys. Rev.* **158**, 450-461 (1967)
16. W. C. Koehler, J. W. Cable, M. K. Wilkinson & E. O. Wollan, *Phys. Rev.* **151**, 414-424 (1966)
17. J. Kwo, E. M. Gyorgy, D. B. McWhan, M. Hong, F. J. DiSalvo, C. Vettier & J. E. Bower, *Phys. Rev. Lett.* **55**, 1402-1405 (1985)
18. F. W. Lipps & H. A. Tolhoek, *Physica (Utrecht)* **XX**, 85-98 & 395-405 (1954)
19. F. E. Low, *Phys. Rev.* **96**, 1428-1432 (1954)
20. K. Namikawa, M. Ando, T. Nakajima & H. Kawata, Submitted to *J. Phys. Soc. Japan*
21. R. Nathans, S. J. Pickart, H. A. Alperin & P. J. Brown, *Phys. Rev.* **136**, A1641-A1647, (1964)
22. Y. Nishina, *Z. Phys.* **52**, 869-877 (1929)
23. P. M. Platzman & N. Tsoar, *Phys. Rev. B* **2**, 3356-3359 (1970)
24. P. M. Platzman & N. Tsoar, *J. Appl. Phys.* **57**, 3623-3625 (1985)
25. M. J. Pechan & C. Stassis, *J. Appl. Phys.* **55**, 1900-1902 (1984)
26 N. Sakai & K. Ono, *Phys. Rev. Lett.* **37**, 351-353 (1976)
27. J. Schwinger, *Phys. Rev.* **73**, 407-409 (1948)
28. S. Ramaseshan, T. G. Ramesh & G. S. Ranganath, *Anomalous scattering* , edited by S. Ramaseshan & S. C. Abrahams, pp. 149-152, Munskgaard, Copenhagen (1975)
29. C. G. Shull, W. A. Strauser & E. O. Wollan, *Phys. Rev.* **83**, 333-345 (1951)
29a. O. Steinsvoll, G. Shirane, R. Nathans, M. Blume, H. A. Alperin & S. J. Pickart, *Phys. Rev.* **161**, 499-506 (1967)
30. R. M. Stephen & H. Fraunfelder, *Polarization of Radiation following β-Decay* , *in Alpha, Beta, Gamma Ray spectroscopy* , Edited by K. Siegbahn, ch. 4, pp. 1453-1490, North-Holland, Amsterdam (1965)
31. T. Suzuki, A. Tanokura, K. Onoue, *Jap. J. Appl. Phys.* **17** (Sup. 17-2), 320 (1978)
32. H. A. Tolhoek, *Rev. Mod. Phys.* **28**, 277-298 (1956)
33. C. Vettier, D. B. Mc Whan, E. M. Gyorgy, J. Kwo, B. M. Buntschuh & B. W. Batterman, submitted to *Phys. Rev. Lett.*
34. O. L. Zhizhimov & I. B. Khriplovitch, *Sov. Phys. J. E. T. P.* **60**, 313-319 (1984), transl. from *Zh. Eksp. Theor. Fiz.* **87**, 547-557 (1984)

A NEW EXPECTATION VALUE : THE COMPTON DEFECT, AND ITS APPLICATIONS TO THE STUDY OF ATOMIC AND MOLECULAR ELECTRONIC STRUCTURES

C. Tavard and F. Gasser
C.M.S.R., Faculté des Sciences
Université de Metz, Ile du Saulcy
57045 Metz Cedex
France

ABSTRACT. The concept of Compton profile allows an interpretation of X-Ray and electron impact spectroscopic studies for atomic, molecular and solid targets and hence gives a detailed information on their corresponding electronic structures.

However some characteristic differences, either of positive or negative types, have been observed between Compton profile calculations and very precise experimental measurements. The main purpose of this article consists here to explain the physical origin of these differences, or Compton defects. As a conclusion, such an observable function will be shown to provide useful additional informations on the target system.

1 . INTRODUCTION AND GENERAL REFERENCES

A wide part of our detailed experimental knowledge in electronic structures comes from X-Ray or electron differential scattering cross sections. The Waller-Hartree theory [1] explains their angular behaviour in connection to the first and second order density matrices [2] of the scatterers in their real space.

A spectral analysis at large scattering angles exhibits also a dominant contribution, discovered by A.H. Compton [3] for the case of photon-electron interactions. A Döppler broadening of the Compton peak is finally due to the velocity (or momentum) density distribution of the target electrons. Consequently such experiments give access to momentum electronic densities.

With the extent of new techniques and unceasing progress in experimental resolution, both angular and spectral investigations in differential scattering cross sections have been performed during the past twenty years. All major scientific contributions to these fields can be found reviewed and analyzed in several reference textbooks [4,5]. Furthermore, a number of very recent works on X-Ray studies of solid systems are discussed in a former article of these series [6].

R. Daudel et al. (eds.), Structure and Dynamics of Molecular Systems – II, 87–108.

This article deals with recent progress in scattering processes and a special interest is brought to Compton scattering studies beyond the impulse approximation, or Compton defect studies. Fondamental concepts and approximations are emphasized here in a condensed self consistent presentation.

The Hartree atomic units ($\hbar=1$, $m_e=1$, $a_0=1$) are used throughout the manuscript. Some basic properties and approximations are summarized in the next paragraphs.

1.1. Classical kinematics of a collision (Fig. 1) :

An incoming "structureless" particle (photon, electron...) of selected momentum $\vec{k}_i$ (and hence of energy E_i) impinges upon a given target, usually at rest in the laboratory frame.

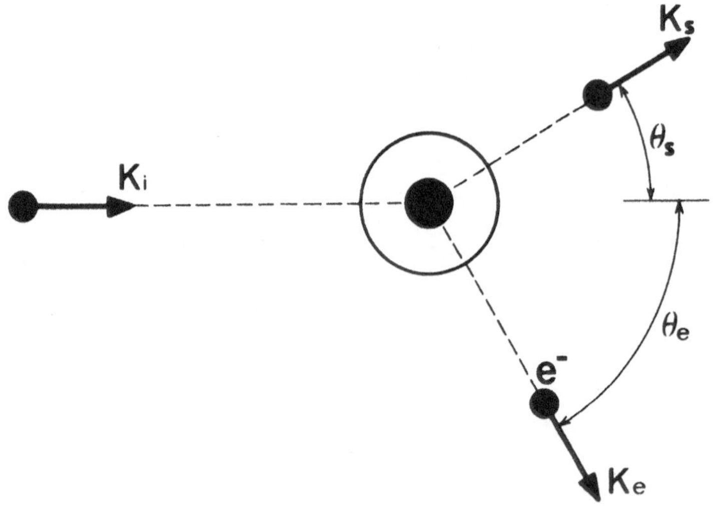

Fig. 1 - Kinematics of an ionizing collision.

After collision and according to the asymptotic values of its energy E_s and its momentum $\vec{k}_s$, the scattered particle can be detected with an appropriate experimental device (photographic plate, X-Ray spectrometer, Wien lens, 127° electrostatic analyzer, scintillator-photomultiplier or channeltron detectors...).

The following difference $\vec{K} = \vec{k}_i - \vec{k}_s$, or momentum transfer, is left to the target. Both scattering angle θ_s and energy loss $\Delta E = E_i - E_s$ can be shown to be ajustable parameters of such experiments. Theories will rather use E and k as independent parameters, with

$$k^2 = k_i^{\ 2} + k_s^{\ 2} + 2\ k_i\ k_s \cos\theta_s. \qquad (1)$$

The energy-loss spectrum contains usually an elastic peak ($\Delta E = 0$), in addition to inelastic contributions ($\Delta E > 0$).

1.2. Binary encounter collision :

This approximation assumes that the collision takes place between an incoming particle and only one target electron, neglecting collective effects such as quantum mechanical exchanges inside the target. It can be shown to be valid especially at large momentum transfers.

1.3. Compton effect :

The Compton effect is a particular incoherent (or inelastic) scattering process where a photon interacts with a free (or quasi-free) electron [3] at rest. This electron is ejected with an asymptotic moment $\vec{k}_e$ (Fig. 1). The energy-momentum classical conservation conditions allow to write

$$\vec{k}_i = \vec{k}_s + \vec{k}_e, \quad E_i = E_s + E_e \qquad (2)$$

and $E_e = k_e^2/2$, or alternatively

$$\vec{k}_e = \vec{k}, \quad \Delta E = k_e^2/2 = k^2/2 \qquad (3)$$

This situation is illustrated in Fig. 2. Later, a quite similar effect was predicted [7] and observed [8] in the case of fast incident electrons.

1.4. First-Born approximation :

The simplest way to represent both incoming and scattered particles considered here in states of uniform motion and described respectively by the following plane waves exp $(i\vec{k}_i.\vec{r}_o)$ and exp $(i\vec{k}_s.\vec{r}_o)$.

This description assumes frozen states of the target before and after an instantaneous collision. With respect to the target ionization energy I_i of a given ionized subshell, it can be shown to be valid only at sufficiently high energies (E_i and $E_s \gg I_i$).

1.5. Impulse approximation (IA) :

A quite similar plane wave description exp $(ik_e.r)$ of the ejected electron is found to be valid when $E_e \gg I_i$. IA supposes tacitly that first-Born validity conditions are also fulfilled. Then a Compton peak $\delta(\Delta E - k^2/2)$ takes place at $\Delta E = k^2/2$. Furthermore the observed Döppler broadening effects on the Compton peak (Fig. 2) can be taken into account in a IA theoretical frame.

Deviations to this plane wave behaviour are obviously due to the electrostatic field of the remaining ion, hence responsible of observed Compton defects.

1.6. Compton profile :

A quantum·mechanical concept associated to the target electron momentum density $\rho(\vec{p})$ and first introduced by Du Mond [9]. For a free moving electron, the energy-momentum conservation conditions must be rewritten

$$\vec{k}_i + \vec{p} = \vec{k}_s + \vec{k}_e \;, \; E_i + p^2/2 = E_s + k_e^2/2 \qquad (4)$$

or

$$\vec{k}_e = \vec{k} + \vec{p} \;, \; \Delta E = k^2/2 + \vec{k}.\vec{p} \; . \qquad (5)$$

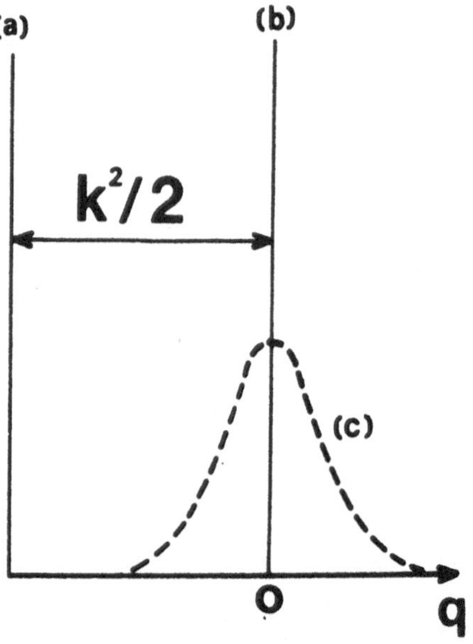

Fig. 2 - Main characteristics of the energy-loss spectra at large momentum transfers (notations in atomic units).
(a) elastic peak, (b) inelastic Compton peak, (c) Döppler broadening.

A simple classical approach to describe the Döppler broadening effects (Fig. 2) consists to perform a statistical average on the momentum density distribution. Under assumption of energy conservation (Eq. 5) the energy-loss spectrum (Fig. 2) is proportional to

$$\int d\vec{p} \; \rho(\vec{p}) \; \delta(\Delta E - k^2/2 - \vec{k}.\vec{p}) = k^{-1} \int d\vec{p} \; \rho(\vec{p}) \; \delta(p_z - q) = k^{-1} \; J(q) . \qquad (6)$$

$J(q)$, an even function of the <u>Compton parameter</u> $q = \Delta E / k - k/2$, represents the target <u>Compton profile</u>. $J(q)$ depends only on the target electronic structure. Amongst properties of $J(q)$, the following theorem [10,11]

$$\int_{-\infty}^{\infty} q^{2n} \; J(q) \quad dq = \int d\vec{p} \; \rho(\vec{p}) \; p_z^{\;2n} = \langle p^{2n} \rangle / \; 2n{+}1 \qquad (7)$$

can be established for spherical targets, with interesting applications for n=0 or n=1 (and uses of the virial theorem $\langle p^2 \rangle = - E$).

1.7. Sum rules for scattering cross sections :

Eq. (7) corresponds also to a number of integral properties [12] established for the double differential scattering cross section

$$\sigma_s^{(2)} = d^2\sigma/d\Omega_s.d\,\Delta E\ . \tag{8}$$

Under first-Born conditions and after integration on the energy loss-spectrum, the most interesting results are respectively

$$\int \sigma_s^{(2)}\ d(\Delta E) \simeq \sigma_o.N\ , \tag{9}$$

$$\int \sigma_s^{(2)}\ \Delta E\ d(\Delta E) = \sigma_o.\ N.\ k^2/2\ . \tag{10}$$

σ_o is a constant factor representing the basic interaction process, related to Thomson or Klein-Nishina formulas in the case of X-Rays, and alternatively to Rutherford or Mott-exchange formulas with incident electrons. Eq. (9) is valid only at large k under binary encounter assumptions, but Eq. (10) is exact. This result is known as the Bethe sum rule [13] and generalizes the Thomas-Reiche-Kuhn sum rule. Thus Eq. (10) provides a simple normalization procedure [14] of experimental results.

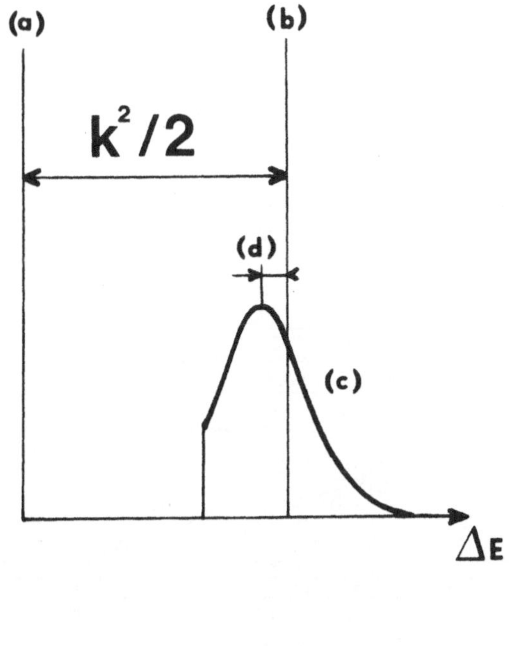

Fig. 3 - Energy-loss spectrum near ionization thresholds.
(a) elastic peak, (b) average energy-loss, (c) Döppler broadening, (d) Compton shift.

1.8. General conclusions :

The average energy-loss $\langle \Delta E \rangle$ is equal to $k^2/2$. That corresponds strictly to the situation found for collisions with a free electron at rest (Eq. 2), a result completely independent of any scattering structure. From Eq. (10) and Fig. 3 one can conclude that the energy-loss spectra must present an enhanced contribution in the neighbourhood of $\Delta E = k^2/2$ (or $q = 0$) . This situation suggests to plot experimental results and theoretical models as fonctions of the Compton parameter q in order to generalize the concept of Compton profile.

Another remark may concern the occurence of collective phenomenons which are known to produce a specific and important contribution to the Waller-Hartree incoherent scattering factors [1] at small scattering angles. From Eqns. (9) and (10), such contributions are expected in inelastic transitions to discrete states, but do not significantly contribute to the continuous spectrum. They will be neglected in the following sections.

1.9. Bethe surface (Fig. 4) :

A two-dimensional description of double differential scattering cross sections $\sigma_s^{(2)}$ at given values of ΔE (or q) and θ_s (or k).

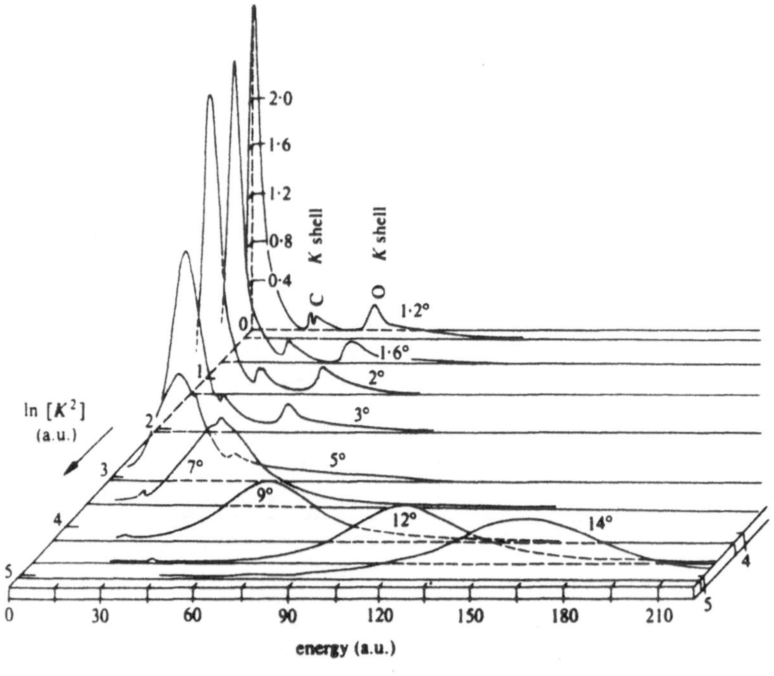

Fig. 4 - The Bethe surface of carbon dioxide with respective carbon and oxygene K-shell ionization thresholds. (courtesy from R.A. Bonham and H.F Wellenstein).

In the case of small momentum transfers, the energy-loss spectrum shows all successive ionisation thresholds as well as tail effects resulting from successive partial Compton profiles. At the contrary a total Compton line shape of the scatterer is observed at large k.

2 . SPECIFICITIES OF COMPTON SPECTRA

The Compton line shapes own typical asymmetries and their maximum is usually found shifted of the amount δE with respect to the average energy-loss $k^2/2$. A negative shift is shown in Fig. 3.

2.1. Compton shifts determinations :

Negative Compton shifts were observed in early experiments [15], but precise measurements have been obtained only during the past ten years. Amongst significant results are those of Weiss and Cooper and al. [16, 17] for several X-Ray measurements on lithium, polyethylene, beryllium and aluminium which indicate a maximum approximately equal to $\delta E = -10 eV \pm 3$ eV, since these shifts are found to vary with the momentum transfer.

At the same time were produced a number of electron scattering studies [18] exhibiting extreme negative shifts for H_2 and helium targets ($\delta E = -5 \pm 1$ eV) but also positive shifts for neon (+ 14 eV).

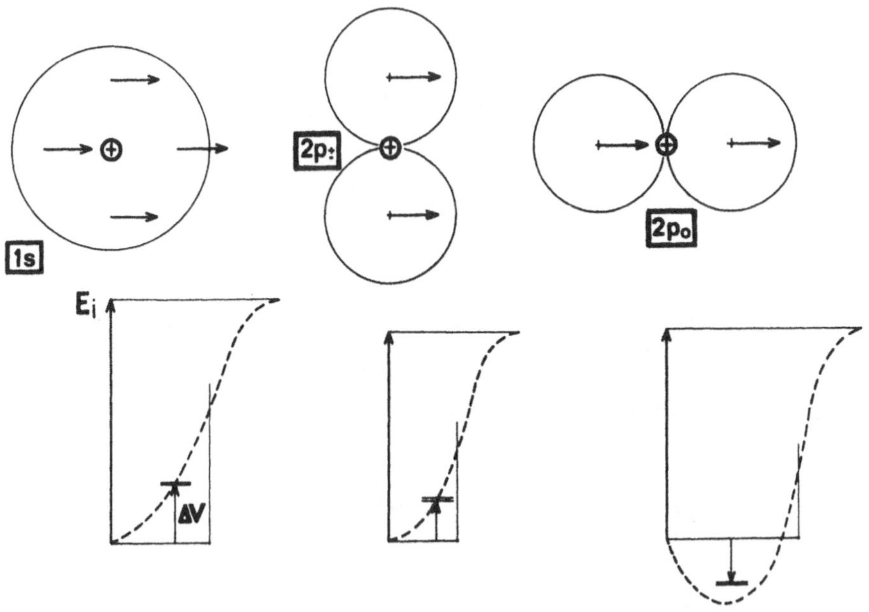

Fig. 5 - Qualitative orders of magnitude of average electrostatic field effects acting upon an electron ejected from various orbitals. Arrows represent the momentum transfer $\vec{K}$.

In fact a direct explanation of such signs and magnitudes is not easy. A preliminary approach [19] is presented in Fig. 5 and 6. It gives qualitative estimates of the average electrostatic field effects ΔV acting upon the ejection trajectories in the respective cases of 1s, $2p_{\pm}$ and 2 p_o atomic orbitals.

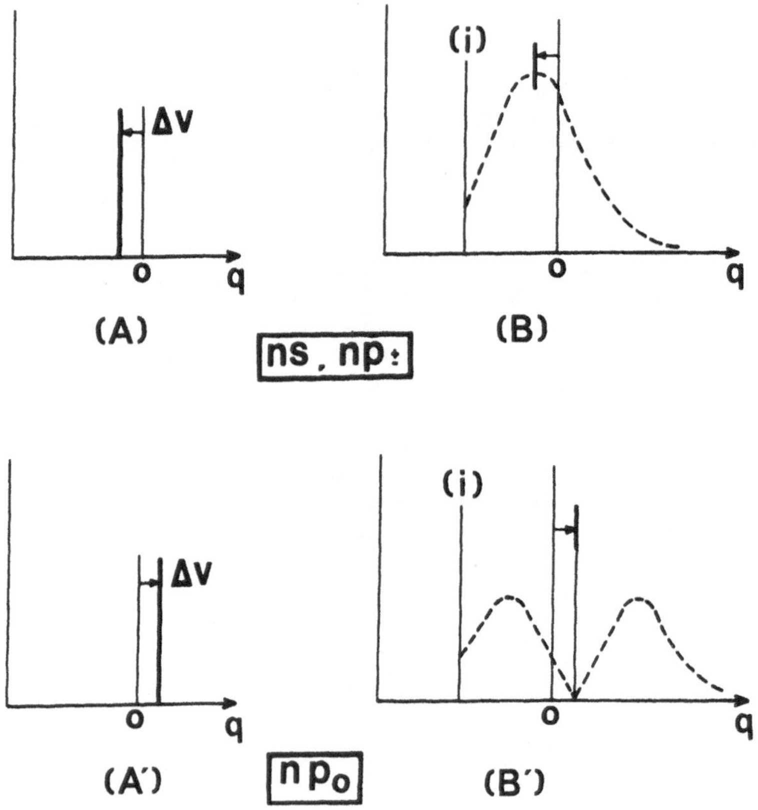

Fig. 6 – Schematic presentation of negative and positive Compton defects. A and A' represent the occurence of both possible ΔV shifts, (i) is the ionisation threshold. B and B' correspond to broadened and shifted peaks whose shapes can be shown to result from a Fourier transform of the orbital self-overlapping function S(R).

Since the energy conservation is rewritten here

$$\Delta E = k^2/2 + \vec{k}.\vec{p} - \Delta V \tag{11}$$

or $$q = p_z - \Delta V/k \quad , \tag{12}$$

this approach is sufficient to predict a correct sign for Compton shifts ($\delta E = - \Delta V$) together with their correct behaviour at large momentum transfers. Similar discussions can be held on other atomic or molecular orbitals in terms of their relative orientation with respect to k. However such cases are less evident to analyze and quantitative estimates of ΔV are not simple to define. Finally, such studies have been forsaken to the benefit of a full determination of Compton line asymmetries [20, 21] and the definition of Compton defects.

2.2. Compton line shape classical description :

According to Eqns. (3), (5) and (11), the simpler descriptions of the double differential scattering cross section $\sigma_s^{(2)}$ make successive uses of :

- a δ-function representation of the collision with N free electrons at rest :

$$\sigma_s^{(2)} = N\,\sigma_o\,\delta(k^2/2 - \Delta E) = N\,\sigma_o\,\delta(kq) = (N\,\sigma_o/k)\,\delta(q). \quad (13)$$

- Döppler broadening effects due to the initial velocity (or momentum) distribution of target electrons which leave the following average :

$$\sigma_s^{(2)} = \sigma_o \int d\vec{p}\; \rho(\vec{p})\; \delta(\vec{k}.\vec{p} - kq) = \sigma_{IA}^{(2)} =$$
$$= (\sigma_o/k) \int d\vec{p}\; \rho(\vec{p})\; \delta(p_z - q) = (\sigma_o/k)\; J(q) . \quad (14)$$

- average electostatic field effects as given by Eqns. (11) or (12) :

$$\sigma_s^{(2)} = (\sigma_o/k) \int d\vec{p}\; \rho(\vec{p})\; \delta(p_z - q - \Delta V/k) . \quad (15)$$

In fact this interpretation suffers from the lack of precise definitions of ΔV , which varies also with k. This important difficulty suggests rather the elaboration of an exact quantum mechanical treatment together with further possible approximations such as perturbation expansions limited to the most significant low-order contributions.

2.3. Quantum theory of Compton line shapes :

Under first-Born assumptions, the following result

$$\sigma_s^{(2)} = \sigma_o \langle \Psi_i | \sum_{\mu} e^{-i\vec{k}\vec{r}_\mu}\, \delta(K - E_i - \Delta E) \sum_{\nu} e^{i\vec{k}\vec{r}_\nu} | \Psi_i \rangle \quad (16)$$

was derived by several authors [10,22]. Ψ_i and E_i are the respective initial (ground) state and energy eigenvalue appearing in the Schrödinger equation $K\,|\Psi_i\rangle = E_i\,|\Psi_i\rangle$ of the target.

For sake of simplicity in the present discussion, the effects of collective contributions will be neglected in Eq. (16), in accordance to the previous discussions on binary encounter approximation, sum rules properties and large k oscillatory behaviour of exponential functions. Uses of the time-dependent δ-function representation

$$\delta(K - E_i - \Delta E) = (2\pi)^{-1} \int_{-\infty}^{\infty} dt \exp[it.(K - E_i - \Delta E)], \qquad (17)$$

leads finally to the calculation of

$$\langle \Psi_i | e^{-i\vec{k}\vec{r}_\mu} e^{itK} e^{i\vec{k}\vec{r}_\mu} e^{-itK} | \Psi_i \rangle =$$
$$= \langle \Psi_i | e^{-i\vec{k}\vec{r}_\mu(0)} . e^{i\vec{k}\vec{r}_\mu(t)} | \Psi_i \rangle. \qquad (18)$$

Eq. (18) is a particular expression appearing in the Van Hove formalism. Since the operators $\vec{r}_\mu$ (0) and $\vec{r}_\mu$ (t) do not commute, various approximations have been proposed in several effective calculations such as :

- statistical treatments with cumulants ;
- quasi-classical theories with time-expansions of the type $\vec{r}(t) = \vec{r}(0) + t\,\vec{k}/2 + \ldots$ which are found, for example, in the description of the inelastic scattering processes of slow neutrons.

In fact a correct calculation of Eq. (18) must follow very closely the physical events. The approximations used for slow incident neutrons do not have to describe the ejection of target particles. Above the first ionization threshold they must account for the ejection of target electrons in the case of X-Ray or fast electon scattering processes. The most convenient approach consists to write

$$\exp(-i\vec{k}\vec{r}_\mu) \exp(itK) \exp(i\vec{k}\vec{r}_\mu) = \exp[it(K + \vec{k}.\vec{p}_\mu^{\,op} + k^2/2)]. \qquad (19)$$

The hamiltonian K of the target system contains a potential energy contribution which does not commute with $\vec{p} = -i\vec{\nabla}$ and Eq. (19) has no simple matrix elements. At large k, Eq. (19) is expected to reproduce the impulse approximation with

$$K - E_i + \vec{k}.\vec{p}^{\,op} \simeq \vec{k}.\vec{p}^{\,op} . \qquad (20)$$

The plane-wave closure theorem

$$(2\pi)^{-3} \int d\vec{p}\; |e^{i\vec{p}\vec{r}}\rangle\langle e^{i\vec{p}\vec{r}}| = 1 \qquad (21)$$

together with the definition of $\rho(\vec{p})$ from p-Fourier transforms of Ψ_i give the following result :

$$\sigma_{IA}^{(2)} = \sigma_o (2\pi)^{-1} \int_{-\infty}^{\infty} dt \exp[it(k^2/2 - \Delta E)] \langle \Psi_i | \sum_\mu \exp(it\vec{k}.\vec{p}_\mu^{\,op}) | \Psi_i \rangle =$$
$$= \sigma_o \int d\vec{p}\; \delta(k^2/2 + \vec{k}.\vec{p} - \Delta E)\, \rho(\vec{p}) = \sigma_o/k.J(q). \qquad (22)$$

At this stage it is important to notice that $\vec{p}^{\,op} = -i\vec{\nabla}$ represents the translation operator whose average values for a given orbital

$$\langle \varphi(r) | \exp(it\,\vec{k}.\vec{p}^{\,op}) | \varphi(r) \rangle = \langle \varphi(\vec{r}) | \varphi(\vec{r} + \vec{k}t) \rangle = S(\vec{k}t) \qquad (23)$$

introduce (Fig.7) the spatial self-overlap function $S(\vec{R}) = S(\vec{k}t)$. In that case the IA result (Eq. 23) becomes, with $R = kt$

$$\sigma_{IA}^{(2)} = \sigma_o (2\pi)^{-1} \int_{-\infty}^{\infty} dt \exp[it(k^2/2 - \Delta E)] \langle \varphi(\vec{r}) | \varphi(\vec{r} + \vec{k}t) \rangle =$$
$$= \sigma_o (2\pi)^{-1} \int_{-\infty}^{\infty} dt \exp[it(k^2/2 - \Delta E)] S(kt) =$$
$$= (\sigma_o/k)(2\pi)^{-1} \int_{-\infty}^{\infty} dR \exp(-iRq) S(R) = (\sigma_o/k) J(q) \quad (24)$$

where $J(q) = \int_{-\infty}^{\infty} dR \exp(-iRq) S(R)$. (25)

J(q) is simply a Fourier transform of the self-overlap fonction S(R). The above equations suggest here that a molecular formalism can be used to solve problems related to Compton line shapes of atomic systems. This description requires only to define atomic or molecular orbital orientations with respect to k and explains the observed anisotropies in the Compton scattering of solids [6]. It suggests to that extent a number of simple pictures (as illustrated in Figs. 7, 8 and 9) to understand Compton defect phenomenons.

2.4. Compton defect theory :

From Eq. (19), one can rewrite exactly [23]

$$\exp[it(C + K - E_i)] =$$
$$= \exp(it\,C).[1 + i \int_0^t dt' \exp(-it'C)[K - E_i] \exp(it'C) + \ldots] \quad (26)$$

with $C = C^{op} = \vec{k}.\vec{p}^{\,op}$, in terms of a perturbation expansion similar to the time-ordered expansion first given by Dyson [24]. The leading term has been shown to reproduce the IA result (Eq. 24), i.e. Döppler broadening effects.
With $\Delta V = K - E_i$, $\Delta V(R) = \Delta V(kt) = \exp(-itC)\, \Delta V \exp(itC)$, the second term has a simple physical interpretation for individual orbitals

$$\int_0^t dt' \langle \varphi(\vec{r}) | \exp[i(t-t')C] (K - E_i) \exp(it'C) | \varphi(\vec{r}) \rangle =$$
$$= \int_0^t dt'' \langle \varphi(\vec{r}) | \Delta V(\vec{k}t'') | \varphi(\vec{r} + \vec{k}t) \rangle =$$
$$= k^{-1} \int_0^R dR' \langle \varphi(\vec{r}) | \Delta V(\vec{R}') | \varphi(\vec{R} + \vec{r}) \rangle = k^{-1} \Delta \mathcal{V}(R). \quad (27)$$

This contribution represent, under impulse assumptions, an estimate of potential energy changes $\Delta\mathcal{V}(R)$ at a given stage of the ejection. When ΔV is reduced only to an electron-nuclear potential interaction, exact analytical expressions are available from the well-known Roothaan methods and tables for molecular orbitals [25] . In this case and according to the quantum mechanical statistics, $\Delta\mathcal{V}(R)$ is the result of an average, on the impulse path $\vec{A'A} = \vec{R} = \vec{k}t$ of the ejected electron, over all intermediate nuclear positions (Fig. 7) .

A final Fourier transform J' (q) of $\Delta\mathcal{V}(R)$ thus represents a first order corrective contribution to the IA J(q), such as

$$\sigma_s^{(2)} = \sigma_o[k^{-1}.J(q) + k^{-2}.J'(q) + k^{-3}.J''(q)....]. \qquad (28)$$

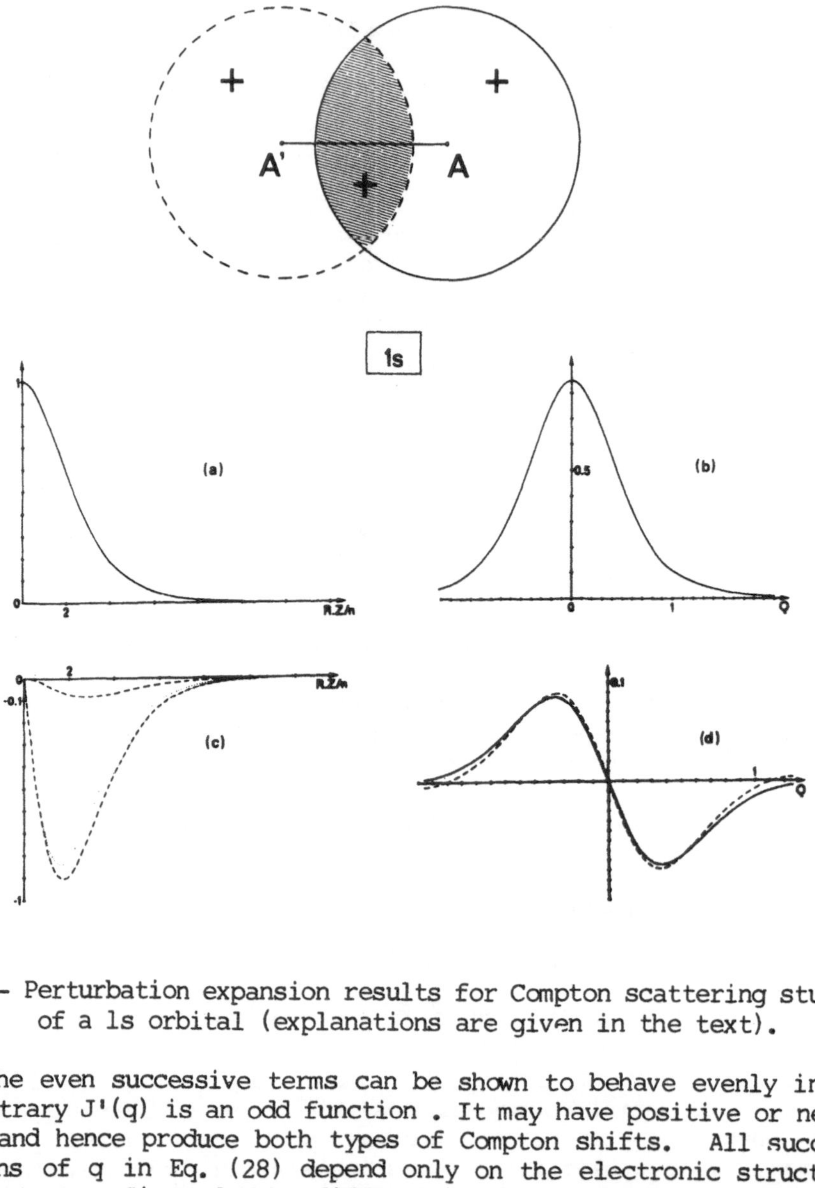

Fig. 7 - Perturbation expansion results for Compton scattering studies of a 1s orbital (explanations are given in the text).

The even successive terms can be shown to behave evenly in q. At the contrary J'(q) is an odd function . It may have positive or negative slopes and hence produce both types of Compton shifts. All successive functions of q in Eq. (28) depend only on the electronic structure of the scatterer. Since double differential scattering cross-sections in the q-scale can be splitted formally into even and odd parts, $k^{-2}J'(q)$ here represents a first order approximation of the Compton defect. In other words, the ejected electron can be considered as a probe which explores the electrostatic field of remaining ion and gives a number of additive informations on the target electronic structure.

2.5 . Compton defect : semi-quantitative pictures :

Figs. 7, 8 and 9 are typical illustrations of exact J(q) and J'(q) calculations [26] for some atomic orbitals whose self-overlap functions S(R) are represented schematically by shadow areas. Every situation is analyzed in terms of universal coordinates ζR for S(R) and $\Delta \mathcal{V}(R)$ or $Q = q/\zeta$ for J(Q) and J'(Q).

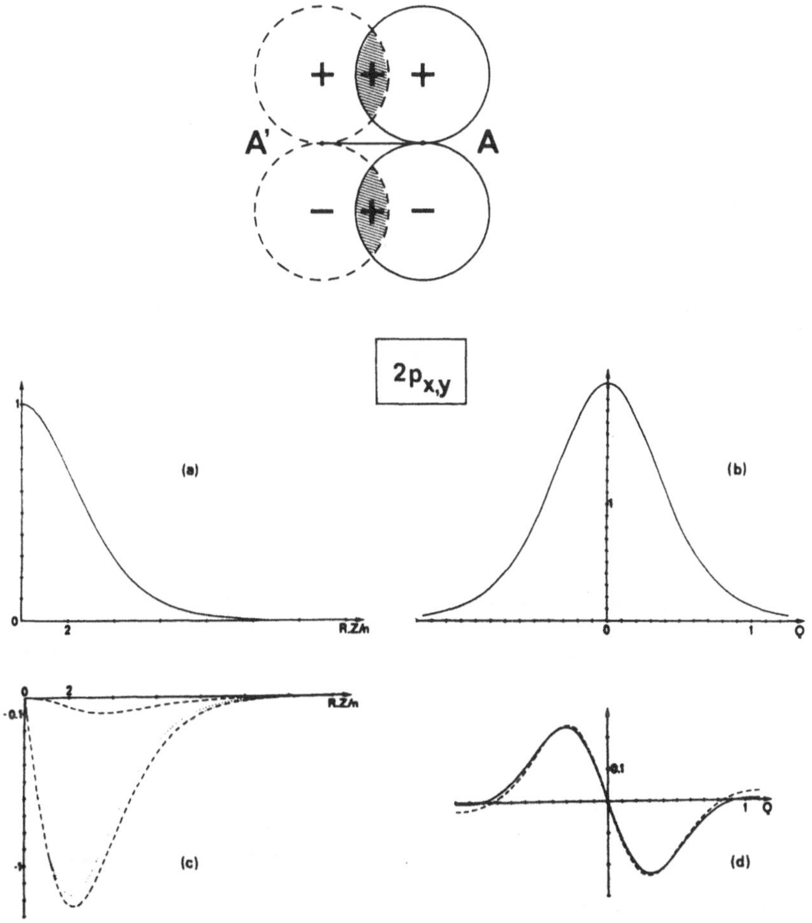

Fig. 8 - Perturbation expansion results for Compton scattering studies of a $2p_{\pm}$ orbital (explanations are given in the text).

Self-overlap functions S(R) are shown in Figs.(a) together with their Fourier transforms J(Q) in Figs.(b). $\Delta \mathcal{V}(R)$ functions in Figs.(c) result from the difference between two potential energy terms, also represented. The final result of their Fourier transforms is given in Figs. (d) (dashed lines) together with the result of an exact first-Born calculation [27] which is in fact only available for hydrogenic systems.

The observed magnitudes appear to be related to the orbital extents rather than the energetic specificities of nucleus-electron interactions, a conclusion in agreement with above discussions on δV shifts. Besides, a comparison beetween the different situations exhibits respective positive ns and $np_{\pm}$ and negative np_o self-overlaps along intermediate nuclear positions, which explains clearly both types of defects.

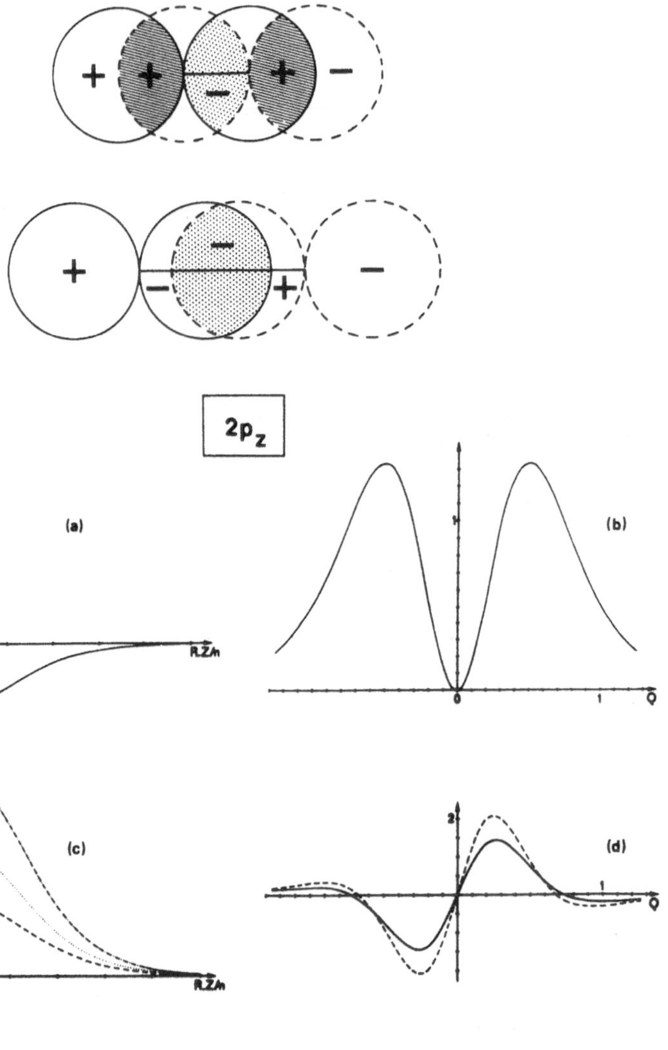

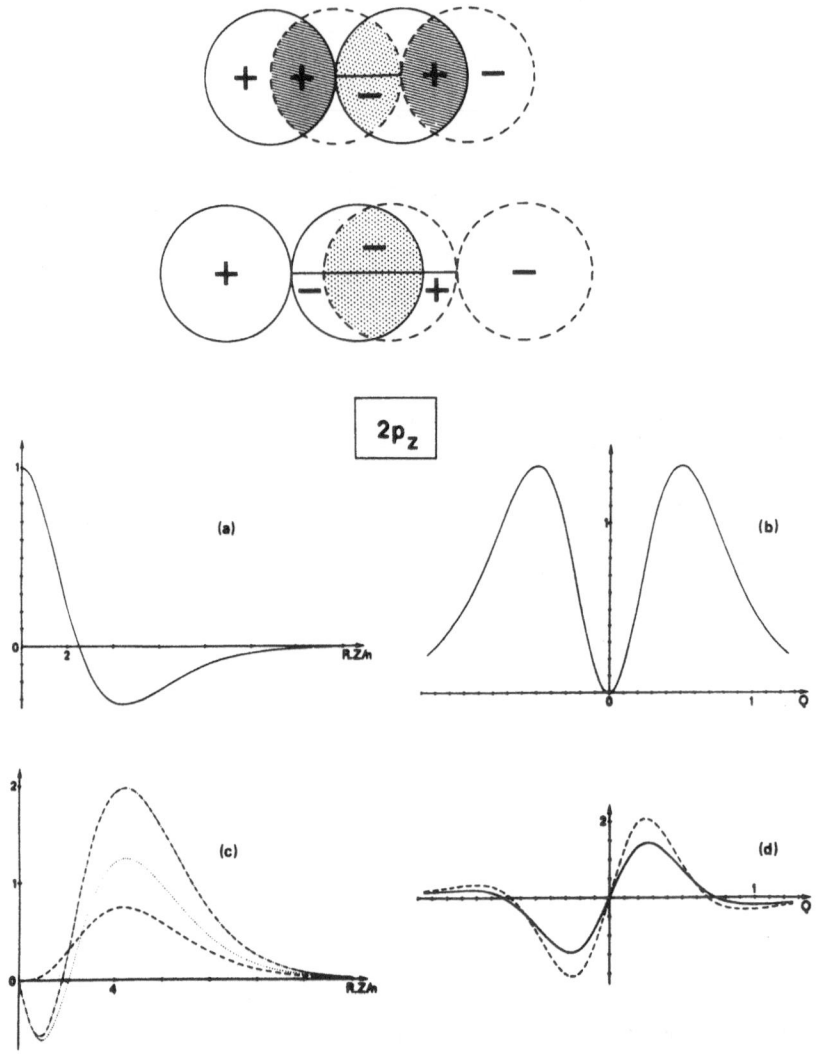

Fig. 9 - Perturbation expansion results for Compton scattering studies of a $2p_o$ orbital (explanations are given in the text).

Similar considerations allow to generalize the above results in predicting the correct sign of individual atomic Compton shifts from the parity of (l + m) quantum numbers. Since magnitudes were seen to depend strongly on the orbital extent along the direction k_0 only two "$2p_0$" electrons may induce a sufficient effect to produce positive shifts in the Compton scattering of neon atoms.

The formalism can be developped to include applications to the case of molecular orbitals as well as the contribution of bielectronic terms coming from the electron-electron interaction effects. As an example Fig. (10) illustrates an exact account of nuclear and electronic electrostatic fields acting upon the ejected electron of an helium target (F. Gasser and C. Tavard, unpublished results), together with the experimental statistics of counting rates.

However Compton defects calculations on molecular systems or solid structures were not attempted yet. Actually, unexplained differences between experimental results and impulse models [6] ask for such works.

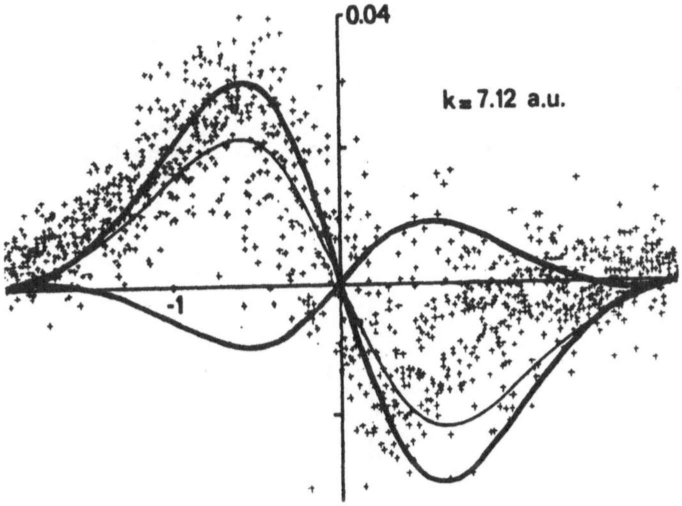

Fig. 10 – Compton defect of an helium target : theory vs. experiments. The crosses correspond to a complete set of experimental measurements from which an exact impulse profile calculated by Benesch has been substracted. The full lines show Compton defect contributions from the nuclear field (Z=2) [28], the electronic field due to the remaining electron and their sum (thin line). This sum would be reproduced approximately by an effective nuclear charge $Z^* \simeq 1.5$. Introduction of second order symmetric corrections [28] would improve also the observed agreemeent.

2.5. Other theories and results :

The first order corrective term $J'(q)$, based upon impulse assumptions to describe the first stage of ejection, gives a lesser agreement when the ejected electron is extracted from an orbital of large self-overlap extent (such as $2p_o$ in Fig. 9-d). A Coulomb wave description [27] is in fact strictly applicable to the formal case of hydrogenic targets. Approximate treatments using an effective screening factor Z* have been developped in several articles [29,30,31]. The main difficulty consists here to formulate a correct definition of Z*, found otherwise to vary with experimental conditions on k.

Another very interesting approach [32, 33] consists, with a generalized polarizability (determined with a complex coordinates method), in attempts to reproduce the final states involved into the transition matrix elements.

Finally, with the advent of new experimental techniques which determine triple differential scattering cross sections by coincidence techniques (see next chapter in this text), a summation over all angles of ejection performed from any accurate model [34] or experiment must also reproduce the corresponding asymmetries of the Compton line shape. While these treatments provide more or less accurate descriptions of observed Compton defects, a physical access to target electronic properties is generally lost. Their lone remaining interest consists to produce estimates of possible corrections to IA results.

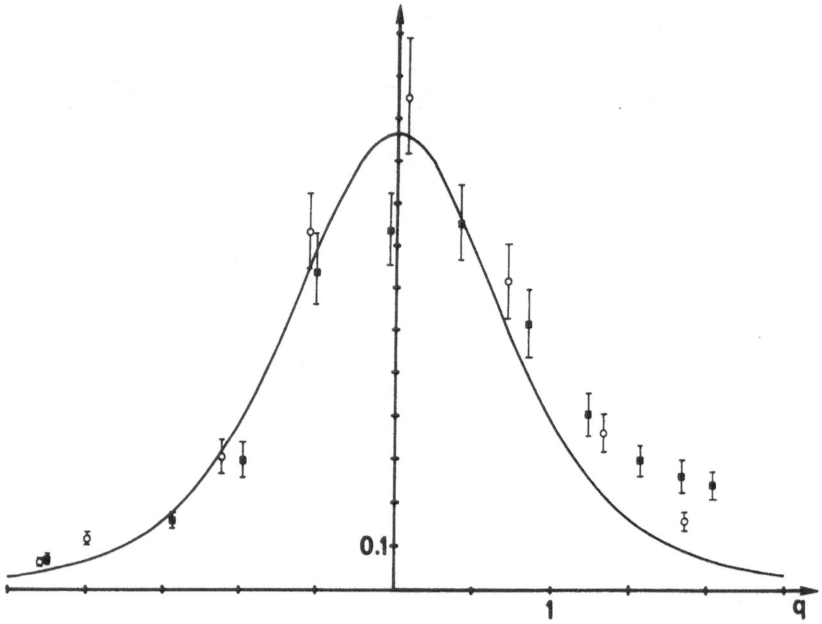

Fig. 11 – J(q) Compton profile and differential cross sections measurements on ejected electrons from an helium target [39]. The full line represents an exact impulse calculation of Benesh, together with experimental data from Opal and Beaty (2 keV incident electrons, white marks) and Toburen (2 Mev incident protons, black squares).

We must also mention here some interesting discussions on the physical interpretation of Compton defects [35,36,37] together with a simple method [38] for such corrections.

3 . OTHER STUDIES IN COMPTON SCATTERING

A number of other techniques also provide informations on momentum electronic structures.

3.1. Measurement of ejected electron double differential cross section :

The concept of Compton profile can be shown very useful to compare and classify [39] a number of second order differential cross section measurements performed on ejected electrons (Fig. 11) under impulse

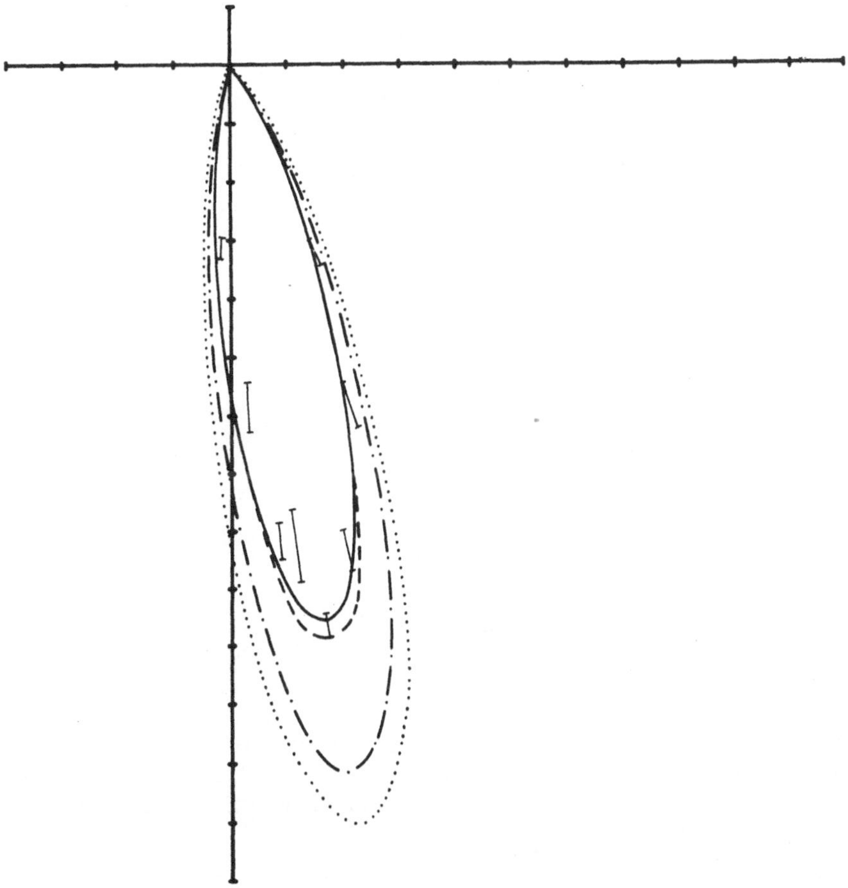

Fig. 12 - Normalized triple differential cross-sections measurements on a neon-2s ionized orbital. Experimental error bars together with the result of several calculations [43].

conditions and with varied projectiles. The observed failures to IA will not be discussed here.

3.2. (e,2e) coincidence studies :

Such measurements consist to detect in coincidence both scattered particle and ejected electron coming from a given collisional event. Until now, a very large number of experiments have been performed with incident electrons under varied kinematical conditions.

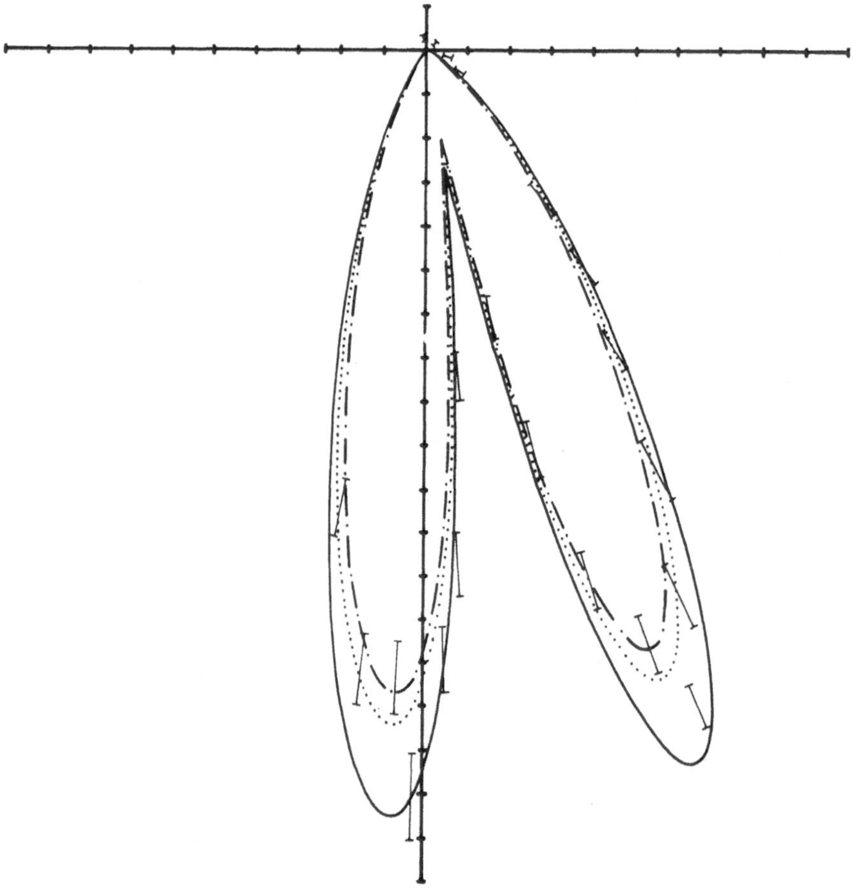

Fig. 13 - Normalized triple differential cross-sections measurements on a neon-2p ionized orbital. Experimental error bars together with the result of several calculations [43]. Evidence of a node (at p = 0) in the radial behaviour of $\rho_{2p}(\vec{p})$.

This new spectroscopic technique covers actually a very large area [40, 41], with applications to ionization mechanism studies as well as (shell by shell) electronic structure determinations in the case of atomic or molecular systems or thin film targets. For such purposes, a number of new devices such as uses of pulsed beams of incident electrons is developped [42].

As an illustration and among some hundreds of interesting results are shown two typical results (Figs. 12 and 13) on neon triple differential (e, 2e) cross sections [43], represented in polar coordinates for fixed values of energy loss ΔE and momentum transfer k . The selection corresponds here to $\Delta E = k^2/2$ in order to satisfy Compton impulse conditions. The horizontal axis represents the $\vec{k}_i$ direction and the results own an axial symmetry along $\vec{k}$.

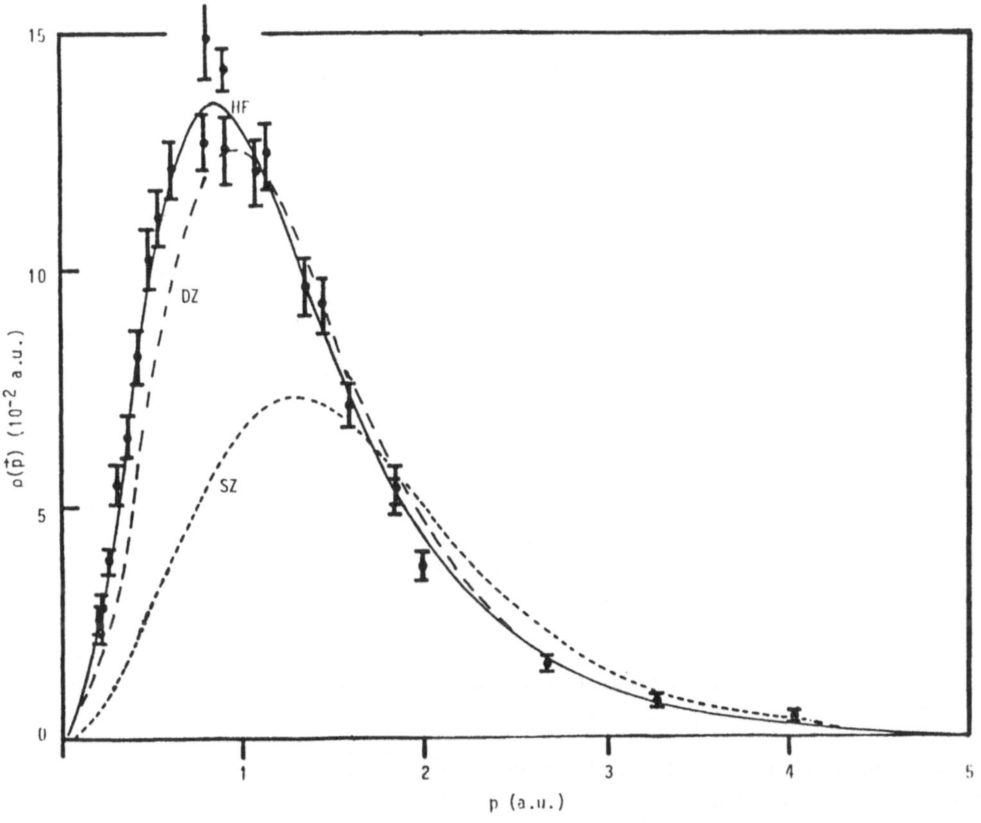

Fig.14 - Neon-2p radial momentum density : (e, 2e) impulse measurements [44] vs. Hartree-Fock calculations (Clementi & Roetti).

An impulse interpretation of these experimental results allows a shell-by-shell direct determination of momentum densities (i.e, Fig. 14). This last result reveals clearly the fundamental interest of (e,2e) techniques in physical chemistry studies. However, complete investigations of all experimental conditions indicate some very strong deviations found with respect to the IA model. Thus, one must take care to a blind use of such impulse models.

It is suggested here that highly intense X-Ray beams produced by synchrotron-radiation new machines may be used to develop similar (X,X+e) experiments for chemical investigations in solid structures.

4 . CONCLUSIONS

Departures from impulse assumptions are now generally explained and several models have been proposed to represent the observed Compton line shapes. Furthermore a suggested treatment yields additional informations on target electronic structures. However, effective calculations of Compton defects remain due in the case of molecular or solid systems.

Also, double differential cross-sections are related to both collisional and structural informations. It can easily be concluded that a one-dimensional information (such as $J(q)$ or $\rho(p)$) is provided here. Then triple differential cross section determinations ask for the dimensiondity of structural informations becoming available from such studies.

According to convenient selection of momentum transfers, a semi-classical discussion reveals that two-dimensional structural information is in fact available. A simple investigation in phase space parameters of $\rho(\vec{r},\vec{p}) = \rho(r,p,\theta_{rp})$ shows that the definition of following projections such as

$$\rho(p, \theta_{rp}) = \int_0^\infty dr\, r^2 \quad \rho(\vec{r},\vec{p}) \qquad (29)$$

and their observability do not contradict the Heisenberg uncertainty relationships, and appears to be related to the above triple differential cross sections. Yet, preliminary investigations performed on helium results indicate that models involving a radial electronic motion near the nucleus give clearly the best agreement with experiments realized at small momentum transfers and high asymptotic energies $k_e^2/2$ of ejected electrons, i.e. classically when the collision takes place in the close neighbourhood of the nucleus [45].

Although some possible definition of $\rho(\vec{r}, \vec{p})$ may rely on the well-known Wigner's isomorphism

$$\rho(\vec{r},\vec{p}) = (2\pi)^{-3} \int \psi^*(\vec{r}-\vec{u}/2) \exp(i\vec{p}\vec{u}) \quad (\vec{r}+\vec{u}/2)\, d\vec{u}\,, \qquad (30)$$

attempts to produce a quantum model using a phase-space description of the target were unsuccessful yet. Nevertheless some (e,2e) experiments performed under asymmetric conditions on atomic hydrogen targets are actually prepared to test IA extreme validity conditions.

A special acknowledgement is dedicated here to Professor R. Daudel and Doctor J. Maruani that kindly accepted the present contribution in these series on "Structure and Dynamics of Molecular Systems". Thanks are also due to Mr P. Senot and Mrs B. Chéry for their collaboration to the final typing.

REFERENCES

[1] - I. WALLER & D.R. HARTREE, Proc. Roy. Soc. London, 124 (1929),119.
[2] - P.O. LOWDIN, Phys. Rev., 97 (1955), 1474.
[3] - A.H. COMPTON, Phys. Rev., 21 (1923), 207 and 483.
[4] - R.A. BONHAM & M. FINK, High Energy Electron Scattering , Van Nostrand Reinhold Co, 1974.
[5] - Compton Scattering, B. WILLIAMS Editor, Mc Graw-Hill Inc, N.Y. 1977.
[6] - G.LOUPIAS in Structure and Dynamics of Molecular Systems, D. Reidel Publishing Co., Dordrecht, 1986, vol. II.
[7] - G.E.M. JAUNCEY, Phys. Rev., 50 (1936), 326.
[8] - A.L. HUGHES & M.M. MANN, Phys. Rev., 53 (1938), 50.
[9] - J.W.M. DU MOND, Phys. Rev., 33 (1929), 643.
[10] - C. TAVARD and R.A. BONHAM, J. Chem. Phys., 50 (1969), 1736.
[11] - I.R. EPSTEIN, Phys. Rev. A, 8 (1973), 160.
[12] - M. INOKUTI, Rev. Mod. Phys., 43 (1971), 297.
[13] - H. BETHE, Ann. Physik, 5 (1930), 325.
[14] - H.F. WELLENSTEIN & R.A. BONHAM, Phys. Rev. A, 7 (1973), 1568.
[15] - P.A. ROSS & P.KIRKPATRICK, Phys. Rev., 46 (1934), 668.
[16] - R.J. WEISS, Phil. Mag., 32 (1975), 247.
[17] - R.J. WEISS, M.J. COOPER & RS. HOLT, Phil. Mag., 36 (1977), 193.
[18] - W.H. RUECKNER, A.D. BARLAS & H.F. WELLENSTEIN, Phys. Rev. A, 18 (1978), 895.
[19] - C. TAVARD & F. GASSER, C.R. Acad. Sc. Paris, série II, 297 (1983), 101.
[20] - A. LAHMAM-BENNANI, A. DUGUET & H.F. WELLENSTEIN, Chem. Phys. Lettrs, 60 (1979), 405.
[21] - A. LAHMAM-BENNANI, A. DUGUET, H.F. WELLENSTEIN & ROUAULT, J. Chem. Phys., 72 (1980), 6398.
[22] - P.EISENBERGER & P.M. PLATZMAN, Phys. Rev. A, 2 (1970), 415.
[23] - F. GASSER & C. TAVARD, Phys. Rev. A, 27 (1983), 117.
[24] - F.J. DYSON, Phys. Rev., 75 (1949), 486.
[25] - C.C.J. ROOTHAN, J. Chem. Phys., 19 (1951), 1445.
[26] - F. GASSER & C. TAVARD, J. Chim. Phys., 78 (1981), 341 and 487 ; 79 (1982), 771.
[27] - B.J. BLOCH & L.B. MENDELSOHN, Phys. Rev. A, 9 (1974), 129.
[28] - C. TAVARD, M.C. DAL CAPPELLO, F. GASSER, C. DAL CAPPELLO & H.F. WELLENSTEIN, Phys. Rev. A, 27 (1983), 199.
[29] - L.B. MENDELSOHN & B.J. BLOCH, Phys. Rev. A, 12 (1975), 551.
[30] - T.C. WONG, L.B. MENDELSOHN, H. GROSSMAN & H.F. WELLENSTEIN, Phys. Rev. A, 26 (1982), 181.
[31] - S. MUKHOPADHYAYA, S.N. RAY & B. TALUKDAR, J. Chem. Phys., 76 (1982), 2484.

[32] - P. FROELICH & W. WEYRICH, J. Chem. Phys., 80 (1984), 5669.
[33] - P. FROELICH & W. WEYRICH, J. Chem. Phys., 82 (1985), 2305.
[34] - C. DAL CAPPELLO, C. TAVARD & A. DUGUET, J. Physique, in press (1986).
[35] - R.J. WEISS, Phil. Mag., 36 (1977), 193.
[36] - P. PATTISON & S. MANNINEN, Phil. Mag., 36 (1977), 1265.
[37] - T. SUZUKI & H. NAGASAWA, J. Phys. C, 14 (1981), 783.
[38] - B.G. WILLIAMS & R.F. EGERTON, Chem. Phys. Lettrs, 88 (1982), 95.
[39] - C. DAL CAPPELLO, C. TAVARD & A. LAHMAM- BENNANI, Phys. Rev. A, 26 (1982), 2249.
[40] - Wave functions and Mechanisms from Electron Scattering Processes, F.A. GIANTURCO & G. STEFANI Editors, Lecture Notes in Chemistry, Springer- Verlag, Berlin, 35 (1984).
[41] - K.T. LEUNG & C.E. BRION, J. Electron Spectrosc. Relat. Phenom., 35 (1985), 327.
[42] - R.A. BONHAM, private communication.
[43] - C. DAL CAPPELLO, C. TAVARD, A. LAHMAN-BENNANI & M.C. DAL CAPPELLO J. Phys. B., 17 (1984), 4557.
[44] - A. DAOUD, A. LAHMAM-BENNANI, A. DUGUET, C. DAL CAPPELLO & C. TAVARD, J. Phys. B, 18 (1985), 141.
[45] - C. TAVARD, C. DAL CAPPELLO, F. GASSER & M.C. DAL CAPPELLO, Lecture Notes in Chemistry, Springer-Verlag, Berlin, 35 (1984), 264.

NONLINEAR OSCILLATIONS IN MODEL SYMMETRIC TRIATOMIC MOLECULES. INTRAMOLECULAR RELAXATION. EFFECT OF A LASER FIELD

S. BESNAINOU
Centre de Mécanique Ondulatoire Appliquée du C.N.R.S.
23, rue du Maroc, 75940 PARIS CEDEX 19 - FRANCE
D.F. ESCANDE
Laboratoire de Physique des Milieux Ionisés
Ecole Polytechnique, 91128 PALAISEAU - FRANCE

ABSTRACT. The transition to stochasticity for model symmetric triatomics has been studied by classical mechanics methods. For the first time it is demonstrated that the onset of stochasticity is due to the blow-up of the (1,1) stochastic layer in the local coordinate phase space, and the stochasticity threshold E_c of various molecules is calculated by means of an entirely analytical approximate formula. It is shown that E_c depends on the ratio (m_2/m_1) of the central atom mass to the terminal atom mass. The Lagrangian energy can be partitioned into sub-components whose variations indicate quasi periodicity of the energy flow below E_c and vibrational dephasing above E_c. In the test molecule CO_2 the normal mode behavior predominates within a rather large range of energies and dissociation by a laser field is not favored when exciting a local mode. The dissociation threshold is very high and not much lowered when there is an initial energy content. Results of similar studies in the literature are discussed.

1. INTRODUCTION

The methods of classical mechanics have been used in the past, mainly by astrophysicists[1], but also by physicists[2] to study non linear oscillations of Hamiltonian systems. The extension of these methods to investigate the dynamics of molecules[3,4], is rather recent. It is known that, at low energies, classical trajectories are quasi periodic and that the motion is stable until the energy reaches a critical value E_c which could be the threshold of irreversible intramolecular relaxation and beyond which the behaviour of the trajectories is stochastic. By sending a laser pulse on a molecule one can hope to induce, through multiphoton absorption, highly excited vibrational states and hence change the motion from stable to chaotic. Symmetrical triatomics belong to simple systems having more than one degree of freedom. When the bending motion is ignored they may be modelled as two kinetically coupled Morse oscillators. Under this assumption some studies on the stability of the oscillations[5] or on normal and local mode behavior[6] have already been published. In what follows we investigate again the dynamics of such symmetrical triatomics.

R. Daudel et al. (eds.), Structure and Dynamics of Molecular Systems – II, 109–128.

Studying the motion in both normal and local coordinate spaces we show that the onset of stochasticity is induced by the blow-up of the stochastic layer of the local (1,1) resonance and that the normal mode description is the most adequate over a rather large range of energies below E_c.

The stochasticity threshold E_c of various symmetrical triatomic molecules is calculated using an analytical approximate method[7]; its dependence on the mass factor is demonstrated. For such systems the transition to stochasticity is based on semi qualitative criteria such as a) the appearance of stochastic trajectories on Poincaré's surfaces of section [8] ; b) the beginning of irregular variations with time of some energy components which will be defined further ; c) the broadening of the spectral lines in the power spectrum of the classical trajectories discussed by NOID et al. [9]. The CO_2 molecule is chosen as a test molecule and the stochasticity threshold is determined following these criteria. We investigate also the effect of a driving force such a laser field on the dynamics of CO_2.

2. METHOD AND MODEL

The XZ_2 molecule is considered as a two degree of freedom system. Each XZ bond is pictured as a Morse oscillator whose potential energy is equal to : $D\,(1-e^{-ar})^2$ where D is the bond dissociation energy, a the anharmonicity parameter, r the internal or local coordinate (stretching vibration). The total energy is then given by the relation

$$E = \frac{m_1 m_2}{2(m_1^2\sin^2\alpha + m_2^2 + 2m_1m_2)}\left[(m_1+m_2)(\dot r_1^2+\dot r_2^2)\;\; 2m_1(\cos\alpha)\dot r_1\dot r_2\right]$$

$$+ D\,(1-e^{-ar_1})^2 + D\,(1-e^{-ar_2})^2$$

m_1 and m_2 are the masses of Z and X, α the bond angle. The kinetic energy contains a coupling term which for given m_1 and m_2 is maximum when $\alpha = \pi$.

The dynamics of the molecule will be studied by means of two complementary approaches.

1) <u>Lagrange equations of motion</u> are written in terms of the dimensionless symmetry coordinates Y_1 and Y_2 defined by the equations

$$Y_1 = a(r_1+r_2)$$

$$Y_2 = a(r_1-r_2)$$

Y_1 and Y_2 are also unnormalized normal coordinates when the potential is harmonic. Y_1 and Y_2, diagonalize the kinetic energy. The total energy may be written as follows in terms of D units, which will be used throughout this paper,

$$E = \frac{\dot{Y}_1^2}{2\omega_1^2} + \frac{\dot{Y}_2^2}{2\omega_2^2} + [1-e^{-\frac{Y_1+Y_2}{2}}]^2 + [1-e^{-\frac{Y_1-Y_2}{2}}]^2 \qquad (1)$$

with

$$\omega_1^2 = 2\,\frac{m_2+2m_1\cos^2(\alpha/2)}{m_1 m_2}\,Da^2$$

$$\omega_2^2 = 2\,\frac{2m_1\sin^2(\alpha/2) + m_2}{m_1 m_2}\,Da^2$$

Lagrange equations are solved numerically for the CO_2 molecule. The Runge -Kutta - Mc Gill [10] integrator (RKGS) with constant step is used when there is no driving force. Test calculations have been done for various trajectories. The chosen step is such that smaller steps do not change the results. The trajectories are integrated up to the time 1.25×10^{-12} sec., ($\simeq$ time 4000 in scaled units) which ensures moderate rounding off errors. If necessary the integration time could be increased. When a laser field is applied, the variable step Hybrid-Gear [11] integrator is used. The step value varies around the value kept for R.K.G.S. with error parameters $0.66\ 10^{-8}$ and $0.86\ 10^{-10}$. It has been checked that the two methods give comparable results when there is no laser field. In this latter case the total energy is constant but it is possible to define a symmetrical energy [5] satisfying the relation,

$$E_{SYM} = \frac{1}{2}\,\frac{\dot{Y}_1^2}{\omega_1^2} + 2\,[1-e^{-\frac{Y_1}{2}}]^2 \qquad (2)$$

and an antisymmetrical energy [5] such that

$$E_{ASYM} = \frac{1}{2}\,\frac{\dot{Y}_2^2}{\omega_2^2} + [1-e^{-\frac{Y_2}{2}}]^2 + [1-e^{\frac{Y_2}{2}}]^2 \qquad (3)$$

The coupling energy is then

$$E_{COUP} = E - E_{SYM} - E_{ASYM} \qquad (4)$$

As long as the total energy is low the normal coordinates are appropriate to describe the motion. Each subsystem (symmetric and antisymmetric) is nearly conservative and the variation in time of E_{SYM}, E_{ASYM}, E_{COUP} are expected to be periodic, oscillating around the initial values with a low "coupling" energy. When the total energy increases reaching the critical energy and going beyond, then the motion can no longer be described in terms of normal modes ; there is mode mixing. We then propose another partition of the energy according to

$$E'_{SYM} = E_{SYM} + \frac{1}{2} E_{COUP} \tag{5}$$

$$E'_{ASYM} = E_{ASYM} + \frac{1}{2} E_{COUP} \tag{6}$$

Since their sum remains constant E'_{SYM} and E'_{ASYM} will have opposite variations.

If the initial conditions are such that a small fraction of the total energy is in the antisymmetric mode, a linear approximation to the differential equations can be used. The problem is then separable ; Y_1 obeys a Morse oscillator differential equation whose analytical solution is known. The equation of motion of Y_2 still depends on Y_1 ; once Y_1 is replaced by its analytical form, after some rearrangements Y_2 is shown to be the solution of a Mathieu's equation[5].

When there is a driving force such as a laser the variations in time of the total energy E as defined in (1) are most significant. The interaction energy of the laser with the molecule is expressed in the classical form as : $\mu_1 \varepsilon_1 Y_1 \cos \omega_1 t$ or $\mu_2 \varepsilon_2 Y_2 \cos \omega_2 t$ following the excited mode ; $\mu_1 Y_1$, $\mu_2 Y_2$ are the transition dipoles. In a linear molecule, the symmetric mode is not infrared active so that μ_1 is the derivative of a polarizability component whereas μ_2 is the derivative of a dipole moment component.

The electrical anharmonicity is not taken into account. ε_1 and ε_2, ω_1 and ω_2 are respectively the laser field amplitudes and frequencies.

2) Hamilton's method applied to the conservative system Hamiltonian using local coordinates.

If p_i is the conjugate momentum to the local coordinate r_i then the Hamiltonian is given by the equation

$$H = \sum_{i=1}^{2} \frac{1}{2} G_{ii} p_i^2 + \sum_{i=1}^{2} \sum_{j>i} G_{ij} p_i p_j + \sum_{i=1}^{2} D_i (1-e^{-ar_i})^2$$

the G_{ij} are Wilson's[12] G matrix elements depending on the masses of the atoms and on the geometry of the molecule. H is considered[3,6] as the sum of an integrable Hamiltonian H_o with

$$H_o = \sum_{i=1}^{2} \frac{1}{2} G_{ii} p_i^2 + \sum_{i=1}^{2} D_i (1-e^{-ar_i})^2$$

and a perturbation term V with

$$V = \sum_{i=1}^{2} \sum_{j>i} G_{ij} p_i p_j = \sum_{i=1}^{2} \sum_{j>i} V^{ij}$$

V is the kinetic couling. We apply to H_o a canonical transformation choosing as new variables action angle variables J_i and ϕ_i (13,14). H_o changes into $\mathcal{H}_o$

$$\text{with } \mathcal{H}_o = \sum_{i=1}^{2} (J_i \Omega_i - J_i^2 \frac{\Omega i^2}{4D_i}) \tag{7}$$

Ω_i is the harmonic frequency of the local oscillator and is such that

$$\Omega_i^2 = 2D_i a_i^2 G_{ii} \tag{8}$$

The zero frequency ω_i satisfy the equality

$$\omega_i = \frac{\partial H_0}{\partial J_i} = \Omega_i (1 - \frac{J_i \Omega_i}{2D_i}) = \Omega_i \lambda_i \tag{9}$$

or
$$\omega_i = \Omega_i \sqrt{1 - \frac{E_i}{D_i}} \tag{10}$$

Where E_i is the initial energy of the XZ bond ; ω_i being a frequency associated to a physical system has to be positivê (or exceptionally zero), this imposes :

$$0 \leqslant J_i \leqslant \frac{2Di}{\Omega i} \tag{11}$$

Also we should have $E_i \leqslant D_i$. The perturbation V^{ij} is Fourier analysed (3) to give

$$V^{ij} = \sum_{mn} \frac{1}{2} V^{ij}_{mn} (J_i, J_j)\, e^{(m\phi_i - n\phi_j)} + \text{c. c.}$$

After some algebraic manipulation the Hamiltonian becomes

$$\mathcal{H} = \mathcal{H}_o + \sum_{i=1}^{2} \sum_{j>i} \sum_{mn} - 8G_{ij} \sqrt{\frac{D_i D_j}{G_{ii}\, G_{jj}}}\, \lambda_i \lambda_j \sqrt{1-\lambda_i^2}$$

$$\times \left\{ \sqrt{1-\lambda_j^2} \left[\frac{1-\lambda i}{1+\lambda i}\right]^{\frac{m}{2}} \left[\frac{1-\lambda j}{1+\lambda j}\right]^{\frac{n}{2}} \right\} \cos(m\phi_i - n\phi_j)$$

and in D units

$$H = E = \sum_{i=1}^{2} \left(\frac{J_i \Omega_i}{D_i} - \frac{J_i^2 \Omega_i^2}{4D_i^2}\right) + \sum_{i=1}^{2} \sum_{j>i} \sum_{mn} \frac{-8G_{ij}}{\sqrt{G_{ii} G_{jj}}} \lambda_i \lambda_j$$

$$\times \left\{ \sqrt{1-\lambda_i^2} \sqrt{-\lambda_j^2} \left[\frac{1-\lambda_i}{1+\lambda_i}\right]^{\frac{m}{2}} \left[\frac{1-\lambda_j}{1+\lambda_j}\right]^{\frac{n}{2}} \right\} \cos(m\phi_i - n\phi_j) \qquad (12)$$

This Hamiltonian is valid for any triatomic symmetric molecule as long as the bending motion is not taken into account, and the bond potentials are of the Morse type.

3. STUDY OF THE ISOLATED MOLECULE

We are now going to analyse theoretically the Hamiltonian H in order to determine the conditions under which various resonances become active. As an illustration, we shall study some Poincaré surfaces of section for the CO_2 molecule in the r, p space (the values for r and p are obtained through appropriate linear transformations of the Lagrange equations solutions : Y_1, Y_2 and their time derivatives $\dot{Y}_1$, $\dot{Y}_2$). The perturbation term in H is maximum when $\frac{d}{dt}(m\phi_i - n\phi_j) = 0$

or $\quad m\omega_1 - n\omega_2 = 0 \qquad (13)$

which is the well-known resonance condition. Using this relation we can calculate approximate values for J_1, J_2 from $H_0 \simeq E$. Let us assume $\frac{\Omega_i}{D_i} = 1$ which amounts to change the J_i by a scaling factor ; then (8) becomes

$$0 \leq J_i \leq 2 \qquad (14)$$

this gives the range where resonances should have their centers to be active, for given m,n, E. Although $\frac{\Omega_i}{D_i}$ is in general different for different molecules, condition (14) could be used for any symmetric triatomic to determine approximately the character of the resonances (real or virtual)[15a,15b] at zero order. Since the perturbation term depends on the amplitude V_r of the resonance (term between parentheses) and on the mass factor $\frac{G_{ij}}{\sqrt{G_{ii}G_{jj}}} = \frac{G_{ij}}{G_{ii}}$ characteristic of each individual molecule, it will affect to a large extent zero order virtual resonances so that they become real. Owing to the symmetry of the Hamiltonian we can expect the (1,1) resonance to have an important effect on the dynamics. We are going to demonstrate it. For various values of m,n, E we have calculated the corresponding J_i^r and V_r. Our results show that the main active resonance of H_o is the (1,1) resonance with m=n=1 as it is the only one satisfying condition (14) over a large range of energies. It gives rise to the symmetrical and antisymmetrical vibration modes even at low energy when the resonance width is not very large. The width of the resonance is given by the approximate formula[3]

$$(\Delta J_i)_{mn} = (J_i - J_i^r)_{mn} = 4\sqrt{\frac{|V_{mn}^{ij}|}{\dot{\omega}_i + \left(\frac{\partial J_j}{\partial J_i}\right)^2 \dot{\omega}_j}}$$

and $\dot{\omega}_i = \left(\frac{\partial \omega_i}{\partial J_i}\right)_{J_i = J_i^r}$.

We have calculated this width $(\Delta J_1 = \Delta J_2)$ at E=0.01 for H_2O, where the resonance (1,1) has been shown [6] to be predominant, and for CO_2. We find respectively 6 10^{-4} and 19.10^{-4} (apart from identical constant factors), this suggests that the stochasticity threshold is probably lower for CO_2 than for H_2O.

Up to the rather high value of E=0.7, at zero order the (1,2) resonance remains virtual but some higher order resonances are real with a very small amplitude. In any cases under the effect of the perturbation the (1,2) resonance is expected to become real at a smaller E value. To illustrate this point we show some Poincaré surfaces of section[1] for CO_2 at E = 0.5 in the (r,p) space. In figure (1a) we see the curves $r_1 = f(p_1)$ $(r_2=0\ p_2>0)$ and in figure (1b) the curves $r_2=f(p_2)$ $(r_1=0\ p_1>0)$. Fourteen trajectories have been integrated over nearly forty periods of the symmetric vibration with initial conditions such that the energy in the symmetric mode goes from 98% to 2% of the total energy, with no coupling energy at time t = 0. This sampling is not refined enough to set out regions where some trajectories could be stochastic as was demonstrated recently[16], but it shows anyway the location of the resonances in the internal coordinate phase space for given initial conditions on the normal modes. When the energy in the symmetric mode is 98% of the total energy we can see the crescent shaped (1,1) resonance curve on the right of the drawing. When the energy in the symmetric mode is decreased, the trajectories are distorted, some of them open and finally the (1,2) resonance is present only when the energy in the symmetric mode is 15% of the total energy and less. It appears on the

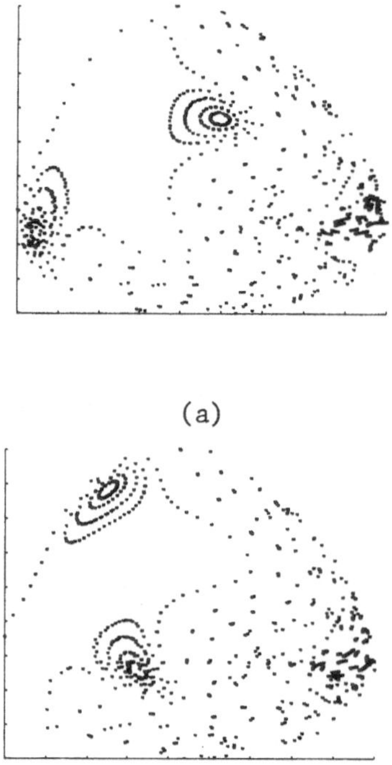

(a)

(b)

Figure 1. Poincaré surfaces of section for CO_2 in local coordinates phase space : a) $r_1 = f(p_1)$ b) $r_2 = f(p_2)$.

Poincaré surfaces of sections as pairs of concentric islands. Thus due to the interaction of various resonances (even virtual) which are a manifestation of the kinetic coupling the (1,2) resonance has become active.

We have also drawn Poincaré surfaces of sections in the $(Y,\dot{Y})$ space, equivalent to the (Y,P) space if P is the conjugate momentum to Y. They are in general good agreement with the more complete ones given in the literature [16].

It must be pointed out that in this case the (1,1) local resonance cannot be seen as it is "killed" through the transformation from local to normal coordinates. Assuming that the (1,2) resonance is the nearest to affect the (1,1) main resonance, we have calculated the stochasticity threshold of several symmetric triatomics by means of an approximate analytical formula [7] applicable to a variety of Hamiltonians. Near the resonance the Hamiltonian is reduced to a time dependent Hamiltonian having one degree of freedom. Let M and P be the energy dependent amplitudes of the relevant resonances (here (1,1) and (1,2)). It is shown that the resonance width w (ρ) is proportional to

P, M and ρ where $\rho = \frac{2M^{1/2}}{\pi k}$, $\rho \ll 1$, $k = \frac{\vec{r}.\vec{p}}{\vec{r}.\vec{m}}$

(here $\vec{r} \begin{cases} \omega_{2r} = \omega_r \\ -\omega_{1r} = -\omega_r \end{cases}$ $\vec{p} \begin{cases} 1 \\ -2 \end{cases}$ $\vec{m} \begin{cases} 1 \\ -1 \end{cases}$).

If $\frac{P}{M}$ is of the order of ρ^s, the value ρ_s of ρ at the threshold of the stochastic layer blowup is estimated from the variation curve of $w(\rho)$ as $\rho_s = [u+2 - 2(u+1)^{1/2}] / u^2$ with $u = 2k+1-s$. Equating ρ_s with ρ_s allows to calculate E_c, M being approximated by $|V_{11}^{12}|$.

The molecules were supposed to be linear, this increases the coupling. The results are displayed in table 1, where we also give when available results obtained by other authors.

A general inspection of the table shows the mass dependence of the stochasticity threshold (17) within the present model. In the XZ_2 series the stochasticity threshold increases with the central atom mass m_2. Indeed the mass factor of the coupling term which could be written as $\left(\frac{1}{1+\frac{m_2}{m_1}}\right)$ decreases and so does the coupling and the transfer of energy from one moiety of the molecule to the other is less easy. On the contrary if it is the terminal atom mass m_1 which increases, then the stochasticity threshold decreases. In this case, one can see that the mass factor of the coupling term increases and so does the coupling *

The variation observed when going from CO_2 to H_2O is in agreement with the results of other authors[6], using a similar model it is based on the appearance of a few bifurcated or stochastic trajectories on Poincaré surfaces of section. In contrast there is a noticeable discrepancy between our calculation and the results of Farantos et al.[18] for the molecule SO_2 whereas for O_3 (third colum) the orders of magnitude are more comparable. Two factors could be invoked to account for the discrepancies. One is the fact that the quoted authors consider that the molecule has three degrees of freedom. In this case the transition to stochasticity is determined by following the variation of the maximal Lyapunov characteristic number and the evolution of the spectral lines in the Fourier transforms of the trajectories. The other factor is the use of a very different type of potential which is a rather accurate quartic expansion fitted on spectroscopic data with correct behavior of the various diatoms in the molecule at all internuclear distances. Indeed, the third degree of freedom probably lowers the stochasticity threshold as it introduces new resonances, but it is our feeling that most of the discrepancy comes from the different anharmonicity character of this potential. A similar calculation has been made by Hänsel[19] (fourth colum) in the case of O_3 with a less sophisticated quartic potential. Here the criterion for the onset of stochasticity is the growing complexity and broadening of the spectral lines in the Fourier transform of the trajectories. The rough agreement observed between the values calculated with such different models (including ours) is probably fortuitous. We shall now analyse in greater details the transition to sto-

* In hydrides where m_1 is equal to unity the coupling takes its minimum value which explains why the description in terms of local coordinates could be successful.

Table I - Stochasticity Threshold E_c of Symmetric Triatomics in D Units

The numbers in the second column are calculated using the approximate formula worked out by D. Escande (ref.7).

Molecule	E_c			
	This work	Other works		
		(i)	(ii)	(iii)
CO_2	0.404	0.37 ∿ 0.38	0.51	0.6 0.48
NO_2	0.42			
SO_2	0.52	0.26		
ClO_2	0.53			
CS_2	0.36			
OH_2	1.1	0.9 ∿ 0.99		
SH_2	1.42			
ON_2	0.44			
OO_2 (O_3)	0.43	0.31	0.39	
OCl_2	0.37			

chasticity for the CO_2 molecule. We shall apply the tests enumerated in the introduction ; this molecule has been studied by several authors (5,16,20) as it is shown in the first row of table 1. Each of the approximate values of the stochasticity threshold has been obtained through different methods but using the same basic model. They are in reasonable agreement : (i) is estimated from Poincaré surfaces of section at various energies (16), (ii) is deduced from the values of local entropies for different energies (20) (the value is smaller than unity when the trajectories are stable and larger when irregular regions appear on Poincaré surfaces of section). Finally (iii) are results inferred from studies of the symmetric mode orbital stability when the initial energy in the antisymmetric mode is a small fraction of the total energy (5,21). As was pointed out (6) this condition restricts the study to a single family of trajectories in phase space. But we have shown that this family belongs to the (1,1) resonance at E = 0.5 and this remains probably true for a larger range of E values. As shown before there is no other large primary resonance for creating large scale stochasticity (15,22). In agreement with reference (7) it is the broadening and the blowing up of the stochastic layer of the (1,1) main resonance which determines the onset of instability. So studying the orbital stability of the symmetric normal mode under these specific initial conditions amounts to study the behavior of the (1,1) resonance and this is less restricted than it could appear. In the recent paper (21) using Floquet theory analysis for the general case of symmetric triatomics, it was demonstrated that depending on the ratio $2(\omega_2/\omega_1)$, there are E values for which stable motion can reemerge even if instability occurs for lower E values. In the case of CO_2, $2(\omega_2/\omega_1) \simeq 3.83$; from the stability maps (fig. 3, reference (21)) one can see that such eventuality does not occur and from $E \simeq 0.48$ the motion is continuously unstable so that there is no ambiguity concerning the onset of stochasticity. In figure 2 the sub-components of the energy defined in part II, are plotted as functions of time. The initial conditions on the trajectories are such that a small fraction of the total energy is in the antisymmetric mode. This does not affect the character of the curves (quasi periodic or chaotic). Other calculations were performed with different initial conditions but the curves are not shown here *. For a total energy E = 0.2, the variations with time of the energy components are very regular (figures 2a, 2b, 2c). The small amount of energy which flows in and out of the symmetric mode seems to be compensated in an intricate way by both the "asymmetric" and "coupling" energy variations. This remains true for different initial conditions. For E = 0.5 the curves are still quasi periodic with greater amplitudes

---

* The stochasticity threshold E_c is difficult to determine accurately. It could happen that for some value E below E_c but close to it some trajectories become erratic. As is known they do fill a very small area in phase space (1a). In this case the motion could be said to have quasi periodic character. Inversely some periodic trajectories could exist even when the motion is stochastic. This seems typical of two degrees of freedom systems (22).

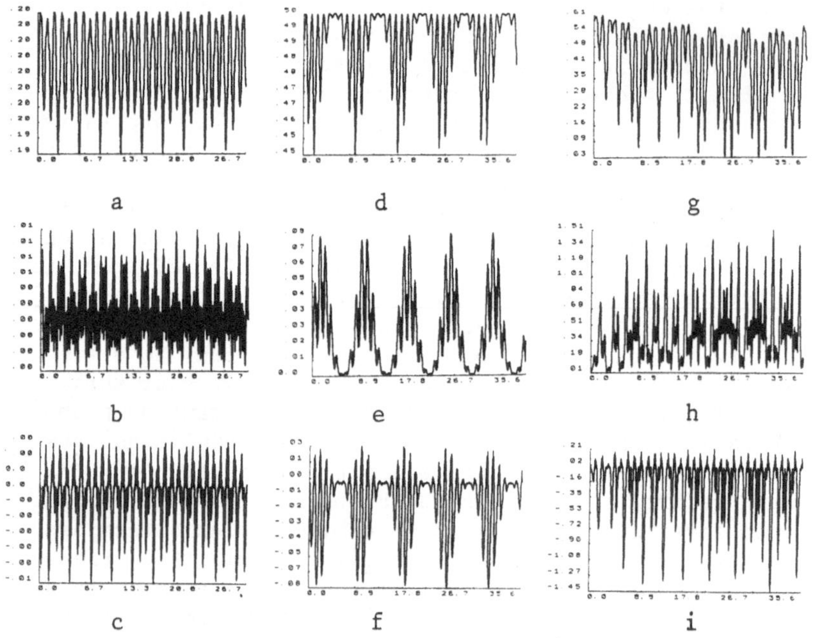

Figure 2. Variations with time of the CO_2 molecule energy subcomponents E_{SYM}, E_{ASYM}, E_{COUP} ; a,b,c, for a total energy of 0.2 ; d,e,f for a total energy of 0.5 ; g,h,i for a total energy of 0.6. Time is expressed in 10^{-2} scaled units (4000 units = 1.25 10^{-12} seconds).

of the oscillations (figures 2d, 2e, 2f) ; Although it appears now that the chosen energy partitioning is approximate the normal mode behavior is nevertheless still predominant. We give also in figures 2g, 2h, 2i the results for E = 0.6. Figures 2e and 2g have to be compared with figures 3a and 3b where the other partitioning which we proposed was used. In any case the variations in 2g and 3b are now chaotic. Figure 3b corresponds to a calculation where the time scale has been changed. From time zero to time 2000 (equal now to 1.25 x 10^{-12} sec.) it shows an irregular transfer of energy from one moiety of the molecule to the other. There is intramolecular relaxation due to the dephasing of the oscillations. The transfer is not irreversible as could be expected for a two degree of freedom system. This is confirmed by inspecting figure 3b from time 2000 to time 4000 ; one can see some periodicity in the chaos [23]. In the light of these results the value of the stochasticity threshold given in reference [20] seems the best estimate. The properties described here are typical of the model and do not depend on the particular choice of CO_2 only E_c is different from one molecule to the other and indicates the borderline between quasi periodic and chaotic behavior, as well as between reversible vibrational energy flow, and vibrational dephasing. The power spectra of trajectories $Y_1(t)$, $Y_2(t)$ for various values of E and the same initial conditions as above are reported on figure 4. The resolution is about 4% of the symmetric mode frequency. We have indicated by a vertical bar the fundamental frequency ω_S of the

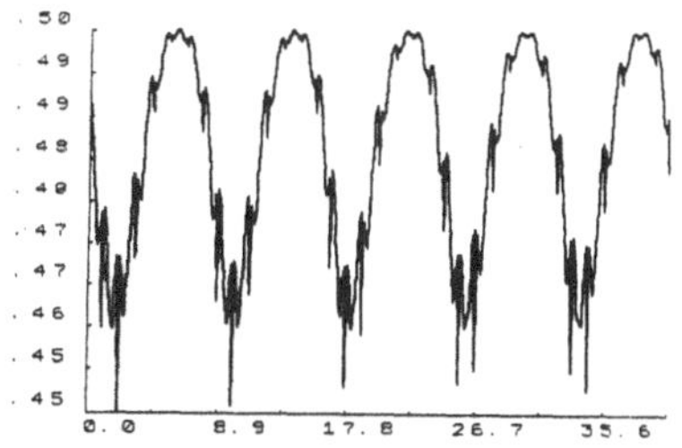

a

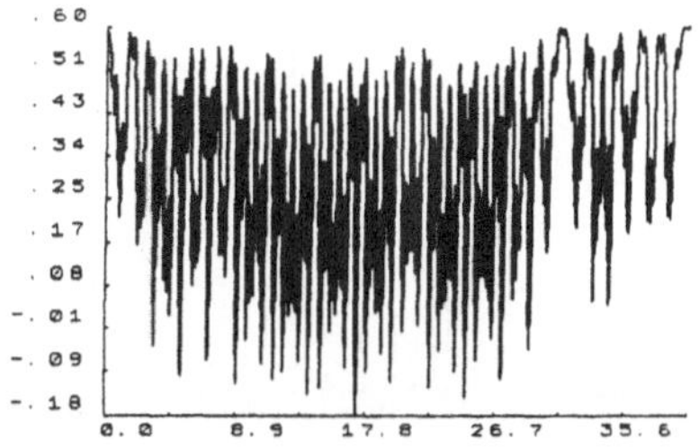

b

Figure 3. Variations with time of the energy subcomponent E'_{SYM} in the CO_2 molecule a) total energy of 0.5 Time expressed in 10^{-2} scaled units ; b) total energy of 0.6 ; time expressed in 0.5 10^{-2} scaled units.

symmetric mode having E energy (the time scale is such that the frequency of an isolated CO bond is 0.1). On the left hand side of the figure (4a, 4c, 4e, 4g) the Fourier transform of Y_1 are plotted for energies E respectively equal to 0.1, 0.2, 0.5, 0.6. Up to E = 0.5 the spectrum consists in a few sharp lines corresponding to the fundamental frequency and some overtones. Only for E = 0.6 do the lines broaden due to intramolecular relaxation. On the right hand side of the figure (4b, 4d, 4f, 4h) the spectra of Y_2 for the same energy values are reproduced. For E = 0.1 and E = 0.2 several lines appear, the strongest one is not very far from the fundamental anharmonic frequency of the asymmetric mode ω_{AS}, the others are combination bands of the type : $|\omega_{AS} \pm n\omega_S|$ where n is an integer. They correspond to the frequencies of approximate solutions for Mathieu's equation. At E = 0.5 which is near the border between stable and unstable motion the spectrum becomes complex, the lines being split into doublets. At E = 0.6 the lines show a large broadening, new lines appear at very small frequencies, indicating a stochastic behaviour of the trajectory.

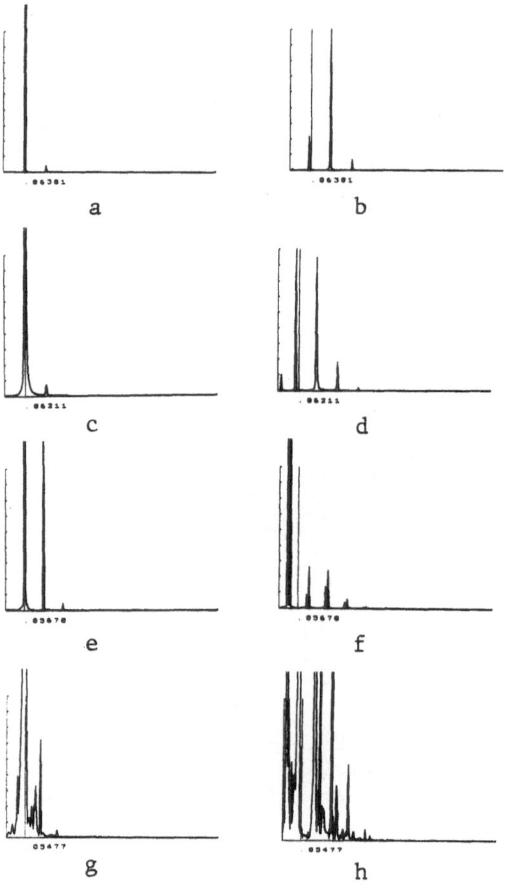

Figure 4. Fourier transform of CO_2 classical trajectories on the left : $Y_1(t)$ transform ; on the right : $Y_2(t)$ transform from top to bottom ; total energy of 0.1 ; 0.2 ; 0.5 ; 0.6.

4. EFFECT OF A LASER FIELD

The effect of an external oscillating force on a one degree of freedom [15,24] or two degrees of freedom [25] system has already been studied either by perturbation theory as in (15) or by classical trajectory analysis as in (24) and (25). In this latter case special attention was paid to the dependence of the trajectories types on the laser intensity and frequency. In what follows we rather intend to determine the intensity threshold I for dissociating the test molecule CO_2 given an initial energy content for different excitations(normal or local). We are interested in early dissociations as the trajectories are integrated over a time length of 1.25×10^{-12} sec., but this time could be lengthened in some cases. We are first sonsidering a molecule having an initial energy content E = 0.5 near the stochasticity threshold with most of the initial energy in the symmetric mode Y_1. We have shown in the preceeding paragraph that in the absence of a laser field, the symmetric mode is still stable. The frequency of the laser is tuned on the frequency $\omega^o_{AS}(1-\frac{E}{2})^{1/4}$ *, where ω^o_{AS} is the harmonic frequency of the asymmetric mode Y_2 and the field is in the direction of Y_2. The molecule dissociates into atoms at $I = 10^3 TW/cm^2$ ** (figure 5a). Inversely if one puts most of the initial energy in the asymmetric mode the threshold intensity is lower but still of the order of hundreds of TW/cm^2, inducing the break up of one bond (figure 5b). Putting again most of the energy in the symmetric mode and exciting it with the frequency $\omega^o_S(1-\frac{E}{2})^{1/2}$ does not lower much the intensity necessary to remove one atom which is now $\simeq 10^2$ TW/cm^2 (figure 5c) (ω^o_s is the harmonic frequency of Y_1). Under the same initial conditions the laser frequency was changed to $2\omega^o_S(1-\frac{E}{2})^{1/2}$ which is the first strong overtone appearing in the Fourier transform of Y_1 at E = 0.5 (figure 5e). A one-bond dissociation occurs for an intensity of 10^3 TW/cm^2 and at a time of $\simeq$ 3psec whereas in all the other cases, the molecule dissociated at early times ($\simeq$ 1.25 psec). The initial energy content of the molecule is then increased with energies much higher than the stochasticity threshold (0.7 and 0.8). When most of the initial energy is concentrated in Y_1, if Y_2 is excited with the frequency $\omega^o_{AS}(1-\frac{E}{2})^{1/4}$ and E = 0.7 one atom is removed for I = 50 TW/cm. But normal modes are known to be inadequate to describe the molecule at high energy, so one of the local coordinate was excited (which could be done by exciting both normal modes) with the frequency $\omega_o(1-\frac{E}{2})^{1/2}$ where ω_o is the harmonic frequency of the standard CO bond. The intensity of each laser must be of some TW/cm^2 in order to succeed in splitting the molecule into one atom and one diatom.

For an energy content of 0.8, the results are similar. Whatever the initial conditions are the intensity thresholdsremain high although they vary in magnitude. When the motion becomes chaotic it is not easier to dissociate the molecule through excitation of a local mode rather than a normal mode, this could be explained by considering the spectral analysis of the trajectories where the lines broaden. They contain probably several components. This is confirmed by inspecting the Fourier transforms of the local coordinates trajectories r_1 and r_2 at high energies. These results are in agreement with those of a recent paper (27) in which the dynamics of the driven unsymmetric mode is studied. It was shown that

---

* This value is obtained by putting q = 0 in Mathieu's equation :

$$\ddot{y} + (a - 2q\cos 2t)y = 0.$$

** To estimate I the following assumptions were made : μ_1 and μ_2 were given one of the values assumed for the HF molecule (24) [1.559(ev x bohr radius)$^{1/2}$]. The dissociation energy of CO in CO_2 was taken equal to 6.ev (26) A harmonic force constant $2Da^2 = 12.10^5$ dynes/cm was chosen for the CO bond in CO_2.

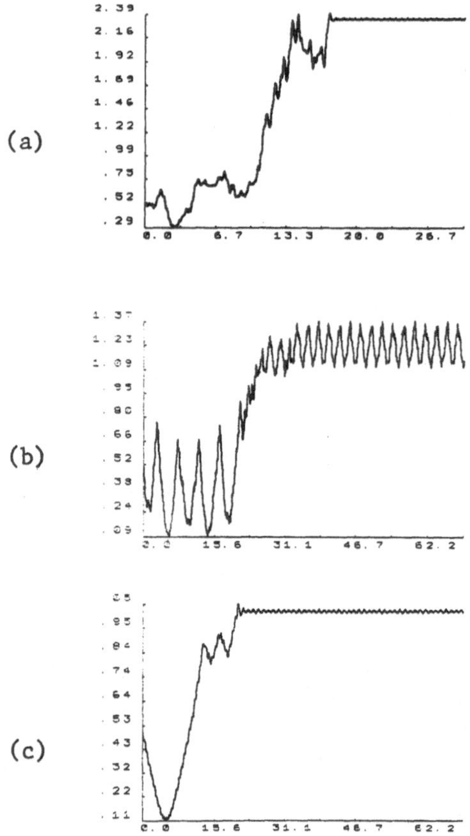

Figure 5. Total energy of a laser driven CO_2 molecule as a function of time. The initial energy content is 0.5. ; 5a) 98% of the initial energy is in the symmetric mode ; excitation along Y_2 with frequency 0.93 ω^0_{AS} ; 5b) 98% of the initial energy is in the asymmetric mode ; same conditions of excitation as in 5a) ; 5c) 98% of the initial energy is in the symmetric mode ; excitation along Y_1 with frequency $0.87\omega_s^0$ Time expressed in 10^{-2} scaled units. Oscillations around the dissociation plateau come from the numerical integration method.

in the unstable domain (here E >0.48) there are no frequencies of the driving term which can give rise to a resonant response.

If one starts with the molecule at classical equilibrium E = 0.0, one bond could be broken with an intensity of a few TW/cm^2 in various way either with the laser field along the asymmetric mode with frequencies : $0.93\omega^0_{AS}$, ω^0_{AS} or along the symmetric mode with frequencies $0.87\omega^0_s$, $0.93\omega^0_s$. The corresponding dissociation curves are reported on figure 6. Some of them show late dissociation (6b, 6d) with an early quasi periodic behaviour corresponding to multiphoton absorption. An energy of $\simeq$ 0.8 can thus be reached which represents 32 photons of the symmetric mode (about 16 of the asymmetric mode). Energetically it is not more advantageous to dissociate a molecule having already a rather high energy content. On the contrary starting from equilibrium allows a more flexible relaxation. The very high intensities necessary for dissociation are a consequence of the constant detuning of the laser with respect to the proper frequencies of the system which vary in time as they depend on the total energy. In the classical model we use this

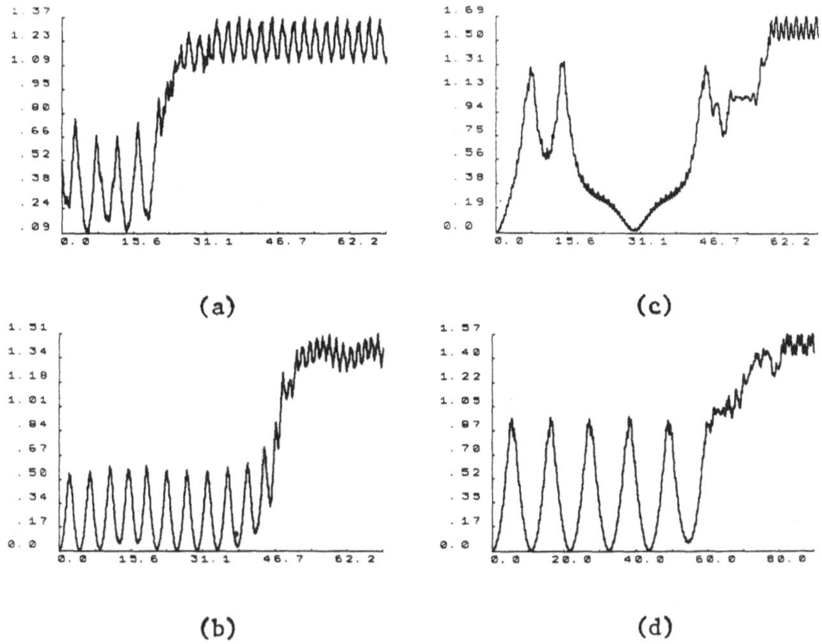

Figure 6. Total energy of a laser driven CO_2 molecule as a function of time with no initial energy content ; a) excitation along Y_2 with frequency $0.93\omega^0_{AS}$; b) excitation along Y_2 with frequency ω^0_{AS} ; c) excitation along Y_1 with frequency $0.87\omega^0_S$; d) excitation along Y_1 with frequency $0.93\omega^0_S$.

detuning cannot be compensated by quantum effects such as the Stark shift or the power broadening. Also the system molecule plus laser field can be considered as a conservative system where the dissociation occurs at a given constant energy. This imposes an initial energy content much greater than the dissociation energy. In this case one can expect the laser intensity to be of the order of TW/cm^2 (28). Figure 7 show the spectral components of Y_1, Y_2 when one of the normal modes is excited by a laser. From top to bottom the initial energies go from 0.7 to 0.0 ; the laser frequency is indicated by an arrow. All the spectra have discrete components embedded in a continuum. Figures 7e, 7f should be associated to curve (6a).

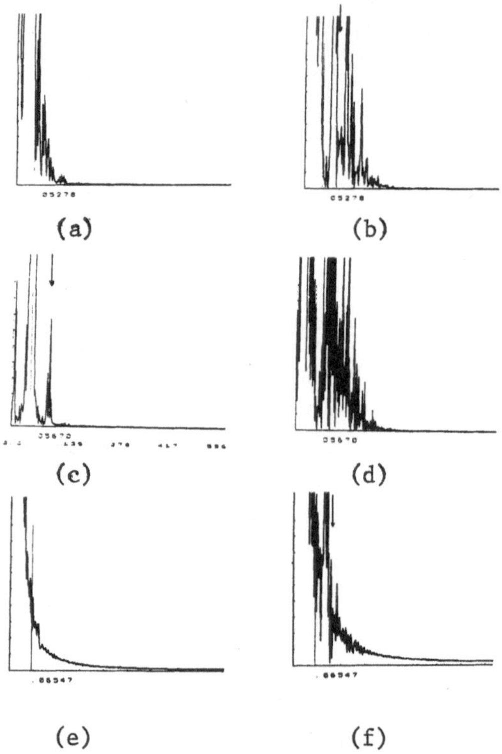

Figure 7. Fourier transforms of the CO_2 classical trajectories with a driving force ; On the left $Y_1(t)$ transform ; On the right $Y_2(t)$ transform. A black arrow indicates the position of the laser frequency for the corresponding excited mode.a,b)Total initial energy 0.7 ; frequency of the laser $0.87\omega^o_{AS}$; c,d)Total initial energy 0.5 ; frequency of the laser $0.87\omega^o_S$; e,f)No initial energy ; frequency of the laser $0.93\omega^o_{AS}$.

CONCLUSIONS

Classical Mechanics proves to be very useful in the study of the dynamics of molecular systems. By modelling symmetric triatomics as two kinetically coupled Morse oscillators we have shown new properties of these systems. We have demonstrated that the leading factor to the onset of stochasticity is the blow-up of the stochastic layer of the (1,1) resonance, while the overlaps of secondary resonances have a small effect. The stochasticity threshold E_c has been calculated using an analytical formula ; it depends on the ratio m_2/m_1. These properties would not necessarily hold for a different model potential. The type of potential governing the oscillations of the molecule determines their behavior at energies where small oscillations are no longer possible.

For the CO_2 molecule it appeared appropriate to partition the energy into sub-components (symmetric and antisymmetric) whose variations with time are not only signifiant of the quasi periodic or chaotic behavior of the motion, but illustrate also intramolecular relaxation. Such a partitioning is generally possible depending on the normal or local character of the vibrations [23,29] even for a greater number of degrees of freedom. In CO_2 the normal mode behavior is predominant. The Fourier analysis of the classical trajectories also help in the understanding of the dynamics. When a laser field is applied, the intensity necessary to dissociate a molecule is rather high. It does not depend substantially on the initial energy content or on the character of the excitation (normal or local).

Finally we would like to point out, although it is beyond the scope of this paper, that classical mechanics studies may be linked to quantum mechanical treatments through semi-classical methods.

ACKNOWLEDGEMENTS

One of us (S.B.) would like to thank Dr. D.A. DOWS (Department of Chemistry, U.S.C., Los Angeles) and Dr. M.F. GOODMAN (Department of Biological Sciences, U.S.C., Los Angeles) for their hospitality during the Summer 1982, and many enlightening discussions. This study was initiated thanks to numerous and active debates with Dr. E. THIELE and Dr. J. STONE (Department of Biological Sciences and Center for Laser Studies, U.S.C., Los Angeles) ; communication of some of their unpublished work is greatly acknowledged. Dr. J.-P. KORB (C.M.O.A.) and Dr. LEFORESTIER (Centre de Chimie Théorique, Orsay) are thanked for interesting discussions, and Mrs G. GIORGI (C.M.O.A.) for her technical assistance. Dr. I. HAMILTON's (University of TORONTO, CANADA) helpful cooperation during his stay at the C.M.O.A. was greatly appreciated.

REFERENCES

1. a) M. HENON and C. HEILES, Astron. J. 69 (1964), 73
   b) B. BARBANIS, Astron. J. 71 (1966), 415
2. G.H. WALKER and J. FORD, Phys. Rev. 188 (1969), 418
3. D.W. OXTOBY and S. RICE, J. Chem. Phys. 65 (1976), 1676
4. K.D. HÄNSEL, J. Chem. Phys. 70 (1979), 1830
5. E. THIELE and D.J. WILSON, J. Chem. Phys. 35 (1961), 1256
6. C. JAFFÉ and P. BRUMER, J. Chem. Phys. 73 (1980), 5646
7. J.P. CODACCIONI, F. DOVEIL and D.F. ESCANDE, Phys. Rev. Lett. 49 (1982), 1879
8. H. POINCARÉ, Les Méthodes Nouvelles de la Mécanique Céleste, DOVER Publications Inc., New York (1957)
9. D.W. NOID, M.L. KOSZYKOWSKI and R.A. MARCUS, J. Chem. Phys. 67 (1977), 404
10. S. GILL, Proc. Camb. Phil. Soc. 47 (1951), 96
11. C.W. GEAR, J. SIAM Numer. Anal. 2B (1964), 69

12. E.B. WILSON, J.C. DECIUS and P.C. CROSS, Molecular Vibrations, Mc GRAW HILL Books Cy Inc. (1955)
13. M. BORN, The Mechanics of the Atom, F. UNGER Publishing Co. (1960)
14. H. GOLDSTEIN, Classical Mechanics, ADDISON-WESLEY (Reading, Mass.) (1965)
15. a) G.M. ZASLAVSKII and B.V. CHIRIKOV, Soviet Physics Uspekhi (American Translation) 14 (1972), 549
b) B.V. CHIRIKOV Phys. Rep. 52 (1979), 265
16. T. MATSUSHITA, A. NARITA and T. TERASAKA, Chem. Phys. Lett. 95 (1983), 129
17. T. MATSUSHITA and T. TERASAKA, Chem. Phys. Lett. 100 (1983), 138
18. S.C. FARANTOS and J.N. MURELL, Chem. Phys. 55 (1981), 205.
19. K.D. HÄNSEL, Laser induced processes in molecules, Springer Series Chem. Phys. 6 (1979), 145
20. I. HAMILTON and P. BRUMER, Phys. Rev. A 23 (1981), 1941
21. E. THIELE, M.F. GOODMAN and J. STONE, J. Chem. Phys. 82 (1985), 2598
22. A.J. LICHTENBERG and M.A. LIEBERMAN, Regular and Stochastic Motion, Springer-Verlag New York Heidelberg Berlin (1983)
23. J.S. HUTCHINSON, W.P. REINHARDT and J.T. HYNES, J. Chem. Phys. 79 (1983), 4247
24. K.M. CHRISTOFFEL and J.M. BOWMAN, J. Phys. Chem. 85 (1981), 2159
25. D.L. MARTIN and R.E. WYATT, Chem. Phys. 64 (1983), 203
26. S.W. BENSON, Thermochemical Kinetics, John Wiley (1968)
27. E. THIELE and J. STONE, J. Chem. Phys. 83 (1985), 312
28. M.J. DAVIS and R.E. WYATT, Chem. Phys. Lett. 86 (1982), 239
29. M.L. SAGE and J. JORTNER, Adv. Chem. Phys. 47 (1981), 293

For more general surveys see
S.A. RICE, Adv. Chem. Phys. 47 (1981), 117
P. BRUMER, Adv. Chem. Phys. 47 (1981), 201

SPECTROSCOPIC INVESTIGATION OF LOCAL MOLECULAR MOTIONS IN POLYMERS

Lucien MONNERIE and Françoise LAUPRETRE
Laboratoire de Physico-Chimie Structurale et
Macromoléculaire associé au C.N.R.S.
Ecole Supérieure de Physique et de Chimie Industrielles
de la Ville de Paris
10, rue Vauquelin
75231 Paris Cedex 05
France

ABSTRACT. This paper deals with the study of local dynamics both in dilute solutions and in bulk polymers. The description of motions in macromolecular chains is first recalled, together with the orientation autocorrelation functions that are associated to the different models of local dynamics. The test of the motional laws by the various spectroscopic techniques is then examined. It now appears that, in spite of its complexity, the local polymer dynamics can be described by orientation autocorrelation functions derived from fluorescence anisotropy decay experiments, which take into account the specific character of the polymer chains, i.e. the orientation diffusion along the chemical sequence and the orientation loss terms arising either from the damping of the diffusion or from isotropic reorientations of some parts of the chain. These functions are able to account for experimental results obtained for polymers in dilute solutions as well as for polymers in bulk at temperatures well above the glass-transition temperature. The specific contributions of the Carbon-13 N.M.R. technique are discussed through several examples: Carbon-13 N.M.R is shown to provide a detailed description of the motions of the different parts of the chain. Such information can be obtained both in solution and in bulk states even at temperatures below the glass-transition temperature when using the high-resolution solid-state techniques.

From the very discovery of macromolecular chains, it has been recognized that, in these compounds which are in most cases built of σ bonds, each bond can adopt different positions on its valence cone and therefore the polymer chain is susceptible to afford a very large number of conformational equilibrium geometries (of the order of 3^N for a chain of N bonds). However, either from the point of view of global motions or from that of localized modes involving only a few segments, the chain dynamics is not completely elucidated and

R. Daudel et al. (eds.), Structure and Dynamics of Molecular Systems – II, 129–154.

is still the subject of a number of investigations.

From the variety of its techniques, Nuclear Magnetic Resonance is able to provide information on both types of motions. For example, the study of the diffusion coefficient in the presence of a field gradient is related to global motions, whereas the relaxation times are only (T_1, $T_{1\rho}$) or mainly (T_2) determined by local segmental modes.

In this paper, we will restrict our purpose to the study of local dynamics of polymers in solution and in bulk. We will first deal with the description of motions in macromolecular chains, then we will precise the specific contributions and capabilities of ^1H and ^{13}C N.M.R. measurements as compared with those of other spectroscopic techniques.

1. MODELS OF LOCAL MOTIONS

With the exception of quasi-elastic neutron and light scattering, the observed data in all the experiments based on spectroscopic techniques are related to the chain brownian motion mainly through the orientation autocorrelation functions of one or several given vectors belonging to the main chain or directly associated to it.

The orientation autocorrelation function involved in dielectric relaxation is:

$$M_1(t) = \langle \vec{U}_n(0).\vec{U}_n(t) \rangle$$
$$= \langle \cos\Theta(t) \rangle$$

whereas the orientation autocorrelation function involved in N.M.R., E.S.R. and fluorescence anisotropy experiments is:

$$M_2(t) = \langle P_2[\vec{U}_n(0).\vec{U}_n(t)] \rangle$$
$$= \langle 3\cos^2\Theta(t)-1 \rangle/2$$

where $\vec{U}_n$ is the unitary vector associated to the nth bond of the chain, $\Theta(t)$ describes the rotation of the unitary vector during time t=0 and t, and P_2 is the Legendre second polynomial.

1.1. Isotropic rotational diffusional motion

Such a model describes the motion of a vector joining the center of a rigid sphere to a point performing a random brownian diffusion on the surface of that sphere. It leads to the following expressions:

$$M_1(t) = \exp(-t/\tau'_R)$$

$$M_2(t) = \exp(-t/\tau_R)$$

$$\tau_R = (6D)^{-1} \quad \text{and} \quad \tau'_R = \tau_R/3$$

where D is the rotational diffusional coefficient of the sphere.

1.2. Conformational jumps in macromolecular chains

The bond motions in a polymer chain are much more complicated than the brownian motion of a small molecule. Clearly, the connectivity of the chain, which implies that no bond can be displaced independently of its neighbors, has to be taken into account.

A classification of the different types of motions that may occur in an alcane chain has been proposed by HELFAND (1). Let us consider a short sequence located in the middle of the polymer chain and having a well-defined conformation, for example:

PgttQ

where P and Q stand for the chain moieties on each side of the sequence under study. No conformational change is assumed to affect the P and Q polymer tails during the time interval where the above sequence undergoes the transition.

- Type 3 motions: in this type of motion, the orientation of Q relative to that of P is changed. An example of such a motion is the rotation of one bond of the sequence:

$$\text{PtQ} \rightleftarrows \text{PgQ'}$$

as shown in Fig.1. In such a motion, the viscous friction resulting from the orientational change of part Q is very large.

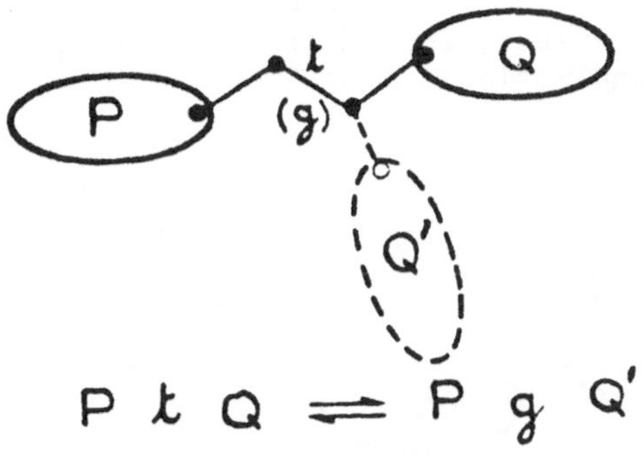

Figure 1. A type 3 transition.

- Type 2 motions: these processes do not alter the relative orientation of P and Q, but they lead to a translation of Q. An example is given in Fig.2. It corresponds to:

PgttQ $\rightleftarrows$ PttgQ'

In such a motion, only one energy barrier height has to be crossed and the viscous friction resulting from the translation of part Q is much less than the one involved in the type 3 processes.

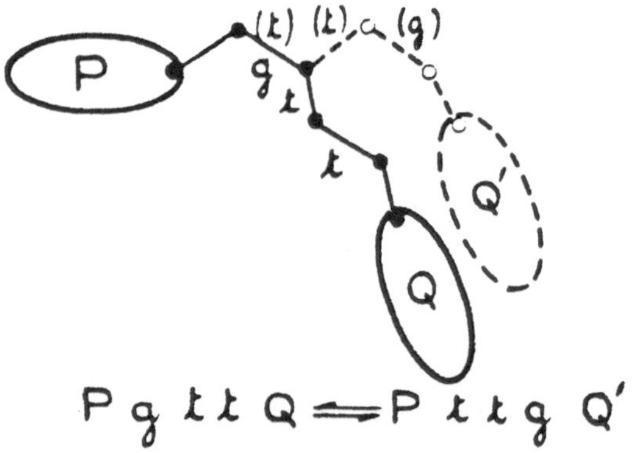

Figure 2. A type 2 transition.

- Type 1 motions: these modes are characterized by the fact that they leave the tails P and Q at the same position at the start and end of the transition; they do not involve any displacement of the ends of the sequence undergoing the motion. Among such motions, there are the 3-bond jump (Fig.3) and the crankshaft (Fig.4) (2). In these cases, the viscous friction is limited to the group of moving atoms and is therefore very weak. However, it must be noted that, due to the simultaneous orientational change of the moving bonds, the motion involves several barrier crossings, for example an average of 2 barrier crossings for a 3-bond motion.

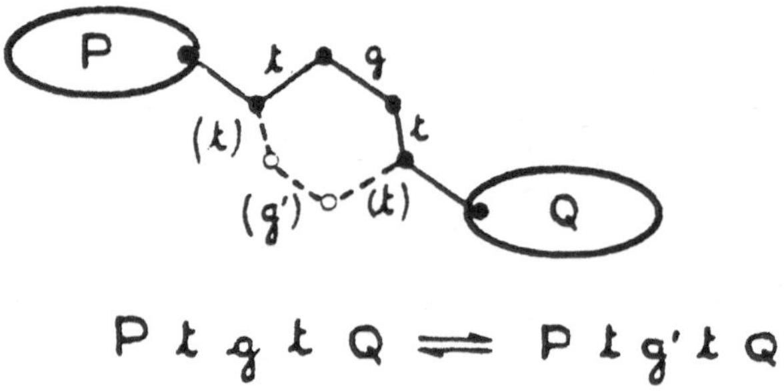

Figure 3. A type 1 transition: the three-bond motion.

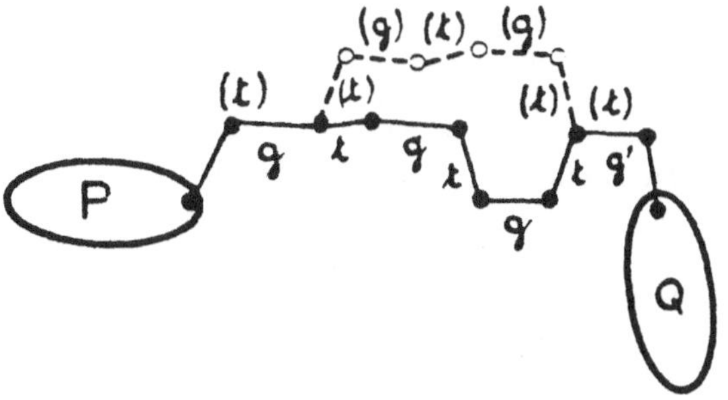

$$Pg(t\,g\,t\,g\,t)g'\,Q \rightleftharpoons Pt(t\,g\,t\,g\,t)\,t\,Q$$

Figure 4. A type 1 transition: the crankshaft transition.

Monte-Carlo brownian dynamics simulation of a polymer chain constrained to a tetrahedral lattice have been carried out based on these conformational jumps which do not affect the rest of the chain (3). As a result of the chain connectivity and of the coincidence of the bonds with the directions of the lattice, the smallest group which can move is made of three bonds and corresponds to the following conformational change: g $\rightleftarrows$ g' (Fig.5).

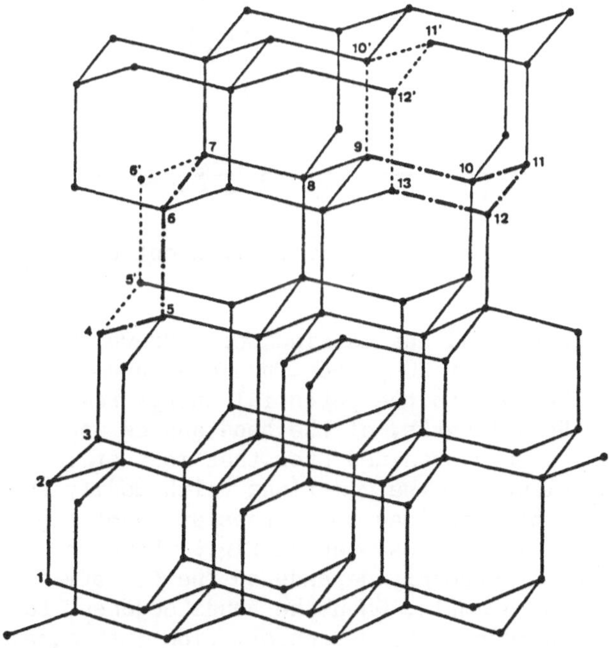

Figure 5. Representation of a chain on a tetrahedral lattice:
* A three-bond motion (g $\rightleftarrows$ g'): 4-5-6-7 $\rightleftarrows$ 4-5'-6'-7
* A four-bond motion (g'g $\rightleftarrows$ gg'): 9-10-11-12-13 $\rightleftarrows$ 9-10'-11'-12'-13

From the point of view of the evolution of the orientations of the vectors associated to the chain, which is our main interest here, it must be noticed that, after such a transition, the orientations of the first and third bonds have been interchanged. No new orientations have been created but orientations have diffused along the chemical sequence. Among all the 4-bond motions that may happen inside the chain, only one cannot result from the combination of two 3-bond jumps: this mode corresponds to the conformational transition: g'g ⇄ gg' (Fig.5). Such a motion leaves the orientations of the second and the third bonds unchanged, on the contrary the first and fourth bonds adopt new orientations after the transition. The 4-bond motions do not cause the orientations to diffuse along the main chain but they destroy the orientations which were originally present.

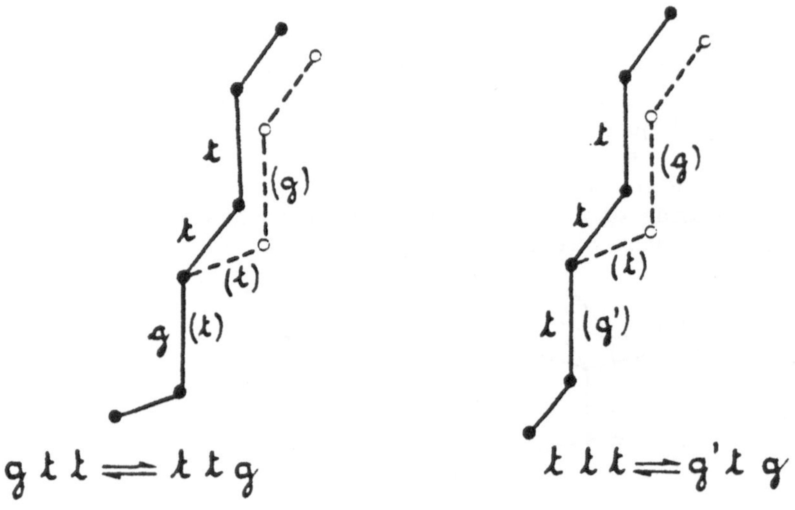

Figure 6. The type 2 transitions which are of frequent occurrence in HELFAND's simulation (4).

More recently, HELFAND (4) has carried out a Brownian dynamics simulation of an alcane chain of 200 bonds in a continuum, taking into account the contributions to the potential energy resulting from the distortions of the bond lengths, of the bond angles and of the torsional angles. Results thus obtained show that most of the occurring transitions belong to type 2 motions which modify the orientation of one bond only and lead to a translation of a part of the chain. Such processes are represented on Fig.6. However, the probability of such motions strongly depends on the conformation of the neighboring bonds. For example, when the bonds adjacent to the one undergoing the motion adopt a trans conformation (ttt case, Fig.6), the second neighbors are parallel to the central bond involved in the transformation and, from the results of the simulation, it seems that this fact promotes the motion to a large extent. HELFAND has noticed that, although type 2 motions are frequent, they do not occur in an isolated way. Indeed, such a motion of a bond, which induces a translation of a part of the chain, and therefore a large dissipated energy resulting from viscous friction, is very often associated to an opposite motion of the second neighboring bond, which allows to accomodate the motion of the first bond, without involving any translation of a part of the chain. As to the orientations of the bonds of the chain, this correlation of individual processes is equivalent to a diffusion of a bond orientation along the chemical sequence.

The fact that all the individual motions are not necessarily coupled is of importance for the bond orientations. Indeed, the isolated modes, which may be eventually associated to particular chain conformations, imply that the diffusional process of a given orientation along the chain will stop and that a new orientation will be created in the chain.

Therefore, from the results of the simulations carried out by HELFAND, it clearly appears that, concerning the orientation of the bonds, the chain motion corresponds to a random diffusion of the orientations along the chain, which is damped by isolated conformational jumps. From that point of view, and although these elementary 3-bond or 4-bond jumps seem to very seldom occur in the simulations carried out on alcane molecules, the conclusions as to the behavior of bond orientations are not modified.

A surprising aspect of the local polymer dynamics concerns the low value of the activation energies which have been determined experimentally or which have been obtained by HELFAND by using his simulations. Indeed, these values are of the order of a few Kcal/mole, which means that they approximately correspond to the barrier height associated to the internal rotation of a single bond in a small molecule. However, as pointed out above, all the motions that may occur in a polymer chain involve either several energy barriers (this is the case for the 3- or 4- bond motions) or the viscous friction of a chain sequence besides the barrier height for one bond. This paradox has been recently elucidated by SKOLNICK and HELFAND (5) by applying the approach of KRAMERS (6) to the trans gauche conformational change for an internal bond in an alcane chain. The basic idea consists in taking into account the distortions of the bond lengths, of the bond angles and of the torsional angles of the units next to the bond undergoing the conformational change. By looking for the path of steepest descent required for the trans gauche conformational change, it appears that the distortion resulting from the bond motion is spread over the neighboring units, without affecting the rest of the chain. This effect is schematically represented on Fig.7. Such an analysis shows that the deformation is mainly accomodated by distortions of the torsional angles of the neighboring bonds,and that some distortions of the bond angles may also occur. These distortions are of opposite sign with respect to the distortion of the bond undergoing the motion, in order to minimize the motion of the more remote parts of the chain.

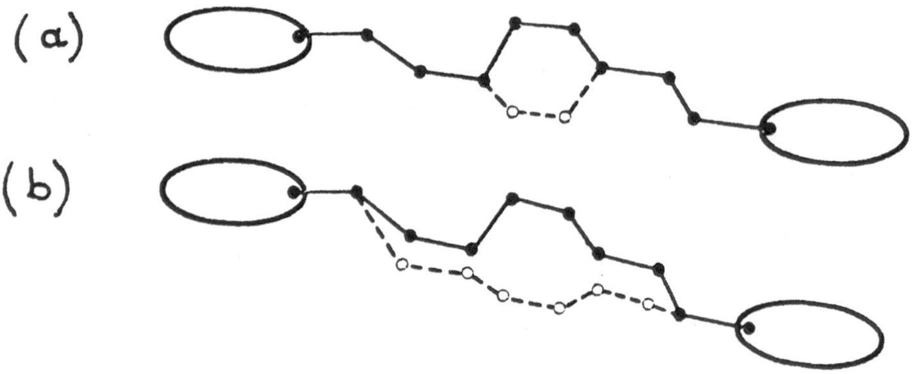

Figure 7. a)A localized transition with fixed ends.
b)The distortion resulting form the bond motion is spread over the neighboring units.

1.3. Motional laws in macromolecular chains

In the case of the spectroscopic techniques under study, the interesting quantity is the orientation autocorrelation function and more precisely $M_2(t)$.

In spite of the complexity and of the variety of motions in a chain, the above results show that the chain connectivity implies that, as to the bond orientation, the chain dynamics corresponds to a damped diffusional propagation of the orientations along the chemical sequence.

Starting from the description of a chain constrained to a tetrahedral lattice, VALEUR, JARRY, GENY and MONNERIE (VJGM) (7) have derived the following analytical expression:

$$M_2(t) = \exp(-t/\Theta) \exp(t/\rho) \operatorname{erfc}(t/\rho)^{1/2}$$

where erfc is the complementary error function, ρ is the characteristic time for a 3-bond jump and Θ describes the damping of the orientation propagation along the chain or the orientation loss processes due, for example, to the overall motion of a part of the chain.

Such a function presents an infinite slope at t=0, which does not correspond to any physical reality.

Starting from the same model, several other analytical expressions (8,9) have been proposed to avoid this defect, but, in some cases (9), it seems difficult to correlate the parameters which are involved in these formula with molecular quantities.

More recently, HALL and HELFAND (10) have derived a model for conformational dynamics based on the conformational correlations observed in their simulations and they have proposed to use the same expression for the orientations:

$$M_2(t) = \exp(-t/\tau_2)\ \exp(-t/\tau_1)\ I_0(t/\tau_1)$$

where I_0 is a modified Bessel function of order 0, τ_1 is the characteristic time responsible for the diffusion of orientations along the chain (similar to ρ in the VJGM model) and τ_2 is the damping term corresponding to diffusion or orientation loss processes.

2. TEST OF POLYMER MOTIONAL LAWS USING THE FLUORESCENCE ANISOTROPY DECAY TECHNIQUE

Tests of the motional models derived for polymer chains can be carried out using the various spectroscopic techniques. However only the fluorescence anisotropy decay is able to provide a quasi-continuous sampling of the second moment orientation autocorrelation function of a vector which is, in the case of this technique, the transition moment of a fluorescent group covalently bound to the chain. From this point of view, this technique is a priviledged tool for a critical discussion of orientational models and, after having summarized the principles on which it is based, we will present the conclusions that can be derived on polymer dynamics both in solution and in the bulk state.

2.1. Principle of fluorescence anisotropy decay technique (11)

The absorption and the emission of a fluorescent molecule can be described by considering the transition moment of the molecule, the direction of which is determined with respect to the geometry of the molecule. As an example, the transition moment of anthracene is shown in Fig.8.

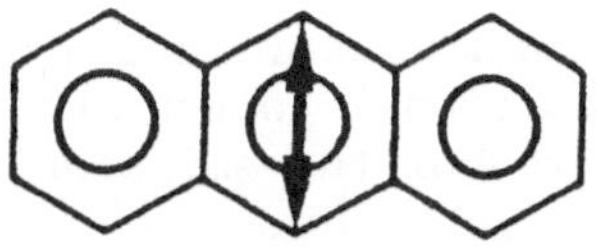

Figure 8. The transition moment (double arrow) of anthracene.

Under the action of a suitable electromagnetic field, polarized along the P direction (Fig.9), the absorption of light is proportional to the scalar product of the incident electric field and of the transition moment. In the same way, the emission of light is proportional to the scalar product of the direction of the analyzer and of the transition moment. Thus, excitation of an isotropic population of fluorescent species by polarized light generally creates a temporary anisotropic population of excited molecules. Molecular motions progressively destroy this anisotropy, and affect the polarization of the reemitted fluorescence light.

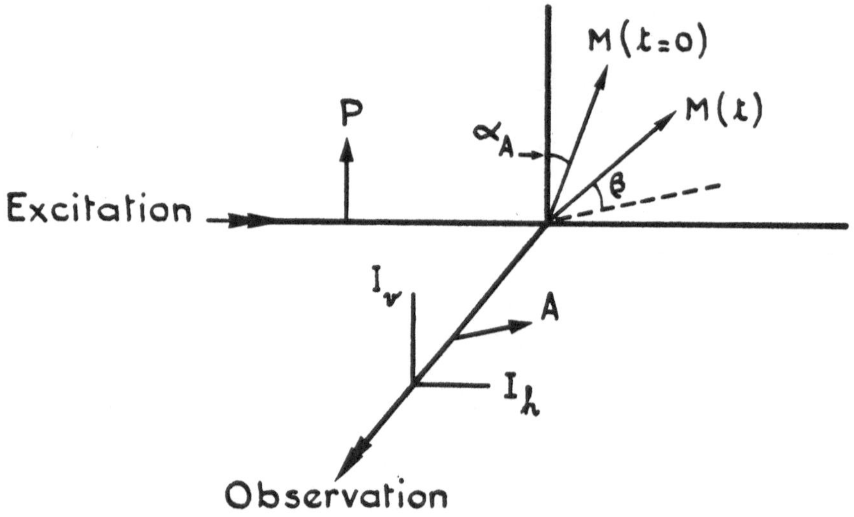

Figure 9. Polarized absorption and fluorescence emission: P, polarizer; A, analyzer.

The interesting quantity, the fluorescence anisotropy, is defined as:

$$r = (I_v - I_h)/(I_v + 2I_h)$$

where I_v and I_h correspond to fluorescence intensities for analyzer direction parallel and perpendicular respectively to the vertical polarization of the incident beam. In this expression (I_v+2I_h) represents the total fluorescence intensity. The fluorescence anisotropy emitted at time t will progressively decrease as a function of time and finally reach a zero value. A complete analysis of the phenomenon shows that the evolution of r(t) as a function of time is directly proportional to the second moment of the orientation autocorrelation function of the emission transition moment:

$$r(t) = M_2(t)$$

The time interval t during which the evolution of $M_2(t)$ can be recorded is directly related to the fluorescence lifetime (10^{-10} to 10^{-8} s); experimentally $M_2(t)$ can be obtained until approximately 10^{-7} s (100ns).

Using such a technique to study polymer dynamics implies that the motion of the fluorescent label reflects the motion of the macromolecular chain. In all the following examples, labelling has been performed in the middle of the chain using an anthracene derivative (12), in such a way that the transition moment of this group lies along the local axis of the chain (Fig.10) and cannot be involved in motional processes independent of the chain ones.

(a)
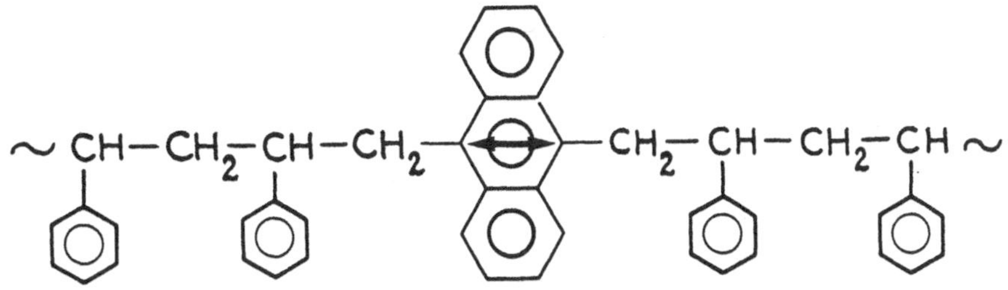

(b)
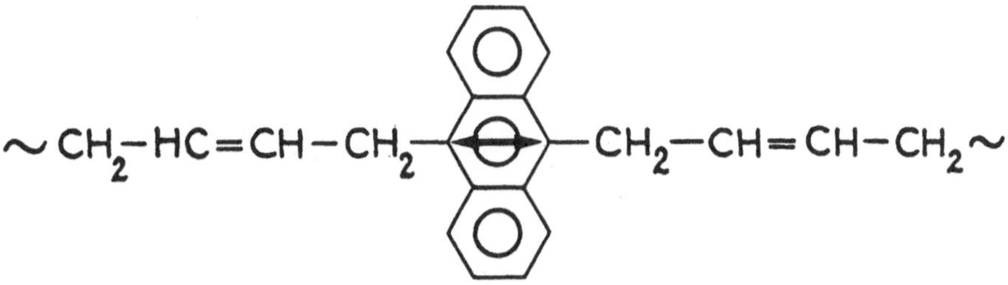

Figure 10. Polymers labelled with anthracene in the middle of the chain: a)polystyrene; b)polybutadiene.

2.2.Polymers in solution

A detailed study of the dynamics in solution of polystyrene chains labelled with anthracene has been carried out using the LURE-ACO (Orsay) synchrotron radiation as the light source (13). The experimental decrease of r(t) as a function of time has been compared with the various mathematical expressions proposed to account for the chain dynamics. For the first time , it has been possible to differentiate the fits obtained with these various functions: for example, although the VJGM function leads to a correct fit at long times, important differences with the experimental data are observed at short times. The best fit is obtained for the HALL-HELFAND function, as shown on Fig.11. For polystyrene at 25°C in dilute solution of ethyl acetate - tripropionin mixtures exhibiting viscosities of 5.40cP, 2.28cP and 0.43cP, the characteristic time τ_1 is equal to 20 ns, 12ns and 2.6 ns respectively.

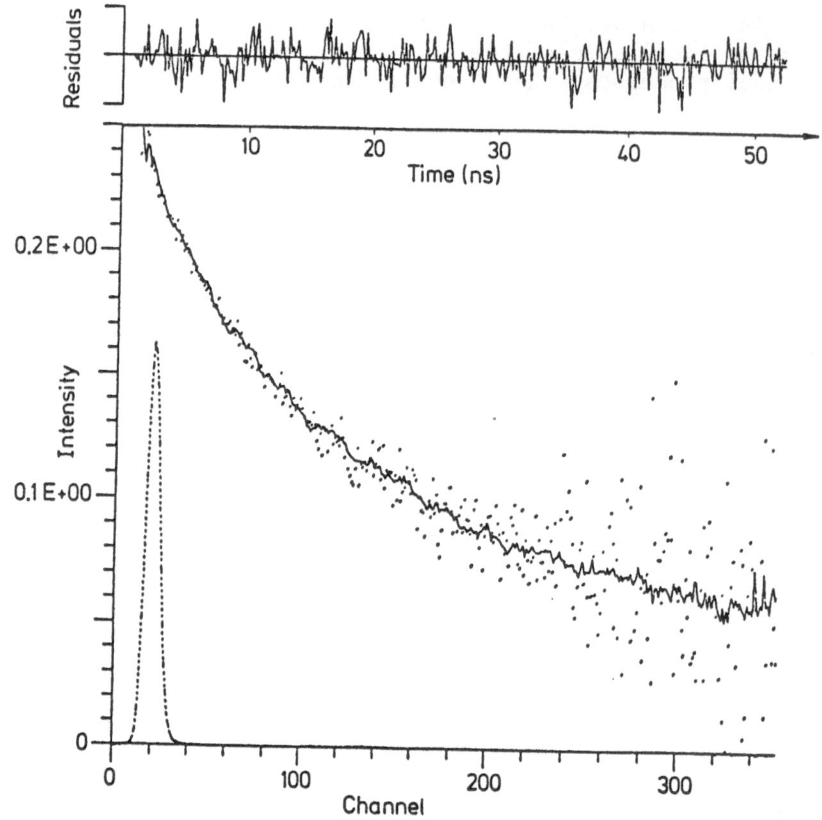

Figure 11. Comparison of the best fit obtained from the HALL-HELFAND expression reconvoluted by the measured instrumental function (exciting pulse) (continuous line) with the experimental anisotropy (dots) of polystyrene labelled with anthracene in the middle of the chain, in tripropionin solution at 25°C (13). The exciting pulse is plotted as a dash-dot line (arbitrary scaled). The upper graph represents the weighted residuals.

2.3.Polymers in bulk at temperatures well above the glass-transition temperature

A similar study has been carried out using anthracene labelled polybutadiene chains dispersed in a polybutadiene matrix (14); experiments have been made in the temperature range -10°C, 80°C. Only the autocorrelation functions which have been derived for chain dynamics are able to account for the experimental data and, among these functions, the best fit is obtained from the HALL-HELFAND expression.

In the case of polymers in bulk, the specificity of the chain dynamics is well evidenced by the study of a series of fluorescent probes (15):

9,10 dialkyl anthracenes: $H-(CH_2)_n-$ 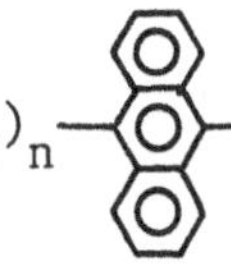 $-(CH_2)_n-H$

When n is equal to 1, the molecule behaves like a small classical anisotropic compound, whereas as n increases, the typical motion of a chain, presenting a diffusive character of the orientation, is observed ($n \geqslant 14$).

Another important result in the field of the dynamics of polymers in bulk concerns the dependence of correlation times towards temperature. Indeed, fluorescence anisotropy experiments carried out on polybutadiene (14) have shown that this dependence obeys the characteristic law (WILLIAM - LANDEL - FERRY expression) which has been derived for the relaxation times deduced from viscoelastic measurements in the frequency range $1\text{-}10^3$ Hz. This expression is known to describe molecular processes associated with the glass-transition phenomenon. Therefore, the motions in which the label is involved in the middle of the chain, although relatively local (of the order of a few monomer units) belong to the processes which are responsible for the glass-transition.

3. TEST OF THE MOTIONAL LAWS FOR POLYMERS USING THE OTHER SPECTROSCOPIC TECHNIQUES

Although the fluorescence anisotropy decay is the only technique able to study directly the autocorrelation function, it is nevertheless possible to verify whether the results obtained from the other spectroscopic techniques are in agreement with the specific motional laws derived for polymers.

From this point of view, N.M.R., and more particularly ^{13}C N.M.R., has an interesting potential: indeed, the relaxation times T_1, T_2, and the nuclear Overhauser effect, are related to the Fourier transforms of the second moment autocorrelation functions and, therefore, these data must be described by the same set of correlation times (ρ and Θ, or τ_1 and τ_2). Moreover, measurements can be carried out at different experimental frequencies, which implies to account for the whole set of data with the same characteristic times.

Such tests have been mainly obtained on polymers in solution. As an example, on table I are shown the experimental and calculated values using the VJGM function for polystyrene in hexachlorobutadiene solution at 60°C (16). The agreement is very satisfying. It must be noted that it is impossible to account for these data using an isotropic diffusional rotational model. In the relaxation resulting

Table I - Comparison between N.M.R. experimental data and calculated values for polystyrene in hexachlorobutadiene solution at 60°C (16).

	T_1 , C_α (25 MHz) (s)	T_1, C_α (15 MHz) (s)	T_1, C_{para} (25 MHz) (s)	T_1, C_{para} (15 MHz) (s)	NOE (25 MHz)	T_2, C_{para} (25 MHz) (s)
Experimental data	0.079	0.042	0.082	0.050	1.0	0.026 0.041
Calculated values from VJGM model $\theta = 2.6\ 10^{-8}$ s $\rho = 9\ .10^{-10}$ s	0.076	0.049	0.082	0.053	0.9	0.033

from the dipolar interaction between two different protons H_i and H_j are involved low-frequency ($\omega_i - \omega_j$) contributions of the spectral densities, typically in the domain of a few hundreds of hertz usually encountered for proton chemical shifts. Such an experiment allows to test motional laws on quite a large frequency range. It has been applied to polystyrene in solution (17). Results obtained on CH_2, CH, meta and para protons can be described using the VJGM expression and a coherent set of correlation times.

N.M.R. studies of polymers in solution have been reviewed (18). From all the results, it appears that only an autocorrelation function specific for chain dynamics can account for the data. However, it must be noted that N.M.R. measurements do not allow to differentiate between the different expressions derived for the autocorrelation function of a polymer (9).

Another technique which can be used as a test for motional laws is the dielectric relaxation. It involves $M_1(t)$ instead of $M_2(t)$, which offers another interesting test. The real (ε') and imaginary (ε'') components of the dielectric constant ε^* are measured as a function of the experimental frequency and results are usually represented by a COLE - COLE diagram:

$$\frac{\varepsilon''}{\varepsilon_0 - \varepsilon_\infty} = f\left(\frac{\varepsilon' - \varepsilon_\infty}{\varepsilon_0 - \varepsilon_\infty}\right)$$

In such a representation, an isotropic motion corresponds to half a circle. On the contrary, results obtained from polymers in solution are characterized by a squewed curve. Such a shape can be reproduced by considering an autocorrelation function of the polymer type (19) and, as shown on Fig.12, the agreement with the experimental results is quite satisfying.

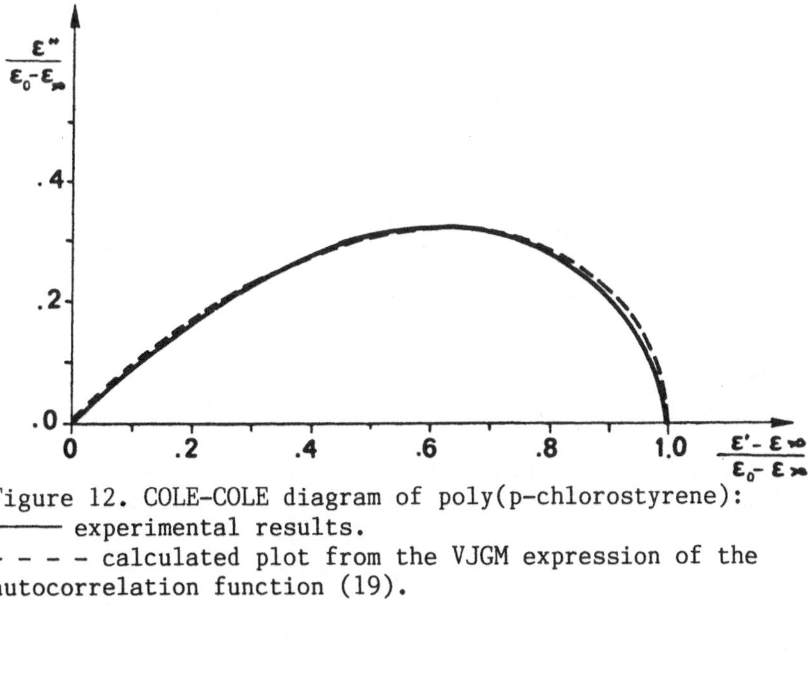

Figure 12. COLE-COLE diagram of poly(p-chlorostyrene):
——— experimental results.
- - - - calculated plot from the VJGM expression of the autocorrelation function (19).

To finish with this confrontation with experimental results, it is of interest to compare the values of the correlation times obtained for the same polymer in solution from the different techniques. Results relative to the polystyrene in solution at different temperatures (20) are shown on Fig.13. The correlation times derived from the different techniques present a parallel evolution towards temperature. Their values are very close for E.S.R., N.M.R. and fluorescence anisotropy. In the case of dielectric relaxation , the observed difference may come from the fact that the involved function is $M_1(t)$ instead of $M_2(t)$.

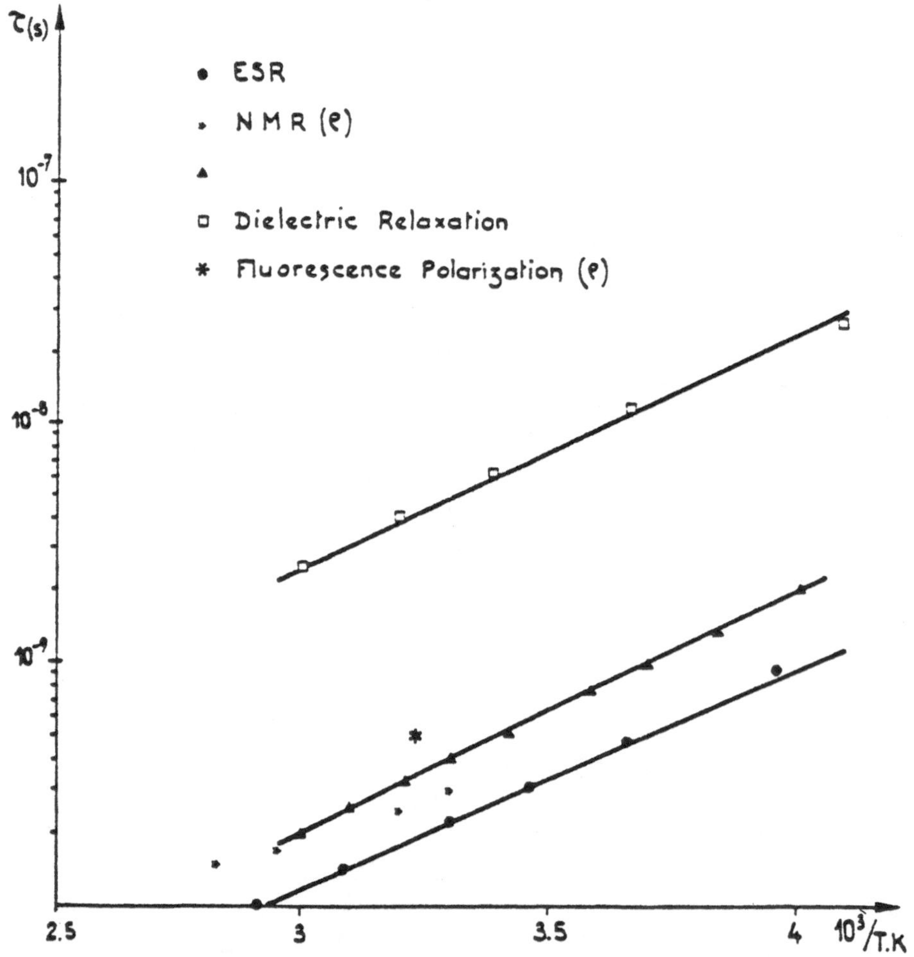

Figure 13. Dependence on absolute temperature T of the correlation times τ derived from different techniques for polystyrene in toluene solutions (20).

4. N.M.R. SPECIFIC CONTRIBUTION TO THE STUDY OF POLYMER LOCAL DYNAMICS

As noted above, N.M.R. does not allow to differentiate the various expressions proposed for the autocorrelation function of a polymer. However, this technique is of great interest for the detailed analysis of polymer dynamics.

First of all, it does not require the introduction of any type of label (fluorescent or spin labels) which usually reflects the general behavior of the chain in a satisfying way, but which may also modify the behavior of the adjacent units. Furthermore, the fixation of the label in a covalent way can only be carried out with certain types of polymerizations (condensation or anionic polymerization), so that any polymer cannot be labelled. From this point of view, ^{1}H and ^{13}C N.M.R. offer a very wide generality; the same holds true in respect of ^{2}D N.M.R. if the deuterated monomer is available.

The specific character of N.M.R. lies in the fact that it allows to observe the dynamics of any group of the chain. It is, therefore, a very unique means for obtaining a detailed description of polymer dynamics. From this point of view ^{13}C N.M.R. is very suitable since, as the ^{13}C relaxation mechanism mainly involves the dipolar interaction with bound protons, the reorientating vector under study is perfectly known. In the same way, the ^{2}D relaxation allows to study the motion of the C-D vector in the molecule. With regard to ^{1}H N.M.R., relaxation comes from ^{1}H-^{1}H dipolar interactions and, in a polymer chain whose flexibility is generally quite high, protons attached to different groups will participate to the relaxation of a given ^{1}H, in such a way that it will not be usually possible to separate the contributions of the different internuclear vectors and follow the dynamics of an unique vector.

It must be noted that the characteristic motional times of the different groups, which are the significant data, can be derived from the macroscopic relaxation times using the orientation autocorrelation function selected by the fluorescence anisotropy experiments on polymers. In the following, we will illustrate the specific potentialities of N.M.R. with examples dealing with polymers in solution and polymers in bulk.

4.1. Local dynamics of polymers in solution

4.1.1. Polystyrene and derivatives. ^{13}C N.M.R. studies on polystyrene and polystyrene derivatives

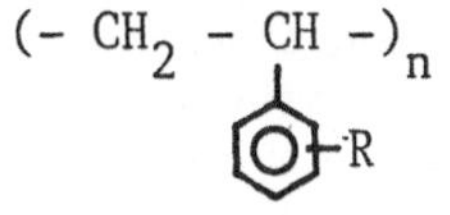

allow to study both the main chain motions using the aliphatic CH and CH_2 data and the side-ring internal rotation by comparing the

C(ortho, meta) behavior to the C(para)one. Experiments carried out on polystyrene (16) (R=H) show that the phenyl ring presents an internal rotation about the C(alpha)-C(para) axis with a correlation time equal to 3.10^{-9} s at 60°C in hexachlorobutadiene. The presence of a substituent R (F, CH_3, Cl, Br) modifies the dynamics of both the chain and the side-chain phenyl ring (21). For example, an ortho substitution induces a significant slowing down of both processes, the effect increasing when going from the fluorine to the bromine.

The role of the organization of the different groups in a chain on its dynamic behavior is well illustrated by the study of head-to-tail (H-T) and head-to-head (H-H) polystyrenes (22):

H-T polystyrene :$(- CH_2 - CH - CH_2 - CH)_n$

H-H polystyrene :$(- CH_2 - CH - CH - CH_2)_n$

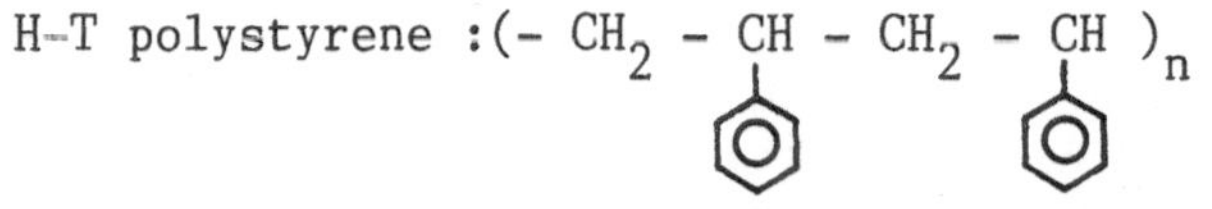

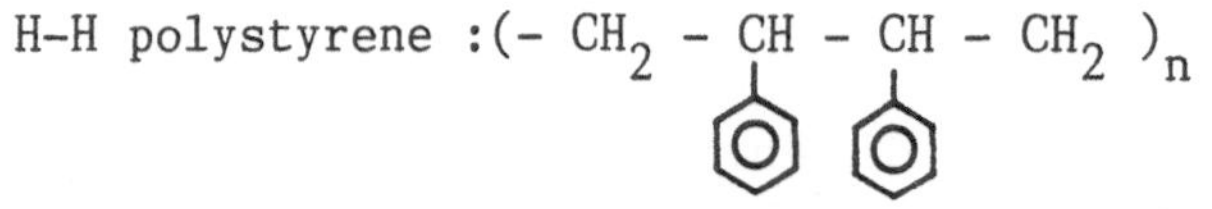

Both compounds present the same correlation time ρ which describes the conformational jumps. On the contrary, the orientation propagation along the chain is damped much more quickly in the H-H polymer which also has a reduced internal mobility of the side-ring owing to the location of these groups on vicinal carbons.

4.1.2. Aryl-aliphatic polyesters. A nice example of the potentiality of ^{13}C N.M.R. in studying the detailed dynamics of a chain in solution is the study, in chloroform solution, of two aryl-aliphatic polyesters differing only by the length of the aliphatic ether sequence separating the terphenyl ester groups (23) :

$[\underset{O}{C}$⬡-⬡-⬡$-\underset{O}{C}O-CH_2-CH_2-O(-CH_2-CH_2-O)_m-CH_2-CH_2-O]_n$
  1 2 3   4   5   6

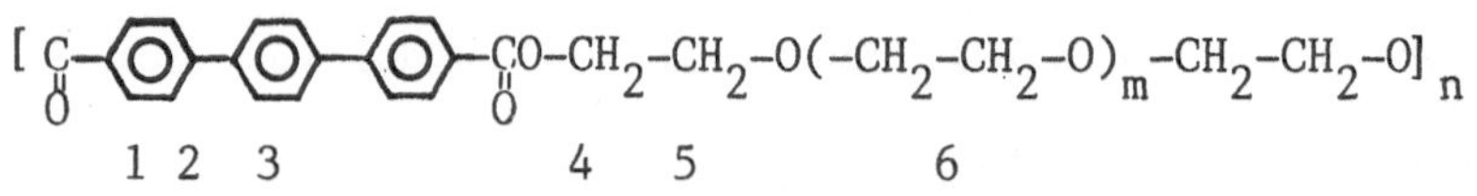

m=2 : polymer A; m=8 : polymer B.

These polymers have large terphenyl groups which may act as anchors for the rapid short-range processes affecting the aliphatic part of the molecule.

Let us consider first the dynamic behavior of the aromatic part. The proton spectrum of the terphenyl groups of polyesters A and B presents three peaks which have all the same relaxation time T_1. This result shows that the relaxation mechanism is a dipolar interaction between pairs of protons attached to vicinal carbons. In the polymers under study, the involved H-H vectors are parallel to the symmetry axis of the terphenyl group. They are not, therefore, sensitive to an internal rotation of the rings about this axis. The relaxation of these protons is well described by an isotropic rotational diffusional motion of the terphenyl unit whose correlation time τ_0 are listed in table II. More detailed information on the ring dynamics are obtained from T_1 and NOE measurements at 20, 25, and 65 MHz on C_1, C_2 and C_3 carbons. Indeed, these data point out the fact that the motion is a combination of an isotropic reorientation of the whole terphenyl group with correlation time τ_0 and of a stochastic rotational process of the phenyl rings about their long symmetry axis with correlation time τ_{ir}. As shown by the τ_{ir} values given in table II, the internal rotations of the rings are faster for the central phenyl ring. The activation energies E_a, which are associated to the τ_{ir} processes and are equal to 5 Kcal/mol for C_1 and C_2 and to 3 Kcal/mol for C_3, are in excellent agreement with results obtained from conformational energy calculations carried out on these compounds (24).

As regards the aliphatic ether part of the chain, three types of carbons can be observed, namely C_4, C_5 and the carbons located in the center of the aliphatic segment designed as C_6. The dynamic analysis of the corresponding data requires an autocorrelation function of the polymer type. Correlation times and activation energies calculated using the VJGM model are reported in table II. Discussion of the results in terms of the HELFAND classification of the different types of elementary motions and of the barrier heights associated with them allows a detailed analysis of the dynamics of the flexible sequence. It appears that in the relaxation of the C_4-H internuclear vector are involved the type 1 or type 2 conformational jumps of the aliphatic part (ρ) and the overall motion of the terphenyl group ($\Theta=\tau_0$). The C_5-H internuclear vector is slightly less sensitive to the overall motion of the terphenyl group and the motional contribution of the aliphatic part is more important. In the case of the central carbons, it must be noted that both correlation times and associated energy barriers E^* depend on the length of the aliphatic sequence. A longer aliphatic ether sequence not only induces an increasing flexibility (ρ), but also leads to a notable decrease of the measured corresponding activation energy. These activation energies are equal to 5.6 Kcal/mol and 2.6 Kcal/mol for polymer A (m=2) and polymer B (m=8) respectively. They are of the order of one barrier crossing for m=8 and of two barrier crossings for m=2. Indeed, type 1 conformational jumps (three-bond motions, crankshaft), which leave the tails of the chain at the same position at the end and start of the transition, require two barrier crossings whereas type 2 motions, which are associated to ulterior counter-rotation processes require slightly more than one. Therefore, owing to the large inertial effects of the terphenyl groups acting as anchors, mainly type 1 motions occur in shorter flexible sequence (m=2) whereas the longer (m=8) flexible one presents a behavior very similar to that of the polyethyleneoxide chain and type 2 motions are therefore very likely.

Table II - Correlation times and activation energies obtained from N.M.R. experiments on $CDCl_3$ solutions of aryl-aliphatic polyesters A and B at 27°C (23).

	Polymer A (m = 2)		Polymer B (m = 8)	
H_1, H_2, H_3	$\tau_o = 1.3\ 10^{-9}$ s	$E_a = 7.5$ Kcal/mol	$\tau_o = 1.3\ 10^{-9}$ s	$E_a = 7.5$ Kcal/mol
C_1, C_2	$\tau_o = 1.3\ 10^{-9}$ s $\tau_{ir} = 1.2\ 10^{-10}$ s	$E_a = 5$ Kcal/mol	$\tau_o = 1.3\ 10^{-9}$ s $\tau_{ir} = 1\ .10^{-10}$ s	$E_a = 4.8$ Kcal/mol
C_3	$\tau_o = 1.3\ 10^{-9}$ s $\tau_{ir}\ 6.8\ 10^{-11}$ s	$E_a = 3.5$ Kcal/mol	$\tau_o = 1.3\ 10^{-9}$ s $\tau_{ir} = 6\ .10^{-11}$ s	$E_a = 3$ Kcal/mol
C_4	$\Theta = 1.3\ 10^{-9}$ s $\rho = 2.4\ 10^{-11}$ s	$E_a = 7.5$ Kcal/mol	$\Theta = 1.3\ 10^{-9}$ s $\rho = 1.7\ 10^{-11}$ s	
C_5	$\Theta = 1.2\ 10^{-9}$ s $\rho = 2\ .10^{-11}$ s		$\Theta = 1.2\ 10^{-9}$ s $\rho = 1.5\ 10^{-11}$ s	
C_6	$\Theta = 1.1\ 10^{-9}$ s $\rho = 2.8 .10^{-12}$ s	$E^{\star} = 5.6$ Kcal/mol	$\Theta = 9\ .10^{-10}$ s $\rho = 2.1\ 10^{-12}$ s	$E^{\star} = 2.6$ Kcal/mol

4.2. Local motions in bulky polymers in the fluid state

As an example of the N.M.R. potentiality in this field, recent results obtained in our laboratory on polyisoprene in bulk in the temperature range -10°C-110°C, i.e at temperatures much higher than the glass-transition temperature of this material (Tg=-60°C) (25), will be described. This polymer of formula

$$(-\ CH_2 - \underset{\displaystyle CH_3}{\underset{|}{C}} = CH - CH_2 -)_n$$

has been studied by Carbon-13 N.M.R. at 25.15 and 62.5 MHz. CH and CH_2 T_1 relaxation times have been determined for the two experimental frequencies. Results reported in Fig.14 present two unusual points: first the $T_1(CH)/T_1(CH_2)$ ratio is notably different from 2, secondly T_1 values at the T_1 minimum are very high. It is of course impossible to account for these behaviors with an isotropic motional model. In the same way a simple autocorrelation function of the polymer type, such as the HALL-HELFAND autocorrelation function, cannot reproduce the whole set of data. Anomalies of the $T_1(CH)/T_1(CH_2)$ ratio have been reported for polybutadiene (26) and interpreted in terms of a local motion which affects the CH and CH_2 internuclear vectors in a different way in these very flexible structures. As the chemical structure of polyisoprene only differs from that of polybutadiene by the presence of a methyl substituent group, such a local mode sketched in Fig.15 is likely to occur in polyisoprene too. Under these conditions, the resulting autocorrelation function can be written as:

$$M_2(t) = a\exp(-t/\tau_0) + (1-a)\exp(-t/\tau_2)\exp(-t/\tau_1)I_0(t/\tau_1)$$

where τ_0 is the correlation time of the specific local process, τ_1 is the diffusive correlation time and τ_2 the damping term.

As shown on Fig.14, a very good fit is obtained with $\tau_2/\tau_1=30$ and $\tau_1/\tau_0=200$.

In this example, N.M.R. has been able to point out the existence of a local mode characteristic of the chains containing double bonds. This very rapid motion ($\tau_0=5.10^{-11}$ s at -10°C) could not have been detected by fluorescence anisotropy studies on anthracene labelled polybutadiene either because of its correlation time much too short for the experimental frequency window of the fluorescence technique or because of the presence of the label which may have affected this local process.

Moreover the dependence of the correlation time of the diffusive motion (τ_1) as a function of temperature obeys a WLF equation, which shows that the associated segmental process is involved in the glass-transition phenomenon of polyisoprene.

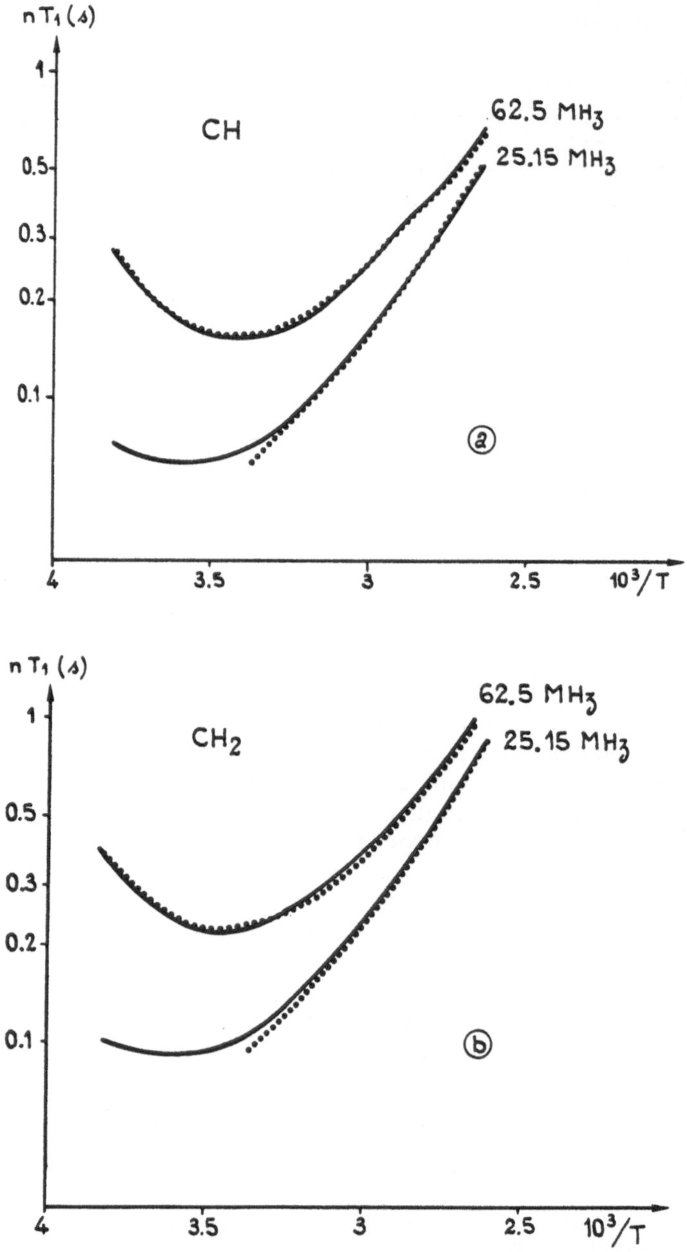

Figure 14. Dependence of nT_1 on absolute temperature T in polyisoprene in bulk (T_1: Carbon-13 spin-lattice relaxation time measured at 25.15 and 62.5 MHz, n: number of protons directly bound to the carbon of interest).

——— calculated plot from the expression of $M_2(t)$ given in paragraph 4.2 of the text:

a/ CH, a = 0.12, $\tau_2/\tau_1 \simeq 30$, $\tau_1/\tau_o > 200$

b/ CH_2, a = 0.40, $\tau_2/\tau_1 \simeq 30$, $\tau_1/\tau_o > 200$

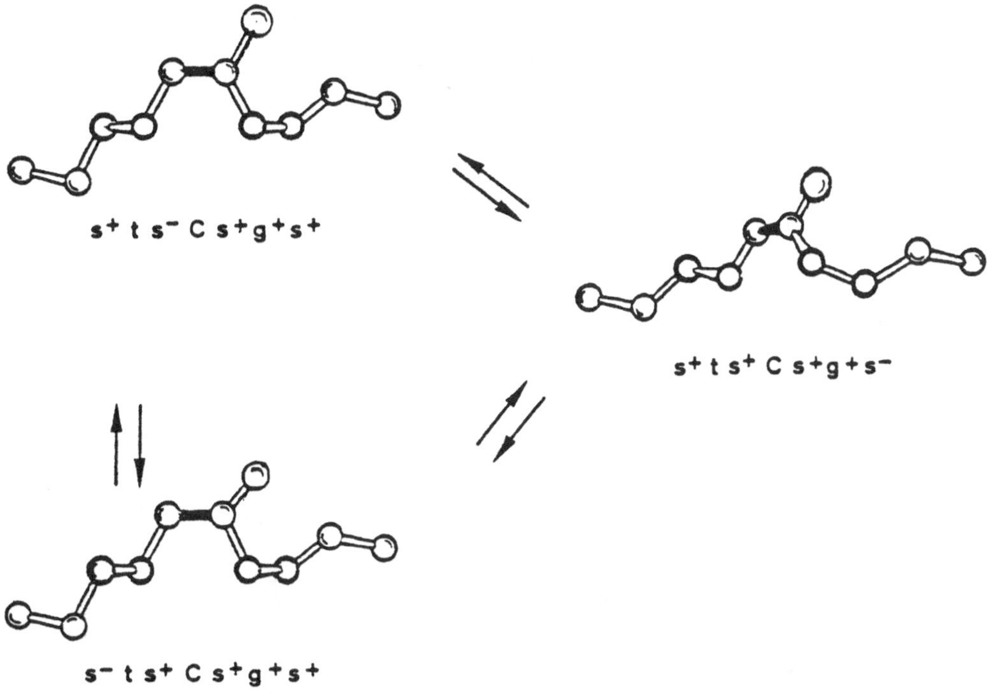

Figure 15. The specific local transitions of the polyisoprene chain.

4.3. Local motions in polymers in the solid state

High-resolution solid-state N.M.R. is a powerful tool for studying polymer dynamics in the solid state. In this case, as polymers under study are examined at temperatures well below the glass-transition temperature, the only involved motions are relative to side-chain groups or to very short sequences of the main chain. It must be noted that cooperative processes diffusing along the chemical structure and leading to autocorrelation functions of the polymer type cannot exist.

To illustrate the N.M.R. potentiality in this field, we will consider two examples:

4.3.1. Polyester ethers. These compounds have the general formula

$$-[(CO-C_6H_4-CO-O-)((CH_2)_4-O-)]_m[(CO-C_6H_4-CO-O-)((CH_2)_4-O-)_{12}]_n-$$

and the case n=0 corresponds to polybutaneterephthalate (PBT). A very detailed study (27) has been performed on these compounds using both Carbon-13 (chemical shift anisotropies, T_1 and $T_{1\rho}$ measurements) and deuterium N.M.R. on solid PBT samples specifically deuterated on the central carbons. In addition to the 180° flip of the phenyl rings about their C_1-C_4 axis, these experiments have shown that the $(CH_2)_4$ sequence performs a three-bond motion which can be likely assigned to the ttt-gtg' conformational transition drawn on Fig.6b. The use of ^{2}D N.M.R. is essential for this assignment since it provides the exact value (103°) of the dihedral angle separating the two equally populated positions of the C-D vectors at the start and end of the transition. A schematic representation of the correlation times at 25°C associated to the motions of the different parts of the polyester ether chain is given in Fig.16.

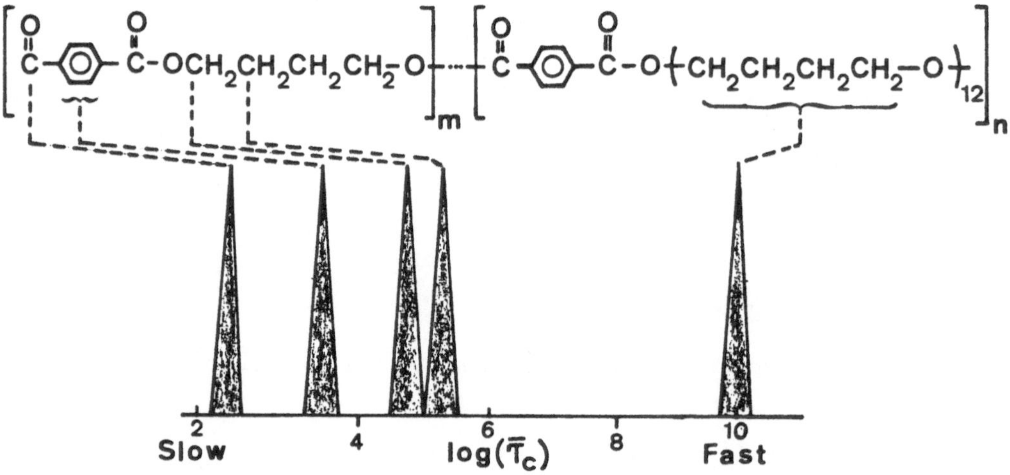

Figure 16. A schematic representation of the dynamics of the different carbons in a polyether ester chain.

4.3.2. Poly(cycloalkyl methacrylates). Linewidth analysis and ^{13}C $T_{1\rho}$ determinations using high-resolution solid-state ^{13}C N.M.R. (28) have been performed on solid poly(cyclopentyl methacrylate), poly(cyclohexyl methacrylate) and poly(cycloheptyl methacrylate) of general formula:

$$(-\ CH_2\ -\ \underset{\displaystyle |\atop \displaystyle COO\ -\ \overbrace{CH\ (CH_2)_m}}{C(CH_3)}\ -)_n$$

In all the compounds under study at 25°C and 12 MHz, the main chain motions cannot be detected by the above experiments, which means that the associated correlation time is higher than 10^{-5} s. On the contrary, the internal motions of the ring are highly dependent on the size of the ring. For the cyclopentyl and the cycloheptyl derivatives, these internal motions are very rapid (of the order of 10^{-12} and 10^{-10} s respectively) and they give rise to narrow spectral lines and high $T_{1\rho}$ values whereas, for the cyclohexyl polymer, the frequency of the motion is of the order of the experimental one (10^6 Hz) which leads to quite broad lines. In the last case, the ring transition is the well-known chair-chair inversion in agreement with the CH and CH_2 $T_{1\rho}$ relative values. Experiments carried out as a function of temperature show that the two conformers present an energy difference of 3 Kcal/mol, higher than the 0.7 Kcal/mol value deduced from conformational energy calculations and likely due to the intermolecular interactions present in the material in bulk. All these results are in agreement with the conclusions obtained from mechanical relaxation experiments carried out on these solid polymers. However, the COO chemical shift anisotropy allows to derive some complementary information: the internal motions of the rings do not affect the orientation of the ester COO group which appears as frozen at the experimental frequency for all methacrylates under study.

5. CONCLUSION

It now appears that, in spite of its complexity, the polymer dynamics can be described by orientation autocorrelation functions which take into account the specific character of the polymer chains, i.e the orientation diffusion along the chemical sequence and the orientation loss terms resulting either from the damping of the diffusion or from isotropic reorientations of some parts of the chain.

These functions are able to account for experimental results obtained for polymers in solution as well as for polymers in bulk at temperatures well above their glass-transition temperature.

On the basis of such functions, it is now possible to calculate, from the N.M.R. relaxation data, the correlation times associated to the motions of the different parts of the chain. Therefore ^{13}C and eventually ^{2}D N.M.R. allows a deeper insight of the molecular analysis of chain dynamics by leading to a detailed description of the involved motions. Some spectacular results have already been obtained. One can expect that, in a new future, and because of the absence of labelling and of the specificity of the measurements, the N.M.R. techniques will be the only ones able to provide a precise interpretation of the polymer dynamics and allow to establish relations between the chemical structure and the dynamic behavior of the chains. From this point of view, studies on polymers in bulk are of decisive importance since they should result in the molecular understanding of the processes which are responsible for the glass-transition phenomenon, which is of main interest for the physical and mechanical properties of polymers.

REFERENCES

1/ E.Helfand, J.Chem.Phys., 54 (1971) 4651.
2/ P.H.Verdier, W.H.Stockmayer, J.Chem.Phys., 36 (1962) 227.

3/ a/ L.Monnerie, F.Gény, J.Chim.Phys. (Fr.), 66 (1969) 1961.
b/ F.Gény, L.Monnerie, J.Polym.Sci., Polym.Phys. Ed., 17 (1979) 131 147 and 173.
4/ E.Helfand, Z.R.Wasserman, T.A.Weber, Macromolecules, 13 (1980) 526.
5/ J.Skolnick, E.Helfand, J.Chem.Phys., 72 (1980) 5489.
6/ H.A.Kramers, Physica, 7 (1940) 284.
7/B.Valeur, J.P.Jarry, F.Gény, L.Monnerie, J.Polym.Sc.,Polym.Phys.Ed., 13 (1975) 667,675 and 2251.
8/ A.A.Jones, W.H.Stockmayer, J.Polym.Sci., Polym.Phys.Ed., 15 (1977) 847.
9/ J.T.Bendler, R.Yaris, Macromolecules, 11 (1978) 650.
10/ C.K.Hall, E.Helfand, J.Chem.Phys. 77 (1982) 3275.
11/ L.Monnerie, p.383 in Static and Dynamic Properties of the Polymeric Solid State, ed. R.A.Pethrick, R.W.Richards, NASI Series, Reidel Publ., 1982.
12/ B.Valeur, L.Monnerie, J.Polym.Sci., Polym.Phys.Ed., 14 (1976) 11, 29.
13/ J.L.Viovy, L.Monnerie, J.C.Brochon, Macromolecules, 16 (1983) 1845.
14/ J.L.Viovy, L.Monnerie, to appear in Macromolecules.
15/ J.L.Viovy, Thèse Dr. Etat, Paris, 1982.
J.L.Viovy, L.Monnerie, C.Franck, submitted to Macromolecules.
16/ F.Lauprêtre, C.Noël, L.Monnerie, J.Polym.Sci., Polym.Phys.Ed., 15 (1977) 2127.
17/ F.Heatley, M.K.Cox, Polymer, 18 (1977) 225.
18/ F.Heatley, Progress in NMR Spectroscopy, 13 (1979) 47.
19/ F.Gény, L.Monnerie, J.Polym.Sci., Polym.Phys.Ed., 15 (1977) 1.
20/ C.Friedich ,F.Lauprêtre, C.Noël, L.Monnerie, Macromolecules, 14 (1981) 1119.
21/ F.Lauprêtre, C.Noël, L.Monnerie, J.Polym.Sci., Polym.Phys.Ed., 15 (1977) 2143.
22/ F.Lauprêtre, L.Monnerie, O.Vogl, European Polymer J., 14 (1978) 981.
23/ P.Tékély, F.Lauprêtre, L.Monnerie, Macromolecules, 16 (1983) 415.
24/ P.Meurisse, F.Lauprêtre, C.Noël, Molecular Crystals and Liquid Crystals, 110 (1984) 41.
25/ R.Dejean, F.Lauprêtre, to be published.
26/ W.Gronski, Makromol.Chem., 178 (1977) 2949.
27/ L.Jelinsky, F.C.Schilling, F.A.Bovey, Macromolecules, 14 (1981) 581.
L.Jelinsky, Macromolecules, 14 (1981) 1341.
L.Jelinsky, J.J.Dumais, A.K.Engel, Macromolecules, 16 (1983) 403.
L.Jelinsky, J.J.Dumais, P.I.Watnick, A.K.Engel, M.D.Sefcik, Macromolecules, 16 (1983) 409.
L.Jelinsky, J.J.Dumais, A.K.Engel, Macromolecules, 16 (1983) 492.
28/ F.Lauprêtre, L.Monnerie, J.Virlet, Macromolecules, 17 (1984) 1397.

NMR APPROACH TO THE OBSERVATION OF COLLECTIVE PROPERTIES IN POLYMER SYSTEMS : GELS AND MELTS

J.P. Cohen-Addad
Laboratoire de Spectrométrie Physique associé au CNRS
Université Scientifique et Médicale de Grenoble
B.P. 87
38402 St Martin d'Hères Cedex
France

ABSTRACT. Dynamic screening effects concerning high frequency monomeric unit motions are known to occur in any polymer system observed above the glass transition temperature. It is shown how these effects may induce specific properties of the transverse nuclear magnetization which can be then applied to the investigation of monomeric unit collective behaviors. The space scale of NMR investigation is determined from the dynamic screening length. Static NMR scaling properties are illustrated from covalent calibrated polymeric gels. Dynamic molecular properties are related to a possible observation of some process involved in the reptational model.

1. INTRODUCTION

This article deals with the interpretation of NMR observations of collective properties of monomeric units in molten polymers ; despite the very short range of magnetic interactions, the scale unit of observation in space may spread over a length about equal to 50 Å, because it is determined by topological constraints in a melt or by cross-links in a gel ; these specific effects will be evoked in next sections. This semi-local space-scale of NMR observation is also commonly known to govern collective properties in polymer systems (1). A two-fold basis underlies the present approach.

i) Collective properties of polymer chains are studied from the relaxation of the transverse magnetization Mx(t) of nuclei attached to macromolecules instead of observing the spin-lattice relaxation process ; this last phenomenon is known to rest upon a quasi-resonant exchange of energy between the spin-system and the surrounding medium ; as a consequence of this quasi-resonance, the relaxational frequency of observation is necessarily locked around the Larmor frequency : $\omega_o/2\pi \simeq 10^8$ Hz ; it is far too high to characterize time-dependent collective properties of macromolecules in a melt, usually described in a time-scale longer than 10^{-4} sec. The relaxation of the transverse magnetization Mx(t) is contrasted to that of the longitudinal one Mz(t) because it mainly depends upon quantum phase coherence properties of wave functions describing nuclear spin motions ; there is

R. Daudel et al. (eds.), Structure and Dynamics of Molecular Systems – II, 155–178.

no resonant condition to be fulfilled ; consequently, the upper limit concerning the time-scale of random motions possibly observed from Mx(t), is given by the spin-lattice relaxation time $T_1 \gtrsim 1$ sec. ; this governs the very life-time of the initial out-of-equilibrium state of the spin-system (2).

ii) The other feature underlying the present approach concerns screening effects now currently considered as governing most static as well as dynamical properties of polymer melts (1)(3). These screening effects are induced by covalent bonds in a polymeric gel or by topological constraints, usually called entanglements, in a melt. Static screening effects have been predicted from an early mean field description of entangled chain systems (4). Then, they were rigorously derived from a general framework of interpretation based upon a critical phenomenon approach providing not only their observed dependence on concentration but also the cross-over curves separating different regimes in the concentration-temperature diagram (1). Short range segmental statistical correlations and long range ones have been extensively characterized from elastic neutron scattering experiments performed on labelled chains (1). Critical or coherence lengths were found to vary from about 20 Å to the average size of a single chain in dilute solution. Dynamic screening effects were formerly introduced from the cut-off observed within the broad relaxational spectrum of molten polymer samples determined from viscoelastic measurements (3). More precisely, high frequency short range segmental motions were accurately shown to be stochastically independent of low frequency long range motions occuring within a scale about equal to the average size of a chain in a melt. This screening effect was assigned to topological constraints fluctuating around polymer chains ; it is well illustrated from the molecular weight independence of short segment diffusional motions ; whereas long range chain motions exhibit a strong molecular weight dependence.

During the process of relaxation of the transverse magnetization starting from an initial time t_o, the frequency cut-off will be supposed to induce a dual behavior of the spin-system response. On the one hand, the spin-system will be found to behave like in a liquid because of high relaxational frequency motions of short segments ; on the other hand, it will behave like in a solid because of very slow long range chain motions.

This paper presents a framework for interpretation of NMR properties observed either in melts characterized by slowly dissociating entanglements or in polymeric covalent gels. The section 2 describes basic properties of submolecules determined by dynamic or static screening effects. General features characterizing NMR observations are presented in section 3. Specific NMR properties associated with static screening effects in a gel are demonstrated in section 4 ; while scaling NMR properties observed in covalent gels are analysed in section 5. Dynamic screening effects in a melt associated with NMR properties are illustrated in section 6 ; finally, the possible NMR observation of a chain migration process in a melt is discussed in section 7.

2. SUBMOLECULE CONCEPT

2.1. Melts–concentrated solutions

Considering a melt or a solution the complete specification of molecular configuration of a chain would require to handle numerous degrees of freedom all together, defined at any time. Difficulties arising from such an approach have not been totally overcome yet. They stress the need to reduce the number of variables to the smallest value compatible with the nature of properties to be observed. Statistical descriptions of polymer systems actually lead to a fictitious partition of a given chain into correlation domains, also called submolecules, and characterized by an average number n_c of monomeric units ; n_c is usually larger than 50. Ignoring short range relationships and attendant behavior at the highest frequencies any submolecule is assumed to be represented by the end-to-end vector $\vec{r}_e$ joining its two fictitious end points separated by n_c monomeric units on the chain backbone. Although $\vec{r}_e$ is a time-dependent vector, it is considered that it fluctuates in a time-scale much longer than relevant correlation times of fluctuations occuring within the corresponding submolecule. In other words, it is considered that the partition function of a whole chain can be calculated in two steps ; starting from a given macromolecule divided into N_s fictitious submolecules all end-to-end vectors $\vec{r}_e(n)$, with $n = 1,2 \ldots, N_s$, are supposed to be fixed in the first step of the calculation ; corresponding partial partition functions $z(\vec{r}_e(n))$, with $n = 1,2\ldots,N_s$, are calculated over all molecular configurations of submolecules compatible with the fixed end-to-end vectors. Then, the partition function of the whole chain corresponding to an end-to-end vector :

$$\vec{R} = \sum_{n=1}^{N_s} \vec{r}_e(n) \qquad [1]$$

is obtained from :

$$Z(\vec{R},N) = \int \mathcal{P}(\vec{r}_e(1),\ldots, \vec{r}_e(N_s)) \prod_{n=1}^{N_s} z(\vec{r}_e(n)) \, d\vec{r}_e(1),\ldots, d\vec{r}_e(N_s) \qquad [2]$$

where N is the total number of monomeric units in a chain ; $\mathcal{P}(\vec{r}_e(1),\ldots, \vec{r}_e(N_s))$ is the probability distribution function of end-to-end vectors of submolecules in a chain. There are two crucial hypothesis underlying the description proposed throughout the present paper :

i) The frequency cut-off of the broad chain relaxation spectrum is wide enough to allow a two-step calculation of the partition function of a given chain. It is like considering that a submolecule is a physical system in a partly out-of-equilibrium state ; variables $\vec{r}_e(n)$ help describing the system in a state close to the equilibrium.

ii) Chain statistics is supposed to be unperturbed within submolecules. Topological constraints induced by the medium in which polymer chains are embedded are solely reflected by the probability distribution function $\mathcal{P}(\vec{r}_e(1),\ldots,\vec{r}_e(N_s))$ which may be a simple product of functions of separated variables $\vec{r}_e(1),\ldots,\vec{r}_e(N_s)$ or a more complicated function of these variables considered as non-independent from one another.

In other words, according to this approach any polymer system is assumed to be described from both an unperturbed individual statistics of submolecules and a constraint induced distribution of collective variables of submolecules instead of building a self-generated statistical description which should start from all monomeric units and should include boundary conditions specifying the physical nature of the polymer system.

Time-dependent fluctuations of $\vec{r}_e(n)$ vectors correspond to the collective motions occuring within a polymer chain and which will be characterized from NMR ; they are involved in the kinetics of disentanglement or more generally in the migration process of a chain in a melt.

2.2. Polymeric Gels

It might appear at a first glance that statistical submolecules contributing to thermodynamic properties of a gel are exactly determined by covalent cross-links. Contrary to this qualitative point of view, physically significant domains have been shown to be determined from screening effects existing before the reaction of gelification. This property was demonstrated from experimental results previously reported (5) ; it will be discussed subsequently from the NMR approach presented in part 5 of this paper. In addition to screening effects trapped during the cross-linking reaction, small coherence domains may be induced by the very overlap of submolecules after the cross-linking reaction ; this property will be also discussed from NMR results reported in part 5. The existence of coherence domains in polymeric gels is not questioned whatever their exact origin ; any submolecule or elementary chain will be described from an end-to-end vector $\vec{r}_e$; there obviously is no long range migration of macromolecules in a polymeric gel ; elementary chain ends are supposed to only fluctuate in space around a mean position ; the situation in such a system is contrasted to that prevailing in a melt.

3. GENERAL NMR APPROACH

One of the main problems encountered in the interpretation of NMR properties observed in polymer systems comes from the lack of quantitative expressions representing the time dependence of relaxation functions of transverse nuclear magnetizations ; in most cases, the response Mx(t) of real spin-systems cannot be calculated as an explicit time expression because molecular time dependent probability distribution functions governing this response are not known exactly. It is not the purpose of the present paper to give an extensive theoretical description of NMR properties observed in polymeric gels ; a model built from two - or three - spin systems will be evoked in this approach. But the most useful experimental characterization of the transverse spin-system response Mx(t) or its Fourier transform $\chi''(\omega)$ is usually based on a single parameter : it is either the time interval Tx associated with a decay of the Mx(t) amplitude equal to e^{-1}, or the width 2δ at half-height of the resonance line $\chi''(\omega)$.

Although, the spin-system response rarely is a single parameter time function, it is considered that Tx or $(2\pi\delta)^{-1}$ give a reasonable experimental estimate of the time scale of the relaxation function Mx(t). A theoretical estimate of the time scale can be obtained according to the following way.

The relaxation function is first expressed as a series expansion :

$$Mx(t) \simeq 1 - M_2\, t^2/2 + M_4\, t^4/4! + \ldots \qquad [3]$$

where M_2 and M_4 are known to be the second and the fourth moments, respectively, of the corresponding resonance line ; Mx(t) is more conveniently expressed as :

$$\mathrm{Log}\,[Mx(t)] \simeq - M_2\, t^2/2 + (M_4 - 3M_2^2)t^4/4! + \ldots \qquad [4]$$

where $M_4 - 3M_2^2$ measures the deviation of Mx(t) from a Gaussian behavior ; the main difficulty encountered in the characterization of Mx(t) arises from the fact that it rarely is a Gaussian function ; consequently, two parameters at least must be used to characterize it ; it is considered that M_2 which is invariant under all molecular motions and M_4 which may reflect molecular motions give a satisfactory approach to the description of Mx(t). A reduced time variable t/τ_a is introduced from what we shall call an apparent correlation time τ_a defined by the relationship :

$$\tau_a^{-2} = (M_4 - 3M_2^2)/M_2 \qquad [5]$$

τ_a goes to infinite values for Gaussian functions ; and :

$$\frac{\mathrm{Log}(Mx(t))}{M_2\ \tau_a^2} \simeq - (t/\tau_a)^2/2 + (t/\tau_a)^4/4! + \ldots \qquad [6]$$

Series expansion [6] illustrates the crucial role played by τ_a in the time behavior of Mx(t) : for t values smaller than τ_a, Mx(t) behaves like a Gaussian function. For t values larger than τ_a, the second term of the series expansion starts governing the spin-system response ; however, formula [6] does not apply at any t value ; it actually is valid within a time interval Δt such that $\Delta t/\tau_a \lesssim 2.4$; it is roughly pictured as an equivalent straight segment defined from an average slope equal to about .6. This is a differential approach to the description of the relaxation function ; it is then recurrently applied over p time intervals ($p = t/\Delta t$) until the time t of observation is finally reached.

Furthermore, the strength of the product $M_2\tau_a^2$ strongly governs the steepness of the amplitude decay of Mx(t) ; for example, let M_2

and τ_a obey the inequality :

$$M_2\tau_a^2 \gg 1 \quad , \qquad [7]$$

then, the Gaussian character of Mx(t) which occurs in the time interval $t \lesssim \tau_a$ is determined by $M_2\tau_a^2(t/\tau_a)^2/2 \simeq M_2t^2/2$ and the amplitude of Mx(t) goes to zero with the time constant $Tx \simeq M_2^{-1/2} \ll \tau_a$; in that case, t values larger than τ_a are not significant. Considering now that M_2 and τ_a obey the inequality : $M_2\tau_a^2 \ll 1$, the behavior of Mx(t) is governed by $0.6\ M_2\tau_a^2(\Delta t/\tau_a) \simeq 0.6\ M_2\tau_a\ \Delta t$, during any time interval Δt such that $0 \lesssim \Delta t/\tau_a \lesssim 2.4$; the apparent time constant of Mx(t) is now :

$$Tx \simeq (0.6\ M_2\tau_a)^{-1} \qquad [8]$$

The last point to be discussed concerns the fourth moment M_4 which is incorporated in the definition of τ_a(5). It is commonly known that in the presence of molecular motions the fourth moment M_4 is expressed as :

$$M_4 = M_4^o - M_2 \left.\frac{d^2\Gamma}{dt^2}\right)_{t=0} \qquad [9]$$

where $M_2\Gamma(t)$ is the correlation function of molecular motions directly participating in the nuclear magnetic relaxation process (2) ; M_4^o is the fourth moment observed in the absence of any molecular motions. In the simplest cases, the second derivative of $\Gamma(t)$ at $t = 0$ is proportional to $-\tau_c^{-2}$, where τ_c is the correlation time of diffusional motions ; then, for random motions fast enough the correlation time τ_c usually obeys the inequality :

$$M_2\ \tau_c^2 \ll 1.$$

Consequently,

$$M_4 \simeq M_2\ \tau_c^{-2} \gg M_2^2$$

and

$$\tau_a \simeq \tau_c$$

the time constant of the relaxation function is expressed as :

$$Tx \simeq (0.6\ M_2\tau_c)^{-1} \qquad [10]$$

The lengthening of the time constant Tx induced by motional averaging is a well-known basic NMR property (2). The foregoing approach clearly shows that a single parameter description of NMR properties does not give any details about the correlation function of random molecular motions ; it only gives an estimate of the second derivative of this function at the initial time.

The submolecule concept is now associated with this NMR approach to define relevant parameters available from relaxation measurements and characterizing a polymer system.

4. POLYMERIC GELS : NMR AND ELEMENTARY CHAINS

We already mentioned that polymeric gels are supposed to be built from statistical structural units also called elementary chains ; all macroscopic properties then result from the collective behavior of this ensemble of structural units. Whatever its exact physical definition, it is considered that any elementary chain can be characterized from an appropriate end-to-end vector $\vec{r}_e$: however, it is assumed that all vector end-points in a gel can fluctuate in space : they are not fixed points. Accordingly, it is more convenient to picture a polymeric gel as a field of forces $\vec{f}$ applied to all elementary chain end-points. For the sake of simplicity, the unperturbed individual statistics of elementary chain will be identified with that of a freely jointed segment made of Ne skeletal bonds of equal length, a, and experiencing a force $\vec{f}$; the corresponding partition function is called : $z[\vec{f}(\nu)]$, where $\nu = 1,2,Ng$, labels elementary chains participating in the gel swelling process. Then, the partition function of a gel considered as a whole is expressed as :

$$Z(Ng,T) = \int \mathcal{P}^*(\vec{f}(1),\ldots\vec{f}(Ng)) \prod_{\nu=1}^{Ng} z[\vec{f}(\nu)]\, d\vec{f}(1)\ldots d\vec{f}(Ng) \qquad [11]$$

The probability distribution function of end-to-end vectors introduced in formula [2] is replaced with the distribution function of forces $\mathcal{P}^*$ defined throughout the polymeric gel. Both statistical treatments are equivalent to each other : one is based on a free-energy approach, while the other one is founded upon a free-enthalpy description.

4.1. Induced Orientational Effect

The mean fluctuation of the end-to-end distance $|r_e|$ of an ideal chain in a melt experiencing a force $\vec{f}$, is given by the well-known state equation :

$$< \vec{r}_e >/aNe = \mathcal{L}(u)\ \vec{f}/|\vec{f}| \qquad [12]$$

$\mathcal{L}(u)$ is the Langevin function and $u = af/kT$. Any skeletal bond obeys a probability distribution function of orientations expressed as :

$$g(\vec{f},\vec{a},Ne) = [I_{1/2}(u)]^{-1} \sum_{\ell,m} I_{\ell+1/2}(u) \; Y_\ell^m(\Omega_{\vec{f}}) \; Y_\ell^{m*}(\Omega(\vec{a})) \qquad [13]$$

where $I_{\ell+1/2}(u)$, $\ell = 0,1,2,...$ are Bessel functions of the second kind ; $\Omega(\vec{f})$ and $\Omega(\vec{a})$ denote angular variables of $\vec{f}$ and $\vec{a}$, respectively. Formula [13] describes the non-isotropic probability of orientations of skeletal bonds within a chain segment experiencing a force $\vec{f}$ applied to its end-points.

4.2. Residual Energy of Spin Interactions

The main variable to which NMR is sensitive, is the angle $\Omega(\vec{a})$ which a given bond $\vec{a}$ makes with the steady magnetic field direction. This angle may correspond to a (CH) bond, or to the axis of rotation of a methyl group or to the vector joining two nuclei ; the angle $\Omega(\vec{a})$ is involved in NMR properties through dipole-dipole interactions or quadrupolar interactions. For two spins (I = 1/2) attached to a skeletal bond $\vec{a}$ which has a fixed orientation, the angle $\Omega(\vec{a})$ induces a broad spectrum structure in the two-spin system response ; the splitting of the corresponding resonance doublet is simply expressed as :

$$\bar{h}\, \varepsilon(\vec{a}) = (3\cos^2[\theta(\vec{a})] - 1)\bar{h}/b^3 \qquad [14]$$

b is the distance between the two nuclei ($\varepsilon(\vec{a})/2\pi = 10^4$ Hz for two protons, with b = 1.8 Å).

The spectrum structure (formula [14]) and the induced orientational effect of skeletal bonds (formula [13]) combine with each other to give rise to a residual energy of spin interactions. Let a given bond $\vec{a}$ belong to a chain segment embedded in a melt, with a fixed force applied to its end-points. Such a bond rapidly fluctuates in space and it is the average value of $\varepsilon(\vec{a})$ which is actually observed from NMR :

$$\varepsilon_e(\vec{f}) = \int \varepsilon(\vec{a})\, g(\vec{f},\vec{a},Ne) d\,\vec{a} \qquad [15]$$

$\varepsilon_e(\vec{f})$ again appears through a spectrum structure ; but the splitting $2|\varepsilon_e(\vec{f})|$ of the resonance doublet is much smaller than $2|\varepsilon(\vec{a})|$ because of the average procedure [15] ; $\varepsilon_e(\vec{f})$ is expressed as :

$$\varepsilon_e(\vec{f}) = (3\cos^2[\theta(f)] - 1)\, I_{5/2}/I_{1/2} \qquad [16]$$

There is a residual energy of spin-interactions induced by the force applied to the observed chain segment ; when $|\vec{f}| = 0$ then $|\varepsilon_e(\vec{f})| = 0$.

4.3. Space-Scale Transfer of NMR Properties

Clearly, the state of stretching of a given elementary chain may be observed from the splitting $\varepsilon_e(\vec{f})$ reflecting the orientational effect of monomeric units induced by the force $\vec{f}$; from [12] and [16] , $\varepsilon_e(\vec{f})$ is easily expressed as a function of the mean fluctuation of the end-to-end distance $<r_e>$. For moderately stretched chains :

$$\varepsilon_e(\vec{f}) \propto (3\cos^2[\theta(f)] - 1)u^2(1 - .1u^2)/15 \qquad [17]$$

or

$$\varepsilon_e(r_e) \propto (3\cos^2[\theta(r_e)] - 1)<r_e>^2/a^2Ne^2 \qquad [18]$$

Formula [18] illustrates a transfer of NMR properties from the local space scale determined by a skeletal bond (≃ 3 Å) to the semi-local scale defined by $<r_e>$ (≃ 30, 50 Å). Details about the structure of monomeric units may be ignored ; these are considered as undistinguishable from one another ; they all are represented by a single physical quantity $\varepsilon_e(\vec{f})$, within a given chain segment.

4.4. Fast Motions of Skeletal Bonds

The full description of the resonance spectrum associated with a given chain segment must take into consideration not only the splitting of the doublet but also the broadening mechanism induced by random motions of skeletal bonds. According to section 3. the spin-system response corresponding to any proton pairs attached to a chain segment experiencing a force $\vec{f}$, is written as :

$$\mathrm{Log}[Mx(t,\vec{f})] = -\mu_2(\vec{f})t^2/2 + \mu_4(f)t^4/4! - 3\mu_2^2(\vec{f})t^4/4! + \dots \qquad [19]$$

$\mu_2(f)$ is the second moment associated with the observed chain segment ; it is invariant under any skeletal bond motion ; the fourth moment $\mu_4(\vec{f})$ includes both the fourth moment $\mu_4^o(\vec{f})$ calculated in the absence of any bond diffusional motion and the additional contribution due to random motions of the chain segment skeleton.

Then, the splitting $\varepsilon_e(\vec{f})$ of the doublet is introduced in the series expansion as follows :

$$\mathrm{Log}\,[Mx(t,\vec{f})] = -(\mu_2 - \varepsilon_e^2)t^2/2 + (\mu_4 - \varepsilon_e^4 - 3(\mu_2-\varepsilon_e^2)^2)t^4/4! - \varepsilon_e^2t^2/2 - 2\varepsilon_e^4t^4/4! - 6\varepsilon_e^2(\mu_2-\varepsilon_e^2)t^4/4! + + \dots \qquad [20]$$

or

$$\mathrm{Log}[Mx(t,\vec{f})] = \mathrm{Log}[\cos(\varepsilon_e t)g_x(\vec{f},t)] \qquad [21]$$

with

$$\mathrm{Log}[g_x(\vec{f},t)] = -(\mu_2 - \varepsilon_e^2)t^2/2 + (\mu_4 - \varepsilon_e^4 - 3(\mu_2-\varepsilon_e^2)^2)t^4/4! + \dots \qquad [22]$$

the two functions $\cos(\varepsilon_e t)$ and $g_x(\vec{f},t)$ are slightly coupled with each other by $- 6\varepsilon_e^2(\mu_2-\varepsilon_e^2)$. The apparent correlation time is now defined by :

$$\tau_a^{-2} = [\mu_4 - \varepsilon_e^4 - 3(\mu_2-\varepsilon_e^2)^2]/(\mu_2-\varepsilon_e^2) \qquad [23]$$

For fast random motions of skeletal bonds described by an appropriate correlation function $\gamma(t)$, under the constraint of a force $\vec{f}$ applied to the chain ends :

$$\tau_a^{-2} \simeq -\left(\frac{d^2\,\gamma(t)}{dt^2}\right)_{t=0} \qquad [24]$$

and the width of individual lines of the resonance doublet is then equal to :

$$T_x^{-1}(\vec{f}) \simeq (\mu_2(\vec{f}) - \varepsilon_e^2(\vec{f}))\ \tau_a(\vec{f}) \qquad [25]$$

Except for specific orientations of the force $\vec{f}$ corresponding to $\varepsilon_e \simeq 0$, the width T_x^{-1} is usually much smaller than the splitting $|\ \varepsilon_e(\vec{f})\ |$; consequently, the resonance doublet has a good resolution, and its splitting may be considered as a direct measure of the stretching state of an elementary chain.

4.5. A Specific NMR Effect

The spectrum actually observed in a polymeric gel is an average of all resonance doublets resulting from all strengths and all isotropic orientations of forces f existing throughout the network structure ; this average gives rise to a single broad line mainly reflecting the ensemble of individual stretching properties of elementary chains ; considering again non-interacting proton pairs on ideal chain segments :

$$\overline{\varepsilon}_e = (Ng^{-1}) \sum_{\nu=1}^{Ng} \int \mathcal{P}^* (\vec{f}(1),\ldots\vec{f}(Ng))$$
$$\varepsilon_e(\vec{f}(\nu))\ d\vec{f}(1),\ldots d\vec{f}(Ng) \qquad [26]$$

$\overline{\varepsilon}_e = 0$ for an isotropic gel ; the mean square value $\overline{\varepsilon_e^2} = \Delta_e^2$ reflects a pure symmetry property namely the deviation of monomeric unit motions from isotropic random rotations ; this represents a collective property of monomeric units within a given elementary chain. The residual energy $\hbar\ \Delta e$ illustrates the average transfer of NMR properties from a local space scale to a semi-local one because it is associated with the overall stretching state of the chain. This is a specific NMR effect induced by the chain structure of any real macromolecule. The residual energy $\hbar\ \Delta e$ mainly concerns quadrupolar interactions or dipole-dipole interactions within proton groups where the distance between nuclei is kept constant. Effects of translational diffusion upon the residual dipole-dipole interactions between nuclei located on different chain segments are negligibly small.

Although the foregoing illustration presented in this section was based upon properties of non-interacting proton pairs located on ideal chain segments, it can be extended to real systems without any major charge brought to the fundamental conclusion of this approach. It must be emphasized that the relaxation process of the magnetization was described within an adiabatic approximation, neglecting quasi-resonant exchanges of energy between the spin-system and the surrounding medium. The approach is based on quantum coherence properties of nuclear spins, only. The width of the resonance line observed on a polymer system, or equivalently, the time constant of the transverse relaxation function may be applied to the characterization of static scaling properties of elementary chains in a polymeric gel.

5. POLYMERIC GELS : SCALING NMR PROPERTIES

In this section, it is shown how the residual energy of spin interactions $\hbar \Delta e$ averaged over the whole sample gel may be used to investigate the static unfolding mechanism of elementary chains induced by external constraints. This will be shown to induce scaling NMR properties as a function of the concentration of the solvent used to swell the gel.

5.1. Introduction

No rigorous description about gel statistics has been proposed until now although it governs static as well as dynamical mechanical responses of commonly used materials such as tires, for example. Although any elementary chain can be described within a reasonable accuracy if it is isolated in a dilute solution, the description of statistical properties of an ensemble of elementary chains in a gel raises theoretical difficulties which have not been overcome, yet (1), (6), (7), (8). The real statistics of an elementary chain embedded in a network structure is not known. Effects of functionality, dangling chains, chain length distribution and screening effects occuring before the cross-linking reaction have been experimentally studied in an extensive way (7). They have been given semi-quantitative interpretations. Some properties are now considered as basic ones ; for example, elementary chains obey a Gaussian statistics in dry gels ; while they exhibit excluded volume properties when gels are fully swollen, using a good solvent ; in addition they obey a packing condition (1), (9).

The purpose of this NMR approach is to attempt to characterize in a semi-local space scale the statistical responses of elementary chains to external constraints. One of these statistical responses can be the static isotropic unfolding of elementary chains induced by swelling a polymeric gel, using a good solvent ; another response may be the elementary chain distortions induced by an uniaxial constraint usually applied by stretching the polymeric gel along one direction. The presence of a residual energy of spin-spin interactions in a polymer network structure is easily experimentally proved (10). The usual order of magnitude of observed line-widths varies from 10^2 Hz to about 10^3 Hz, in most experiments. For an ideal chain segment and proton pairs the stretching parameter u (formula 17) should range from .3 to about 1. ; also the approximate relation :

$$< r_e^2 > \simeq < r_e >^2 \qquad [27]$$

holds on whenever the parameter u has values larger than .3, considering ideal chain segments. For the sake of simplicity, formulae [17], [18] and [27] will be extended to real polymer systems without giving any detailed discussion in the present paper ; it applies except for numerical factors which are not involved in scaling properties. The part $<r_e>^2$ of the residual energy will be used to observe the stretching state of elementary chains while the part N_e^2 will be used to control the effective number of monomeric units actually participating in elementary chain properties.

Since the role of connectivity of the three-dimensional statistical lattice resulting from cross-links is not yet elucidated, the simplest approximation used in the present approach is to consider that the distribution function $\mathcal{P}^*$ can be splitted into a product of identical distribution functions : $p^*(\vec{f})$. The function p* concerns any elementary chain of the polymer network structure. It is a mean field description, neglecting a possible collective behaviour of elementary chains, mainly around covalent cross-links ; $p^*(\vec{f})$ must reflect all symmetry properties of the polymeric gel. Also, this function is supposed to partly compensate the rough approximation proposed by assuming that any elementary chain is similar to a freely jointed segment.

NMR properties are now illustrated from experimental results previously reported in several papers (11)(12). End-linked polydimethylsiloxane (PDMS) networks are well appropriate to such an illustration ; these tetrafunctional covalent gels were made from chains characterized by a number average molecular weight $\overline{Mn} = 10^4$; three polymer volume fractions of synthesis Vc were considered : Vc = 0.46, 0.74 and 0.84 (13). Scaling NMR properties were observed from methyl groups. Three-spin systems (I = 1/2) are easily described from quantum mechanics, most of their properties can be exactly calculated ; the fast rotation of a methyl group around its own c-axis considerably simplifies the expression of the internal dipolar energy. It only depends upon the angle which the c-axis makes with the steady magnetic field direction through a spherical harmonics like in a proton pair.

5.2. Two-Step Swelling of Elementary Chains

In a first approach, small controlled amounts of a good solvent are progressively added to a given dry gel. Although, the volumes of polymer and solvent obey a simple law of addition, the variation of the residual energy $\hbar\,\Delta e$ is found to exhibit a typical behaviour ; this is characterized as a function of the swelling ratio Q = V/Vo (Vo and V are the volumes of the dry gel and the partly swollen gel, respectively). Starting from Q = 1, two well defined swelling ratio ranges may be observed :

i) The first one, hereafter referred to as I goes from Q = 1 to

$Q = Qd \simeq 4$. Within this range, there is a decrease of $\hbar\,\Delta e$, currently observed whenever a good solvent is added to a polymer melt ; Δe varies from about 10^3 rad.sec^{-1} to about 10^2 rad.sec^{-1}.

ii) The second range, hereafter referred to as II, goes from Qd to $Q^* \simeq 12$. Within this range there is an unusual increase of the residual energy ; this is observed until the maximum swelling ratio Q^* is reached, the gel being in equilibrium with the liquid good solvent.

The swelling ratio Qd associated with the minimum of the residual energy $\hbar\,\Delta e$ may be identified with the maximum swelling ratio $Q\theta$ induced by a theta-solvent. For $Q = Qd$, elementary chains obey both a Gaussian statistics and a packing condition, although the gel is partly swollen by a good solvent (12).

5.3. Affine Deformation of End-to-End Distances

For $Q = Qd$, the residual energy $\bar{h}\,\Delta e(d)$ is calculated according to formulae 18 and 27 and by assuming that an affine deformation applies to end-to-end distances of elementary chains :

$$\bar{h}\,\Delta e(d) \propto \langle r_e^2(d)\rangle / Ne^2 \qquad [28]$$

with $\langle r_e^2(d)\rangle \propto Ne$

and $Qd \propto Ne^{-1/2}$;

therefore :

$$\bar{h}\,\Delta e(d) \propto 1/Ne \propto Qd^2 \qquad [28']$$

The product $\Delta e(d)\,Qd^2$ is experimentally found to be a constant (12).

In the concentration range II, the residual energy is found to vary as $Q^{2/3}$. In accordance with formulae 18 and 27 it is predicted to vary as :

$$\bar{h}\,\Delta e(II) \propto Q^{2/3}\,Ne^{-4/3} \qquad [29]$$

provided that both a packing condition and an affine deformation apply to partly swollen chains. Furthermore, the maximum swelling ratio Q^* is expressed as a simple function of the average number of bonds defining an elementary chain : $Q^* \propto Ne^{4/5}$. Consequently, for $Q = Q^*$, the residual energy $\bar{h}\,\Delta e^*$ is predicted to vary as :

$$\Delta e^* \propto 1/Q^* \qquad [29']$$

The product $\Delta e^*\,Q^*$ is experimentally found to be actually a constant (12).

5.4. Trapped Topological Constraints

Screening effects are known to govern statistical properties of overlapping chains in solution (1). The average size of a screening domain is determined from a chain segment containing a number of bonds Ne' given by the concentration Vc of polymer :

$$Ne' \propto Vc^{-5/4}$$

and the correlation length is :

$$\xi \propto Vc^{-3/4}$$

Following the hypothesis proposed by Candau, Bastide et al., it is considered that the actual mesh size of a gel is governed by screening effects occuring before the cross-linking reaction (5). Accordingly, the average number of bonds Ne in any elementary chain must vary as $Ne \propto Ne' \propto Vc^{-5/4}$. Therefore, in the range II, the residual energy $\bar{h}\ \Delta e(II)$ is expressed as :

$$\hbar\ \Delta e(II) \propto Q^{2/3}\ (Vc)^{5/3} \qquad [30]$$

This scaling property has been well observed considering three concentrations of synthesis and two swelling agents (12).

5.5. Disinterspersion of Elementary Chains

In the swelling ratio range I, elementary chains are necessarily swollen by one another. When small amounts of solvent are added to the dry gel, a disinterspersion of elementary chains must occur until they obey a packing condition, corresponding to the swelling ratio Qd. It is postulated that in the range I, elementary chains obey a Gaussien statistics ; their mean dimension is kept constant ; however, there is a screening effect induced by their overlap ; it is associated with small Gaussian domains which are supposed to govern the mesh size of the network actually observed from NMR. Any small domain is assumed to be defined from n bonds (n < Ne) ; from formulae 18 and 27 :

$$\Delta e \propto \langle r_n^2 \rangle / n^2 \propto n^{-1} \qquad [31]$$

Correspondingly, the residual energy $\bar{h}\ \Delta e$ is experimentally found to obey the scaling law :

$$\bar{h}\ \Delta e \propto (Ne)^{-1}\ (Qd/Q)^{2/3} \propto (Q*)^{-5/6}\ Q^{-2/3} \qquad [32]$$

therefore :

$$n^{3/2} \propto Ne^{3/2}\ Q/Qd \qquad [33]$$

Formula [33] describes the statistical pertinent size of space span by the samll screening domains in the range I, until all elementary chains are fully separated from one another at Q = 2d. The present study must be considered as a preliminary approach, only. It stresses the need to confirm the presence of small additional Gaussian screening domains.

5.6. Uniaxial Stretching

The effect of an uniaxial stretching of a polymeric gel on elementary chain properties can be described within the same framework of interpretation. Under stretching, the distribution of forces applied to elementary chain ends is not isotropic anymore. Consequently, the mean value of the residual energy calculated over the whole polymeric gel is not equal to zero : $\overline{\varepsilon e(\vec{f})} = De \neq 0$. Accordingly, a resonance doublet must be observed instead of a single line. Residual quadrupolar interactions instead of dipolar ones were recently observed on polymeric gels. The splitting De of the doublet was shown to vary as :

$$De \propto (Q^*)^{-1} (\lambda^2 - \lambda^{-1}) \qquad [34]$$

where λ is the gel stretching ratio defined in a usual way (14). A simple interpretation of formula [33] can be given by only assuming that the distribution of forces $p^*(\vec{f})$ is shifted toward larger values of fx, where x is the direction of stretching ; correspondingly, $p^*(\vec{f})$ is also shifted toward smaller values of fy and fz, where y and z determine the plane perpendicular to the direction of stretching. The shift factors of p* are set equal to λ and $(\lambda)^{-1/2}$ for the x and y(z) directions, respectively. Also, assuming that $p^*(\vec{f})$ is a homogeneous function of forces $\vec{f}$, mean square values of forces applied to elementary chains are expressed as :

$$\overline{fx^2} = \overline{fy^2} = \overline{fz^2} = \phi_o^2/3 \qquad [35]$$

in the absence of stretching ; and

$$\overline{fx^4} = \overline{fy^4} = \overline{fz^4} = \beta^2(\phi_o^2/3)^2 \qquad [35']$$

where β is a constant specific of the gel system. The residual energy averaged over the whole sample gel is then written as :

$$\Delta e \propto \beta\, \phi_o^2/3$$

Then, in a network structure observed under stretching conditions, the mean square values of forces are written as :

$$\overline{fx^2} = \lambda^2 \phi_o^2/3 \quad ; \quad \overline{fy^2} = \overline{fz^2} = \phi_o^2/3\lambda$$

Considering that formula [17] still applies to stretched gels :

$$\overline{\hbar\, \varepsilon e(\vec{f})} \propto (2\overline{fz^2} - \overline{fx^2} - \overline{fy^2}) \propto (\lambda^2 - \lambda^{-1})\phi_o^2/3$$

The front factor $\phi_o^2/3$ is given by formula [32] for Q = 1 ; it is proportional to $(Q^*)^{-5/6}$; the exponent value .83 is nearly equal to the experimentally determined value about equal to one.

5.7. Conclusion

The reasonable agreement of experimental results with predicted formulae illustrates an unusual application of NMR, leading to the investigation of static scaling properties of polymeric gels. Although the present approach does not claim to bring an answer to any problem encountered in gel physics, it shows the necessity to explore gels properties in a semi-local space scale. Pure asymmetry properties of random segmental motions perceived within individual elementary chains underlie the NMR observation. Experimental results reveal a threshold of affinity which is identified with the average size of actual elementary chains, taking all screening effects into consideration ; these are involved both in the polymer solution before the cross-linking reaction and in the polymeric gel, observed at a low swelling ratio. NMR can be used in addition to measurements of the moduli (shear, osmotic pressure, longitudinal modulus) and to neutron scattering experiments (15)(16). It may be worth emphasizing that NMR properties were analysed according to a model picturing a polymeric gel as a field of forces applies to volumeless effective elementary chain ends ; this model is to some extent analogous to a phantom network (7)(17).

6. POLYMER MELT : NMR AND SUBMOLECULES

Any high molecular weight polymer melt is currently described as a temporary network structure ; this is built from statistical dynamic screening domains. These result from topological constraints, also called entanglements existing in an ensemble of macromolecules. The life-time of the transient network is closely related to the diffusional kinetics of macromolecules, i.e. to the rate of disentanglements of polymer chains. In this section it is shown how dynamic screening effects may define intrinsic sub-systems also called submolecules ; from the NMR point of view they are considered in a state partly out of equilibrium, characterized by a physical quantity derived from NMR measurements : the residual energy of spin-interactions.

6.1. Submolecules. Ephemeral Sub-Systems

Evidence for dynamic screening effects has been obtained from several specific properties generally observed in any polymer melt :
i) The variation of the zero shear-rate viscosity η_o measured as a function of the chain molecular weight M is known to exhibit two well defined ranges, separated by a characteristic molecular weight Mc (3). Below Mc, the viscosity η_o has a linear dependence upon the molecular weight ; above Mc, η_o varies like M^{ε} with $3. \lesssim \varepsilon \lesssim 3.4$; the value 3.4 of the exponent is left unexplained until now whereas the value $\varepsilon = 3$ can be given an interpretation from the reptational model (18). The number of skeletal bonds associated with Mc usually ranges from about 2×10^2 to 5×10^2, considering all polymer melts. The characteristic value Mc is considered as the lowest molecular weight corresponding to the onset of chain entanglements in a given polymer system (19).
ii) Furthermore, above Mc, the relaxation modulus G(t) of any polymer

system is known to present a plateau G_N^o separating the high frequency response to a sudden strain of the melt from the low frequency one. This plateau disappears when the chain molecular weight is lowered below Mc. When it exists, the plateau is equivalently well observed as a frequency cut-off from the shear storage modulus $G'(\omega)$ or from the shear loss modulus $G''(\omega)$ (3).

The striking feature associated with any transient network is that its statistical mesh size does not depend upon the chain molecular weight of the melt, whereas it exhibits a characteristic dependence upon the polymer concentration C (19). The life-time of the network is contrasted to its average mesh size ; it strongly varies with the chain molecular weight with a relaxation time $Tr \propto M^{\varepsilon}$.

The two above specific properties were derived from macroscopic measurements associated with the viscoelastic behaviour of polymer melts.

iii) Another evidence for dynamic screening effects concerns motions occuring in a local space scale and observed from spin-lattice relaxation rates of protons attached to polymer chains in a melt. Measured spin-lattice relaxation rates do not exhibit any chain molecular weight dependence except for very short macromolecules ($M \lesssim 10^4$) when chain ends play the crucial role of a solvent (20)(21). This general result clearly shows that random motions of monomeric units and short segments occuring around the Larmor frequency ($\simeq 10^7$, 10^{10} Hz) are not sensitive to collective motions concerning a whole polymer chain in a melt.

Throughout the present description any polymer melt is considered as a temporary network structure. From property i, the mesh size is supposed to be equal to the mean dimension of any Gaussian chain segment existing between two adjacent entanglements ; this chain segment is also called a submolecule ; its average molecular weight M_e^o is defined by (19) :

$$M_e^o = \rho RT/G_N^o \propto M_c/2 \; ; \qquad [36]$$

ρ is the polymer mass per unit volume, R is the perfect gas constant and T is the temperature ; the plateau modulus value G_N^o is derived from viscoelastic measurements.

Also, from property ii, the relaxational spectrum of a chain in any polymer melt is supposed to have a cut-off ; it consists of two well defined dispersions Ω_1 and Ω_2 (M). The dispersion Ω_1 is associated with high frequency motions occuring within screening domains ; while $\Omega_2(M)$ is associated with collective motions of a chain concerning its random diffusion as a whole ; the dispersion Ω_2 governs dynamical properties of submolecules ; it strongly depends upon the chain molecular weight. Submolecules have not infinite life-times ; they are defined from the initial time of observation t_o ; they will be considered throughout the remaining part of this paper as ephemerical sub-systems to be observed from NMR ; they must vanish within the time interval $Tr \propto M^{\varepsilon}$.

6.2. Submolecules : Superposition NMR Property

Considering a high molecular weight melt it is assumed that the $\Omega_2(M)$ spectrum associated with the life-time of the transient network structure is far too long to be perceived from NMR. It is like observing submolecules with quenched end-to-end vectors ; whereas fast local motions of monomeric units still occur within submolecules since the Ω_1 spectrum associated with them has no chain molecular weight dependence. The NMR approach to the observation of slow chain dynamic processes in a melt is founded upon the existence of sub-systems defined by submolecules and characterized by intrinsic properties. In addition, it is supposed that any state of a given sub-system can be determined from an appropriate variable. Consequently, the relaxation function of the transverse nuclear magnetization Mx(t) is considered as reflecting intrinsic properties of a given sub-system ; while the residual energy of spin-interactions $\hbar\,\Delta e$ which governs the time-scale of variations of the relaxation function $Mx(\Delta et)$ is considered as an appropriate variable characterizing a sub-system. The mathematical structure of $Mx(\Delta et)$ is supposed to be invariant whatever its real complexity, so long as a temporary network structure exists. To prove that a submolecule can be handled like any physical system, concentration effects on the spin-system response were observed. Starting from a pure melt, the decrease of the polymer concentration is known to increase the average chain length Me(c) between two adjacent entanglements, according to the relation (19) :

$$Me(c) = M_e^o\, \rho/c \qquad [37]$$

This property has been generally observed from viscoelastic measurements. Considering formula [18], the residual energy is expressed as a function of the polymer concentration according to the relation :

$$\hbar\,\Delta e(c) = c\,\Delta e^o/\rho \qquad [38]$$

where Δe^o is the residual energy measured in a pure polymer melt. By changing the polymer concentration only, the shift-factor of the time-scale $\rho/c\,\Delta e^o$ of the transverse magnetic relaxation function should have a simple dependence upon the concentration c.

Considering high molecular weight polyisobutylene chains ($\overline{M_w} \simeq 10^6$) in solution in carbon disulfide, the proton transverse magnetizations were recorded as a function of the polymer concentration : c = 0.82, 0.76, 0.68, 0.62 and 0.42 g/cm^3 ; these five concentrations were high enough to let the relaxation functions be purely governed by a residual energy of spin interactions ; life-times of network structures corresponding to these five concentrations were far too long to be observed from NMR ; the shape of the recorded relaxation functions was found to be invariant under polymer dilution ; while a single curve was drawn from the five experimental spin-system responses, using an appropriate shift-factor varing as c^{-1} (22).

The well NMR characterized submolecules will be now used to probe slow chain diffusional processes in a space-scale determined by Me. For polydimethylsiloxane chains, Me ≃ 8800 ; it corresponds to an average chain segment dimension, $\sigma_e \simeq 40$ Å ; correspondingly, the residual energy $\hbar\,\Delta e$ will be used as a low reference frequency $\Delta e/2\pi \simeq 10^2$ Hz.

7. POLYMER MELT : NMR AND DISENTANGLEMENT KINETICS

In this section it is shown how the chain disentanglement kinetics in a melt can be observed from the motional averaging effect of the residual energy of spin interactions ; this effect is induced by shortening the life-time of the studied temporary network structure, using appropriate thermodynamic variables such as the chain molecular weight or the polymer concentration.

The main purpose of the NMR approach is to obtain a deep insight into not only the diffusion process of a chain as a whole but also all internal collective chain motions giving rise to the diffusional mechanism of a chain. For example, using the reptational model proposed to describe any chain migration process in a melt, the mean square displacement of any point R_n located on a chain is expressed as (23) :

$$< (\vec{R}_n(t) - \vec{R}_n(o))^2> \; = \; 6\,D_S t + \sum_{p=1} \frac{4\sigma_e^2}{\pi^2 p^2} \cos^2\left(\frac{\pi\,pn\,\sigma_e^2}{\sigma_T^2}\right) \left(1 - \exp(-p^2 t/T_1)\right) \qquad [39]$$

σ_T^2 is the mean square end-to-end distance of the real chain ; T_1 is the longest chain relaxation time. Several elegant experimental procedures have been proposed to accurately measure the diffusion constant D_S (24),(25). Experimental results were in good agreement with the predicted dependence of D_S upon the chain molecular weight :

$$D_S \propto M^{-2} \qquad [40]$$

Although NMR cannot give a full picture of chain diffusional processes, it will be shown that a partial characterization of a chain relaxation process can be derived from the transverse nuclear magnetization dynamics.

7.1. The NMR Disentanglement Curve

The life-time of any temporary network structure is now supposed to be closely related to the kinetics of chain migration in a melt. It was considered in section 6, that any chain relaxation spectrum consists of two dispersions Ω_1 and $\Omega_2(M)$ well separated by a cut-off. The presence of such a cut-off must give rise to a two-step motional averaging of spin interactions. The first step is induced by local non-isotropic motions associated with the Ω_1 part of the spectrum : it is a partial motional averaging ; the resulting residual energy of spin interactions $\hbar\,\Delta e$ is specific of the studied polymer melt. It has an intrinsic value characterizing the mesh size of the temporary network structure.

The second step of the motional averaging is associated with the $\Omega_2(M)$ part of the chain migration process in a melt ; it is strongly dependent upon the chain molecular weight. Roughly, for high molecular weights ($M \approx 10^6$), the whole Ω_2 spectrum is usually longer than the reference time-scale Δe^{-1} ; consequently, no second step of the motional averaging effect can be observed ; submolecules end-to-end vectors are quenched when observed from NMR ; a solid-like spin-system response is observed. Furthermore, for low molecular weights ($M \approx 10^4$), the whole Ω_2 spectrum is usually shorter than the reference time-scale Δe^{-1} ; consequently, the second step of the motional averaging fully occurs. No residual energy can be measured : a liquid-like response of the spin-system is observed.

A NMR disentanglement curve can be drawn from the resonance line-width or the relaxation time of the transverse magnetization, using polymer samples with decreasing chain molecular weights. The mesh size of any polymer network structure, given by Me, is known to be invariant under changes of the chain molecular weight ; accordingly, $\hbar\,\Delta e$ is kept constant and Δe^{-1} can be used as a reference time-scale. It is clearly seen that a characterization of the $\Omega_2(M)$ relaxation spectrum is obtained by shifting this spectrum toward shorter and shorter values, starting from a high molecular weight melt and going to a low molecular weight one.

The transition of the spin-system response from a solid-like behaviour (also called pseudo-solide response (26)) to a liquid-like behaviour has been well observed considering several polymer melts such as polydimethylsiloxane, polyisobutylene, polystyrene and cis-1,4-polybutadiene. The two-step motional averaging of spin-interactions occuring in a melt is not questioned. NMR disentanglement curves must then be used to characterize the terminal relaxation spectrum $\Omega_2(M)$ (2),(27),(28).

7.2. Chain Migration Models

Two main models have been proposed until now to picture the chain migration process in a melt. According to the Rouse model the memory of organization in space is lost at any time and at any point of a chain divided into submolecules (29). While according to the reptational model proposed by De Gennes, any chain is supposed to more along a fictitious tube built from neighbouring chains ; the memory of organization is lost at chain ends, only (30). Such a mechanism considerably increases the diffusion constant D_S. The dependence of D_S upon the chain molecular weight is predicted to be M^{-1}, using the Rouse model ; the dependence resulting from the reptational model is M^{-2}. Nowadays, the Rouse model is considered as well appropriate to the description of diffusional processes of short chains ($M \lesssim 10^4$) i.e. without topological constraints or entanglements. The reptational model concerns long chains.

Dynamic screening effects are now applied to the characterization of the chain disentanglement kinetics in a melt, in the following way. Any chain is divided into N_S submolecules or sub-systems. A given submolecule is supposed to be fully characterized from a single time dependent vector $\vec{r}_n(t)$, $(n = 1,2, \ldots N_S)$; the residual energy of spin interactions $\hbar\varepsilon e[\vec{r}_n(t)]$ concerning a submolecule is a time dependent function ; however, the mean square value of εe is a constant equal to Δe^2. Correlation functions of the residual energy concerning a given submolecule n :

$$C_n(t_1,t_2) = \overline{\varepsilon_e^n(t_1)\ \varepsilon_e^n(t_2)} \qquad [41]$$

are calculated applying either the Rouse model or the reptation mechanism $(\varepsilon_e^n(t) = \varepsilon_e[\vec{r}_n(t)])$. It is not the purpose of the present paper to describe numerous mathematical details about the calculation of spin-system responses. Main results will be now given.

7.2.1. <u>NMR and the Rouse Model.</u> An exact calculation of the proton pair response can be given, using the Rouse model applied to an ideal chain ; the time scale of the transverse magnetic relaxation function is expressed as :

$$Tx^{-1} \simeq \Delta e^2 \sum_{p,q=1}^{N_S} \tau_{p,q}/N_S^2 \qquad [42]$$

with

$$\tau_{p,q}^{-1} = \tau_p^{-1} + \tau_q^{-1} \qquad [43]$$

and

$$\tau_p^{-1} = \lambda_o \sin^2(\pi p/2N_S+1) \qquad [44]$$

λ_o does not depend upon the chain molecular weight M (i.e. N_S) ; the dependence of Tx^{-1} upon this variable can be exactly computed, assuming that the terminal chain relaxation time τ_1 has a M^2 dependence in accordance with the Rouse model ; the apparent variation of Tx^{-1} is simply written as :

$$Tx^{-1} \propto M^{0.26} \qquad [45]$$

Experimental results lead to a chain molecular weight dependence of Tx^{-1} expressed as :

$$Tx^{-1} \propto M^{1.3} \qquad [46]$$

Therefore, experimental results can fit the predicted dependence of Tx^{-1} upon the chain molecular weight, using the Rouse model in the only case where the terminal time τ_1 is supposed to be proportional to M^3 instead of M^2 $(\lambda_o^{-1} \propto M)$.

7.2.2. <u>NMR and Reptation.</u> There is no time-dependent probability distribution functions concerning the reptational model. Average quantities involved in polymer melt properties are calculated using undirect mathematical procedures (30). In the case of NMR, a single chain relaxation mode is predicted to govern the spin-system response, like in the case of viscoelastic properties ; the time-scale of the transverse magnetization is then predicted to vary as :

$$Tx^{-1} \propto M^3 ; \qquad [47]$$

this dependence is associated with a sharp NMR disentanglement curve which has been never observed. Therefore, experimental results are not in agreement with the reptation model applied to NMR properties in the simplest way. The observed molecular weight dependence should correspond to a mechanism involved in the reptation process in a very different time scale ; the equilibration time Td is supposed to vary as M^2 ; however, it does not correspond to a disorientation of sub-molecules large enough to induce a strong spin-system relaxation effect.

7.2.3. Discussion. The disagreement between experimental results and formulae predicted from the models should not lead to the conclusion that both the Rouse description and the reptational mechanism must be rejected. Two main features come to light from the analysis of NMR measurements. The first one concerns the observed chain molecular weight dependence of the terminal relaxation time τ_1 ; it is found to be proportional to M^3 as it is predicted from the reptation model ; the other one concerns the chain diffusional spectrum ; all relaxational modes are found to have an uniform statistical weight and relaxational frequency vary like $\tau_p^{-1} \propto p^2$ (p = 1,2,...), in accordance with the Rouse model ; this does not mean that the Rouse model is observed from NMR ; it is only shown that the experimentally determined spectrum has a similar structure.

The NMR approach stresses the need to build a reptation model with more details about secondary mechanisms involved in such a motion. The constant length of the end-to-end vector representing a submolecule probably is a too crude hypothesis for the interpretation of NMR measurements, although it is in reasonable agreement with the analysis of viscoelastic properties.

8. CONCLUSION

The purpose of the present paper was to illustrate the way, dynamic screening effects occuring in polymer systems can be applied to the NMR observation of collective properties of monomeric units. NMR measurements are found to obey scaling laws which conveniently characterize these collective properties. In polymeric gels the scaling laws reflect screening effects occuring both before and after the cross-linking reaction. For NMR observations the threshold of affinity is given by the mean size of the effective elementary chain participating in gel properties.

Screening effects of high frequency monomeric motions also apply to the observation of long range fluctuations of polymer chains in a melt. Given a reference time interval of observation determined from both the mesh size of the temporary network structure and the residual energy of spin interactions resulting from dynamic screening effects, the NMR characterization of the chain diffusional process in a melt is based on a criterion of isotropy ; when the translational diffusion of a chain occurs in a time-scale longer than the reference time interval, it is perceived from NMR as a non-isotropic motion ; the spin-system response then has a typical solid-like behaviour. When the time-scale of chain diffusion is shorter than the reference time interval, the motion is perceived as isotropic ; the spin-system response has a liquid-like behaviour.

Finally, it may be worth emphasizing that the present NMR approach may be more generally applied to the observation of fractal polymer systems ; a few experimental approaches appropriate to studies of such systems have been proposed until now.

REFERENCES

1. P.G. De Gennes - Scaling Cocepts in Polymer Physics (Cornell University, Ithaca 1979)

2. A. Abragam - Principles of Nuclear Magnetism (Oxford University Press, 1960)

3. J.D. Ferry - Viscoelastic Properties of Polymer (Wiley, New-York 1983)

4. S.F. Edwards - Proc. Phys. Soc. (London) **88**, 265 (1966)

5. S. Candau, A. Peters and J. Herz - Polymer **22**, 1504 (1981)

6. P.J. Flory - Principles of Polymer Chemistry (Cornell University, Ithaca 1953)

7. J.E. Mark - Adv. Polym. Sci. **44**, 1 (1982)

8. D.S. Pearson - Macromolecules **10**, 696 (1977)

9. S. Candau, J. Bastide and M. Delsanti - Adv. Polym. Sci.**44**, 27 (1982)

10. J.P. Cohen-Addad - J. Chem. Phys. **63**, 4880 (1975)

11. J.P. Cohen-Addad, M. Domard and J. Herz - J. Chem. Phys. **76**, 2744 (1982)

12. J.P. Cohen-Addad, M. Domard, G. Lorentz and J. Herz - J. Physique **45**, 575 (1984)

13. A. Belkebir-Mrani, G. Beinert, J. Herz and P. Rempp - Eur. Polym. J. **13**, 277 (1977)

14. B. Deloche and E.T. Samulski - Macromolecules **14**, 575 (1981)

15. M. Beltzung, J. Herz and C. Picot - Macromolecules **16**, 580 (1983)

16. M. Beltzung, C. Picot, P. Rempp and J. Herz - Macromolecules **15**, 1594 (1982)

17. A.J. Stavermann - Adv. Polym. Sci. **44**, 73 (1982)

18. P.G. de Gennes - J. Chem. Phys. **55**, 572 (1971)

19. W.W. Graessley - Adv. Polym. Sci. **47**, 67 (1982)

20. J.P. Cohen-Addad and J.P. Messa - J. Physique, Lettre **37**, 193 (1976)

21. C. Cunniberti - J. Polym. Sci. Part A-2 **8**, 2051 (1970)

22. J.P. Cohen-Addad and A. Guillermo - J. Polym. Sci. **22**, 931 (1984)

23. P.G. de Gennes - Macromolecules **9**, 587 (1976)

24. P.T. Callagham and D.N.Pinder - Macromolecules **13**, 1085 (1980)

25. P.T. Callagham and D.N. Pinder - Macromolecules **14**, 1334 (1981)

26. J.P. Cohen-Addad - J. Physique **43**, 1509 (1982)

27. J.P. Cohen-Addad and G. Feio - J. Polym. Sci. **22**, 957 (1984)

28. J.P. Cohen-Addad and R. Dupeyre - Macromolecules **18**, 1105 (1985)

29. P.E. Rouse - J. Chem. Phys. **21**, 1272 (1953)

30. M. Doi and S.F. Edwards - J. Chem. Soc. Faraday Trans.2, **74**, 1789 (1978)

DYNAMIC PROPERTIES OF ENTANGLED POLYMERS : THE REPTATION MODEL

L. Léger
Collège de France, Physique de la Matière Condensée
11 Place Marcelin-Berthelot
75231 Paris Cedex 05
France

ABSTRACT. After a brief review of the principal results of the scaling approach of polymer solutions, we describe the main predictions of the reptation model on the dynamic properties of entangled polymer systems (semi-dilute solutions or melts), both in the semi-local and in the global domain (length scale respectively shorter or larger than the chain size). We then compare these predictions to existing experimental data and try to draw a list of still open questions.

INTRODUCTION

Concentrated polymer solutions or polymer melts have quite fascinating dynamic properties : they flow like ordinary liquids if submitted to perturbations slowly varying with time, but respond like elastic solids at high frequencies. This behaviour has early been attributed to the presence of entanglements between chains(1). Indeed, by analogy with polymer gels in which permanent chemical crosslinks between chains are at the origin of the rubber elasticity, it has been argued that entanglements could play the role of temporary crosslinks. It was also thought that entanglements were responsible for the strong molecular weight dependences of all the rheological properties of polymeric liquids, as for example, the viscosity at zero shear which increases like $M_w^{3.4-3.5}$ for long enough chains.

In a typical experiment in which the deformation of the sample under a constant stress applied at time t = 0 is monitored as a function of time, one obtains the S shape curve schematically presented on fig. 1. Three different time domains can be distinguished : at very short times, the system reacts like a glass (region I) ; at intermediate times (region II) the deformation levels off, and a plateau, characteristic of an elastic rubber appears if the polymerization index is larger than a critical value N_e. N_e depends upon the characteristics of the polymer : chemical structure, linear or branched chain, overall polymer concentration for a solution... . In the case of linear chains, to which we shall restrict in all the following, N_e can be interpreted in terms of average number of monomers between two entanglements : the plateau represents

R. Daudel et al. (eds.), Structure and Dynamics of Molecular Systems – II, 179–194.

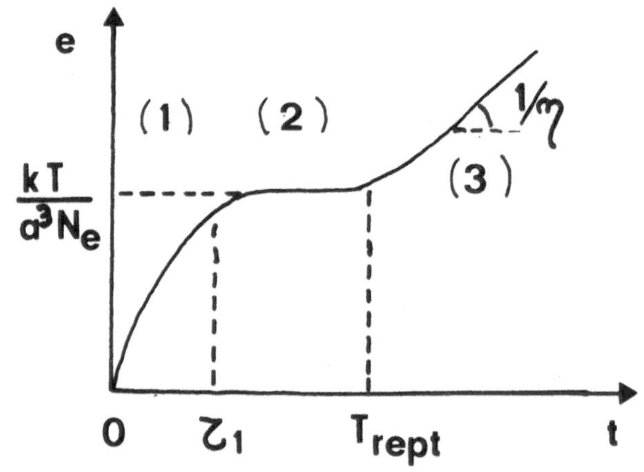

Figure 1. Typical behaviour of the deformation under constant stress of an entangled polymer system displaying three time domains i) glass behaviour at short times, ii) elastic rubber plateau at intermediate times, and iii) viscous fluid at long times. The elastic modulus associated with the intermediate plateau region gives an experimental determination of the average number of monomers between entanglements N_e, assuming an elastic behaviour analogous to that of a true crosslinked network with N_e monomers between crosslinks, i.e. $E = kT/a^3N_e$. N_e depends on the chemical structure of the polymer and usually is of order 10 to 100 (a is the monomer size).

the elastic response of the transient network formed by the chains as long as the entanglements have no time to be undone by Brownian motion. In region III, ($t > T_{rept.}$) the systems behaves like a conventional viscous fluid, and the chains can entirely rearrange with respect to each other.

For short chains ($N < N_e$), region II disappears, and one goes directly from a glass to a liquid like behaviour.

In order to make these notions more quantitative, several theoretical models have been developed in the past decade, in order to 1) take into account both excluded-volume and hydrodynamic interactions in that typical many body problem 2) describe more locally the chains disentanglements. The scaling approach to polymer solutions developed by de Gennes(2) and des Cloizeaux(3) which deals with point 1), and the reptation model developed by de Gennes(4) have revealed quite fruitfull to describe and predict the dynamic properties of entangled linear chains. We shall present briefly their principal features, first for the dynamics at equilibrium (diffusive motions of the chains) and then for a system out of equilibrium (for example relaxation after deformation). In the last part we shall pin point some still open questions which are not quantitatively accounted for by the reptation model.

1. THE DYNAMICAL MODELS : SYSTEM AT EQUILIBRIUM

1.1. The scaling approach to polymer solutions(2)

We just recall here the results which will be used in the following : when in very dilute solutions linear flexible chains adopt a coil configuration, and can be characterized by their radius of gyration $R_G \sim N^{\nu}a$ (a and N are respectively the size and the number of statistical units in the chain, and the Flory exponent $\nu = 3/5$ at three dimensions, for good solvent conditions). Due to excluded volume interactions, R_G is larger than for an ideal chain ($\nu > 1/2$). If the concentration is increased above $c^* = N/R_G^3$, the chains start to overlap. Excluded volume interactions and hydrodynamic interactions are then screened out over a distance $\xi \sim c^{-3/4}$ a : two monomers very far apart along a given chain can no longer come very close to each other to interact through the short range excluded volume interaction and solvent flow induced by one monomer motion is more rapidly damped with distance due to the presence of all the other chains. A semi-dilute polymer solution can thus be considered as made of densely packed hard spheres of size ξ, which behave independently from both static and hydrodynamic point of view. If g is the number of monomers contained in one such subunit, $g = c\xi^3 \sim c^{-5/4}$, i.e., $\xi \sim g^{\nu}a$. This last relation means that inside one subunit all the monomers, on the average, pertain to the same chain. g thus represents the average number of monomers between two entanglements.

If one further increases the concentration, one reaches a regime in which the interactions become completely screened out (concentrated regime). The chains are then ideal at all scales and characterized by $R_{G0} \sim N^{1/2}$ a. The average distance between entanglements is $d \sim N_e^{1/2}$ a.

1.2. Non-entangled regime : $N < N_e$ (case of a concentrated system)

When the chains are short enough, their relaxation under applied stress does not show an elastic response (plateau of region II on fig. 1). This behaviour is interpreted by assuming that there are no entanglements. The chains are thus independent of each other, and their dynamic properties are identical to those of chains in dilute solutions, with a total screening of the excluded volume and hydrodynamic interactions (Rouse chains(7)).

1.2.1. Chain mobility : under an external applied force, $\vec{F}$, the chain takes an average velocity $\vec{V} = \mu\vec{F}$, with μ the global mobility of the chain.

For a chain containing N monomers, the friction is N times larger than for a monomer (for simplicity we assume that the monomer is identical to the statistical unit), i.e. $\mu = \mu_1/N$, with μ_1 the mobility of one monomer. Using an Einstein relation, the diffusion coefficient of the chain is thus

$$D = \mu\, kT = \frac{D_1}{N} \qquad [1]$$

(D_1 is the Brownian diffusion coefficient of one monomer).

1.2.2. Characteristic time of the chain : the spontaneous fluctuations of the chain can be analysed in term of normal modes. Each mode is characterized by one relaxation time(8). The slowest mode corresponds to fluctuation of the chain size with a wavelength comparable to R_G. Its characteristic time or Rouse time, T_{Rouse}, is comparable to the time it takes to the chain to diffuse over its radius :

$$D\,T_{Rouse} \sim R_{G_0}^{\;2} \qquad , \qquad R_{G_0}^{\;2} = N\,a^2$$

$$T_{Rouse} \sim \frac{R_{G_0}}{D} \sim \tau_1\, N^2 \qquad [2]$$

with τ_1 the characteristic time of the motion of one monomer : $D_1\tau_1 \sim a^2$.

1.2.3. Monomer displacement : let us suppose that we are able, by an appropriate labelling technique to follow the displacements of one monomer on the chain (for example by inelastic neutron scattering, with a small sequence of deuterated monomers).
For very long times, and spatial scales larger than R_{G0}, the motion of the labelled monomer is identical to that of the center of mass of the chain : if r is its displacement with respect to the position at t = 0, we have

$$< r^2 > = 6Dt \qquad [3]$$

On the contrary, if one follows the monomer motions over distances smaller than R_{G0} (and consequently during shorter times), they are no longer identical to the center of mass motions, because of the internal modes of the chain. By analogy with eq. [3] , we can suppose that

$$< r^2 >_t \sim t^x \qquad [4]$$

and determine x in order that :

$$< r^2 >_{t\,=\,T_{Rouse}} = R_{G_0}^{\;2}$$

$$< r^2 >_{t\,=\,\tau_1} = a^2$$

These two boundary conditions are fullfilled if x = 1/2 and

$$\underset{\tau_1<t<T_{Rouse}}{< r^2 >_t} = \left(\frac{t}{T_{Rouse}}\right)^{1/2} R_{G_0}^{\;2} \qquad [5]$$

For $r < R_{G0}$, $< r^2 >_t$ is independent of N, as the motion of the labelled monomer does not necessitates that all the chain moves.

The evolution of $< r^2 >$ with time is schematically summarized on fig. 2, on which the different time and length scales for global ($< r^2 > \gg R_{G0}^2$, $t \gg T_{Rouse}$) or semi-local ($< r^2 > < R_{G0}^2$, t_{Rouse}) appear, each with its own behaviour.

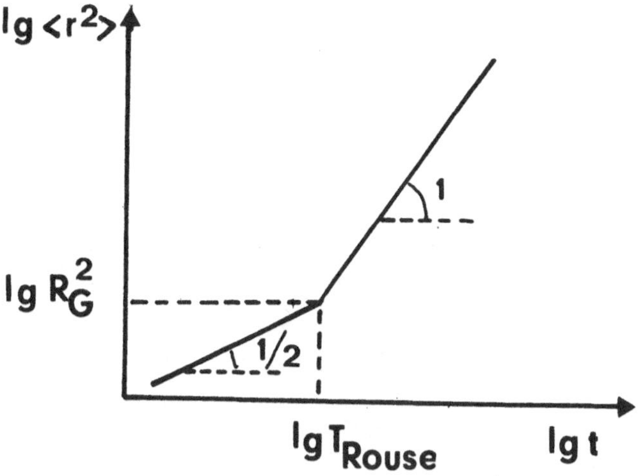

Figure 2. Evolution of the average quadratic distance covered by one monomer as a function of time for a non entangled polymer melt ($N < N_e$). For $t > T_{Rouse}$, the monomer follows the global center of mass motion of the chain.

1.3. Entangled system : $N > N_e$ for a melt
$c > c^*$ for a solution

This case appears a priori much more delicate than the preceeding, because of the interactions between the chains. The more natural approach consists in atempting to describe the structure of the entanglements from topological invariants. However the dynamical properties at times close to T_R are dominated by the fact that the chains are open curves which can modify their knots. This fact is not taken into account in the topological approaches.

On the contrary, the "reptation" model proposed by de Gennes[4] explicitly describes how one chain renews its entanglements with its neighbours, assuming that it reptates like a snake by hindered Brownian motion among obstacles.

1.3.1. Motion of one chain among fixed obstacles : in a polymer melt or in a semi-dilute solution, the chains can change their conformation and rearrange with respect to each other by local Brownian motion, but they cannot cross each other. S. Edwards[9] has proposed to take into account those topological constraints by assuming that each chain is confined inside a virtual tube, envelope of all the constraints imposed by the neighbouring chains (fig. 3). The chain can slip easily along its tube, but it cannot escape laterally over distances larger than the tube diameter d. d is of order of the average distance between entanglements, a few 10 Angströms in a melt or ξ in a semi-dilute solution.

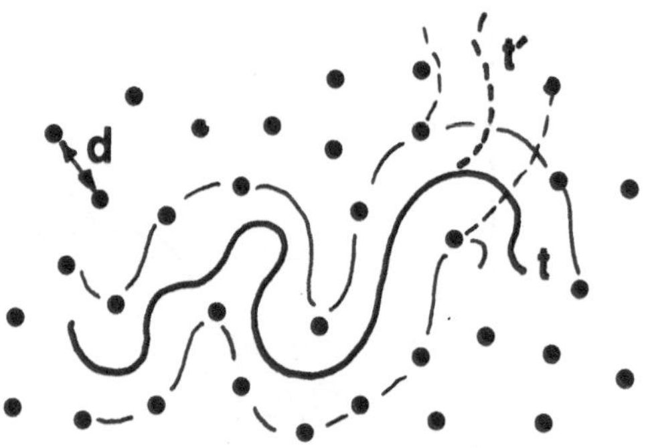

Figure 3. Reptation of one long chain among fixed obstacles. At any time, the chain defines its tube, its extremities choosing their way by diffusion among the obstacles.

1.3.2. Global motion of one chain : for semi-dilute solutions, we have recalled in 1. that both excluded volume and hydrodynamic interactions where screened out over distances ξ. Each chain trapped into its tube can thus be considered as made of N/g independent subunits of size ξ. The tube length is :

$$L_t = N/g\ \xi \qquad [6]$$

The mobility of the chain along its tube is

$$\mu_t \sim \frac{1}{\eta_s \xi}\ \frac{1}{N/g} \qquad [7]$$

with η_s the solvent viscosity.
From [6] and [7] we can deduce the curvilinear diffusion coefficient along the tube :

$$D_t = kT\ \mu_t \sim \frac{kT}{\eta_s \xi N/g} \qquad [8]$$

The time it takes to define a completely new tube is T_R :

$$D_t T_R \sim L_t^2 \qquad [9]$$

i.e. :

$$T_R \sim \frac{\eta_s}{kT} \left(\frac{N}{g}\xi\right)^3 \qquad [10]$$

In the case of a melt we can transpose all these relations, replacing g by N_e and ξ by d the size of the tube ($d \sim N_e^{1/2}$ a), except for relation [7] : when the chain moves, all the monomers experience friction, and we should rather write

$$\mu_{t\ melt} = \frac{1}{\eta\ N\ a} \qquad [7']$$

with now η a monomer-monomer friction coefficient. We thus expect

$$D_{t\ melt} \sim \frac{kT}{\eta\ N\ a} \qquad [8']$$

$$D_t T_R \sim L_t^2 \sim \left(\frac{N}{N_e}\right)^2 d^2 \sim \frac{N^2}{N_e} a^2 \qquad [9']$$

$$T_R \sim \frac{\eta\ a^3}{kT} \frac{N^3}{N_e} \sim \tau_1 \frac{N^3}{N_e} \qquad [10']$$

In a real experiment, one does not measure D_t, as the tube has a Gaussian conformation, but rather D_{self}, with

$$D_{self}\ T_R \sim R_{tube}^2 \sim \left(\frac{N}{g}\right)\xi^2 \quad \text{in a solution}$$

$$\sim \left(\frac{N}{N_e}\right)d^2 \quad \text{in a melt}$$

The self diffusion coefficient of the chains is then

$$D_{self} \sim \frac{kT}{\eta_s \xi (N/g)^2} \quad \text{in a solution} \qquad [11]$$

$$D_{self} \sim \frac{kT}{\eta\ a} \frac{N_e}{N^2} \quad \text{in a melt} \qquad [11']$$

In the semi-dilute solution, with $\xi \sim c^{-3/4}\ a \sim (c/c^*)^{-3/4} R_G$, T_R and D_{self} should follow scaling laws :

$$T_R \sim N^3\ c^{1.5} \qquad [12]$$

$$D_{self} \sim D_0 \left(\frac{N}{g}\right)^{-2} \sim D_0 \, (c/c^*)^{-1.75} \qquad [13]$$

with D_0 the diffusion coefficient at infinite dilution.

1.3.3. Motion of one monomer : in a way similar to the discussion of the non entangled case, we can try to figurate the displacements of one monomer on the chain. We have to distinguish three different time domains :

- $t > T_{rept}$ $(r > R_t)$:

The monomer behaves identically to the center of mass of the chain, with a diffusive motion characterized by D_{self} (eq. [11] , [11']).

- $T_{Rouse} < t < T_{rept}$:

The motion is a global motion of the chain along its tube as we are looking at times longer than the slowest internal mode of the chain T_{Rouse}. Each monomer diffuses along the tube (fig. 4a). We can define the average curvilinear distance covered by this diffusive motion by $< \Delta S >^2 = D_t t$. As the tube is gaussian,

$$< \Delta r^2 > = d < \Delta S > = d\sqrt{D_t t} \qquad [14]$$

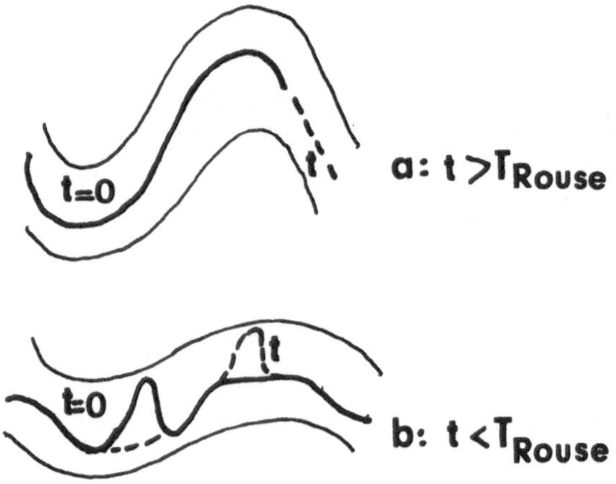

Figure 4. a) Global slipping of the chain along its tube for $t > T_{Rouse}$.
b) Folds diffuse along the chain for $t < T_{Rouse}$.

- $t < T_{Rouse}$:

The different parts of the chain can move with respect to each other, and form folds inside the tube (fig. 4b). These folds diffuses along the chain. De Gennes has proposed a description of such motions in ref. (4), which leads to $< \Delta S > \sim t^{1/4}$ if $\Delta S > d$, i.e.

$$< \Delta r^2 > \sim a\ d\ (t/\tau_1)^{1/4} \qquad [15]$$

We thus expect for the average quadratic distance covered by one monomer three quite distinct spatial and time domains, as summarized on fig. 5.

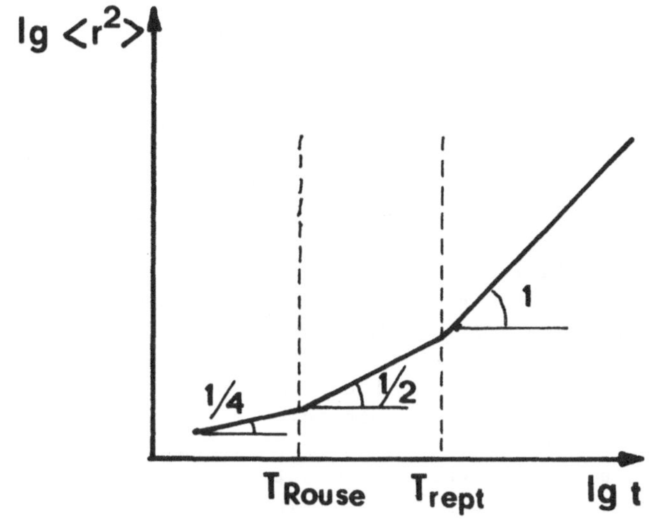

Figure 5. Evolution of the average quadratic distance covered by one monomer as a function of time for entangled chains ($N > N_e$) ; the three domains local, semi-local, and global are easily distinguishable.

We can notice that : i) the different behaviours are continuous at each domain boundary. ii) for long enough chains, the semi-local regime, which corresponds to $T_{Rouse} \sim \tau_1 N^2 < t < T_{rept} \sim \tau_1 (N^3/N_e)$ can be quite large. Typically, for $\tau_1 \sim 10^{-11}$ sec, $N \sim 10^4$ and $N_e \sim 10^2$, $T_{Rouse} \sim 10^{-3}$ sec, and $T_{rept} \sim 10^{-1}$ sec.

2. EXPERIMENTAL TESTS

2.1. Global motion

One has to measure the self diffusion coefficient of the chains. A few chains in the sample have thus to be labelled in order to become distinguishable from the others. If one wants, however, that the motions of the labelled chains remain identical to those of the unlabelled ones, the labelling technique has to be choosen such that it does not introduce specific interactions between chains (labelling by deuteration or by chemical modification of only a few monomers in the chain).

An important difficulty comes from the fact that we are dealing with very slow diffusive processes. From eq. [11], an order of magnitude of D_{self} can easily be evaluated, and ranges in the 10^{-8} to 10^{-13} cm^2/sec for a solution or melt respectively. The time necessary to perform a measurement, τ, is related to the distance over which the diffusion is followed, ℓ by $D_{self}\tau \sim \ell^2$. With the above D_{self} values ℓ has to be small (in the range of a few microns, but still larger than R_G to stay in the global domain) to keep the duration of the experiment within reasonable limits.

A few experimental techniques have allowed to fullfill these two requirements : weakly perturbating labelling and short diffusion length :

- For polymer melts, Klein and Briscoe[11] have used infrared spectroscopy to follow the diffusion of deuterated polyethylene into a matrix of hydrogenated polyethylene, with a spatial resolution of order 100 µm. They have obtained with rather polydisperse samples, but correcting for the polydispersity, $D_{self} \sim N^{-2 \pm 0.1}$.
- For semi-dilute solutions mainly two techniques have been used :

i) pulsed field gradient NMR[12] which realizes a quite ideal labelling of the monomers through their Larmor frequency depending on their position during the pulse of magnetic field, with a resolution imposed by the field gradient. The time window of the NMR spectrometers however does not allow to measure self diffusion coefficients smaller than 10^{-9} cm^2/sec.

ii) Forced Rayleigh light Scattering : the labelling is performed by a few photochromic molecules chemically attached on the chains, which are then locally excited with a pulse of light having a sinusoïdal spatial repartition (interference fringes). An absorption grating is thus printed in the sample, whose amplitude relaxes by diffusion of the labelled chains when the photoexciting pulse has ceased[13]. The spatial resolution is governed by the interfringe spacing and ranges in the .5 to 100 µm. With both techniques, results compatible with eq. [11] have been reported for concentrations smaller than 10 % in weight[12][13].

As an illustration, self-diffusion coefficient data deduced from Forced Rayleigh light Scattering on polystyrene-benzene solutions[13] are reported in reduced units D_{self}/D_0 versus c/c^* on fig. 6. A universal behaviour is observed and self-diffusion coefficients differing by more than two orders of magnitude are all gathered on a unique curve by this choice of reduced units. Moreover, for large enough c/c^* values, the data appear compatible with the prediction of eq. [13] as shown by the straight line which corresponds to the power law $D/D_0 \sim (c/c^*)^{-1.8}$. For c/c^* values larger than 10 slight departure from the predicted behaviour may indicate some additional complications : all the model above developed is only valid if the concentration is small enough to keep monomer-monomer interactions negligible with respect to monomer-solvent interactions. This may no longer be the case for concentrations larger than 15 % by weight, and experiments using very large molecular weights are necessary in order to get a large semi-dilute regime ($c^* \lesssim c \lesssim 15$ %).

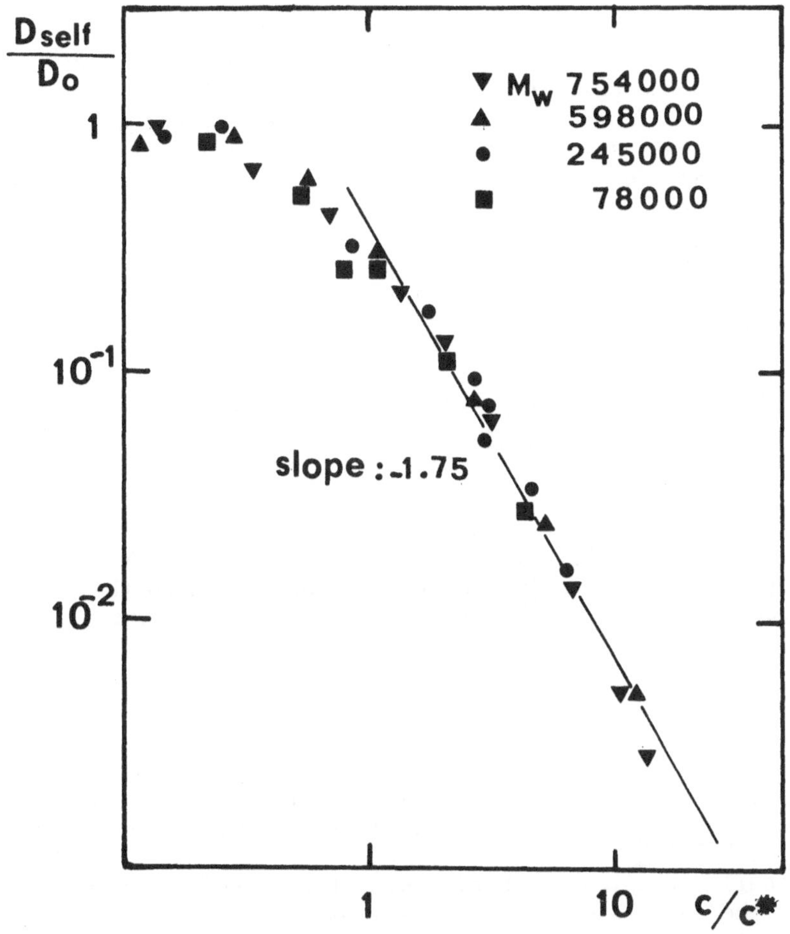

Figure 6. Self-diffusion coefficient deduced from Forced Rayleigh light Scattering on polystyrene-benzene solutions as a function of the concentration. In reduced units (D_0 is the self-diffusion at the zero concentration limit, and c^* the cross-over concentration from dilute to semi-dilute regime) a single curve is observed for molecular weights covering more than two decades, illustrating the validity of the scaling approach to polymer solutions.

2.2. Semi-local scale

This domain is far much delicate to approach experimentally than the previous one.

The more natural idea is to use quasi-elastic neutrons scattering but for most of the polymer melts, the accessible wave vector range corresponds to too small distances with respect to the tube size ($q^{-1} < d$: d is of order 40 to 50 Å for $N_e \sim 200$, and $q^{-1} < 30$ Å).

Quite recent experiments on polytetrahydrofuran melts[14] seem to show a time dependent correlation function S(q,t) characteristic of a constrained motion of the monomers for $q^{-1} \sim d$, giving thus a determination of $d \sim 30$ Å for this particular system.

Several groups try to investigate chains dynamics in the semi-local domain from numerical simulations[15-16-17]. The essential difficulty is to generate long enough chains to get sure that they are entangled, within non prohibitive computation duration. With chains of a few hundred of monomers, Baumgartner et al[16] have obtained a behaviour of $< r^2(t) >$ analogous the one reported on fig. 5, for one mobile chain among frozen chains. Thus, the question of experimental tests of the predictions of the reptation model in the semi-local domain remains open. One has to notice that pulsed field gradient NMR could lead to quite interesting informations as it gives access to the correct time and length windows. When measuring self-diffusion, Callaghan and Pinder have indeed observed a short time behaviour which could well correspond to the penetration in the semi-local regime[10][18].

3. DYNAMICAL MODELS : SYSTEMS OUT OF EQUILIBRIUM

Starting from the basic ideas of the reptation model, Doi and Edwards have established detailed constitutive equations for the rheological properties of entangled polymer systems (flow properties and deformation under stress)[5][19].

We shall emphasize the basic ideas they have developed without entering in the details of the calculations.

Let us start again with the deformation experiment presented on fig. 1 : a constant stress is applied at time t = 0 on a sample of entangled chains.

Suppose that we are now able to follow the evolution with time of the conformation of one particular chain (for example from neutrons small angle scattering on a system containing a few deuterated chains mixed with identical hydrogenated chains). Such an evolution is schematically presented on fig. 7.

Immediately after the application of the stress, both the chain and its tube are deformed in a way similar to a polymer network. Due to the large differences between the three characteristic times of the chain introduced in I (τ_1, T_{Rouse}, T_{rept}) the relaxation takes place in two steps :

1) In the first stage the chain slips rapidly along its tube, and relaxes inside a deformed tube.

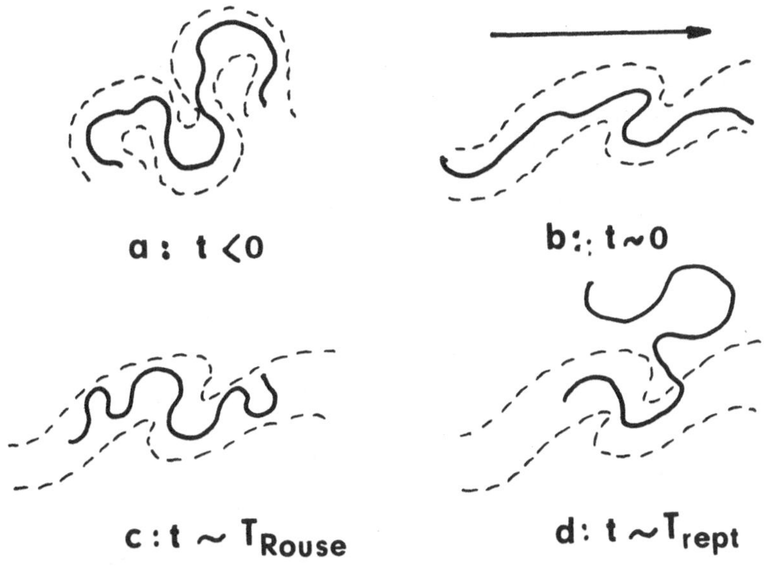

Figure 7. Schematic representation of the different steps of the relaxation of an entangled polymer system after deformation :
a) Chain at equilibrium before the application of the stress.
b) Just after the stress is applied, the chain and its tube are deformed.
c) For $t \sim T_{Rouse}$, the chain has relaxed in its still deformed tube.
d) For $t \sim T_{rept}$ the tube itself relaxes.

2) The tube (or its equivalent in Doi and Edwards language : the primitive chain, a fictive chain following the tube axis) relaxes by reptation of all the chains in the system.

These steps are better and better separated in time when the molecular weight of the chains is increased :
- the slipping of the chain in its tube takes place over a time scale $\tau_d < t < T_{Rouse}$, with $\tau_d = \tau_1 N_e^2$ and $T_{Rouse} = \tau_1 N^2$ the Rouse time of a chain with respectively N_e and N monomers ;

- the relaxation of the tube is completed within a time comparable to $T_{rept} \sim \tau_d(N/N_e)^3$.

When the first step of the relaxation has taken place ($t > T_{Rouse}$), the constraint at each time is proportional to the memory function M(t) which measures the fraction of monomers of the chain still pertaining at time t to the tube defined at time t = 0. M(t) has been calculated by de Gennes(4) and decreases exponentially at long times, like $\exp(-t/T_{rept})$ (fig. 8).

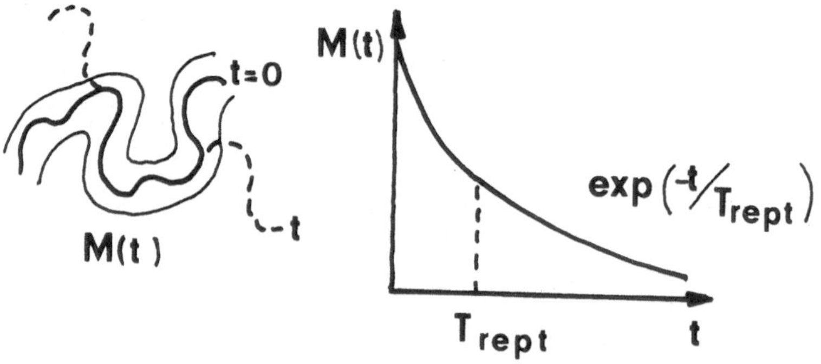

Figure 8. The memory function M(t) measures the fraction of monomers still pertaining at time t to the tube defined by the chain at time 0.

These basic ideas have lead Doi and Edwards to a quite good description of the viscoelastic properties of polymers, both in the linear domain (weak deformations) and in the case of strong deformations, for long enough chains (N/N_e large enough to give rise to well separated relaxation mechanisms) with well defined molecular weights(20)(21)(22).

4. UNSOLVED QUESTIONS

The reptation model thus appears quite successfull, at least qualitatively to explain the dynamic properties of entangled polymer systems. From a more quantitative point of view, however, several questions remain.

One of the most important is the molecular weight dependence of the zero shear viscosity. In a simple approach, one can write $\eta = E\, T_{rept}$ The plateau elastic modulus E is molecular weight independent for $N > N_e$) and proportional to the density of entanglements, while T_{rept}, proportional to the terminal relaxation time, should vary like N^3 (eq. [12]). However, numerous experiments, performed on very different polymer systems, rather give an exponent larger than 3, and closer to 3.4(1). Many ideas have been proposed to explain those differences : Doi(23) has suggested that for not very high N/N_e values, fluctuations of the chain in its tube could lead to an apparent exponent for the viscosity-molecular weight law larger than 3 while the 3 value should be recovered asymptotically at very high N/N_e. Polydispersity effects have also been invoked(24)(25)(26), while the possibility of very long life time tight knots(27) has also been considered. Those different points have not yet been tested experimentally in details.

Another interesting question is risen by the self-diffusion coefficient. A number of available measurements have been performed on systems in which all the chains were mobile, and thus may not correspond to the simple reptation picture in which one mobile chain evolves among fixed obstacles. Klein(28), Daoud and de Gennes(29) and Graessley(30) have described this diffusion in terms of 1) a reptation motion of the test chain in its tube, 2) a tube renewal contribution resulting from the reptation of all the surrounding chains, and they all conclude that tube renewal is a rather slow process leading to only small corrections with respect to pure reptation. Recent Forced Rayleigh light Scattering measurements with labelled and unlabelled chains of different molecular weights(31) have clearly established that for very large molecular weight of the matrix (frozen environment), D_{self} was indeed in very good agreement with the pure reptation prediction, while for comparable matrix and labelled chains molecular weight the chains motions were noticeably accelerated showing an important contribution of the tube renewal to D_{self}. Tube renewal effects have also been reported in strain-stress relaxation experiments performed on systems having a binodal molecular weight distribution(32).

With regards to the semi-local domain, only few experimental investigations are available at the present time. This domain is anyhow quite interesting and should provide more refined tests of the reptation model than the global domain.

We want to emphasize once more that all the above considerations only apply to linear and flexible chains. For branched systems, the reptation should be highly hindered, and de Gennes has predicted for star molecules a reptation time decreasing exponentially with the length of the arms(33). Klein(34) has indeed measured the self-diffusion coefficient of star polyethylene molecules in a melt of linear chains, and obtained a behaviour compatible with de Gennes'prediction but, for very long chains, tube renewal processes become much more efficient than reptation and the exponential dependence is no longer observed. For a melt of rigid rods, Doi and Edwards(35) have developed a complete description the rheological properties taking into account the hindrance of the rotation of each rod molecule due to the presence of the others.

The reptation model has lead to a clear progress in the understanding of the dynamic properties of entangled polymer systems. One of its advantages is to well specify distinct time and spatial domains for the chains behaviour in good qualitative agreement with many experimental observations. Some puzzling questions remain however, and new experiments, using well characterized polymer sample may well lead to a complete draw back of the simple pictures we have presented.

REFERENCES

1. W.W. Graessley, Adv. Polym. Sci., 16, 1 (1974).
2. P.G. de Gennes, Scaling concepts in polymer physics, chap. III and VII, Cornell University Press, Ithaca, N.Y. (1979).
3. J. des Cloizeaux, J. Phys. (Paris), 36, 281 (1975).
4. P.G. de Gennes, J. Chem. Phys., 55, 572 (1971).
5. M. Doi, S.F. Edwards, J. Chem. Soc., Faraday Trans. 2, 74, 1789, 1802, 1817 (1978).
6. P.J. Flory, Principles of polymer chemistry, chap. XII, Cornell University Press, Ithaca, N.Y. (1971).
7. P.E. Rouse, J. Chem. Phys., 21, 1273 (1953).
8. P.G. de Gennes, Scaling concepts in polymer physics, chap. VI, Cornell University Press, Ithaca, N.Y. (1979).
9. S.F. Edwards, Proc. Phys. Soc., 92, 9 (1967).
10. S. Daoudi, F. Brochard, Macromolecules, 8, 804 (1975).
11. J. Klein, B.J. Briscoe, Proc. R. Soc. London, ser. A 365, 53 (1979).
12. P.T. Callaghan, D.N. Pinder, Macromolecules, 14, 1334 (1981).
    Macromolecules, 13, 1085 (1980).
13. L. Léger, H. Hervet, F. Rondelez, Macromolecules, 14, 1732 (1982).
14. J. Higgins, J.E. Roots, presented at the 18th Faraday Symposium 'Molecular basis of viscoelasticity', Cambridge (1983).
15. K.E. Evans, S.F. Edwards, Journ. Chem. Soc., Farad. Trans. 2, 77, 1891 (1981).
16. A. Baumgartner, K. Kremer, K. Binder, Faraday Symp. Chem. Soc., 18
17. J.M. Deutsch, Phys. Rev. Lett., 49, 926 (1982).
18. P.G. de Gennes, L. Léger, Ann. Rev. Phys. Chem., 33, 49 (1982).
19. M. Doi, Journ. Polym. Sc., Polym. Phys. Ed., 18, 1005 (1980).
20. Y. Einaga, K. Osaki, M. Kurata, Polymer J., 2, 550 (1971).
    5, 91 (1973).
21. Y. Einaga, K. Osaki, M. Kurata, J. Polym. Sc., 13, 1963 (1975).
22. C.R. Taylor, R. Greco, O. Kramer, J.D. Ferry, Trans. Soc. Rheol., 20, 141 (1976).
23. M. Doi, Journ. Polym. Sc., Polym. Phys. Ed., 21, 667 (1983).
24. M. Adam, M. Delsanti, J. Physique (Paris), 43, 549 (1982).
25. D.A. Bernard, J. Noolandi, Macromolecules, 15, 1553 (1982).
26. W.W. Graessley, Journ. Polym. Sc., Polym. Phys. Ed., 18, 27 (1980).
27. P.G. de Gennes, Macromolecules, 17, 703 (1984).
28. J. Klein, Macromolecules, 11, 852 (1978).
29. M. Daoud, P.G. de Gennes, Journ. Polym. Sc., Polym. Phys. Ed., 17, 1971 (1979).
30. W.W. Graessley, Adv. Polym. Sci., 47, 67 (1982).
31. M.F. Marmonier, L. Léger, Phys. Rev. Lett., 55, 1078 (1985).
32. J.P. Montfort, G. Marin, Ph. Monge,
33. P.G. de Gennes, Journ. Phys. (Paris), 36, 1199 (1975).
34. J. Klein, Faraday Symp. Chem. Soc., 18,
35. M. Doi, S.F. Edwards, J. Chem. Soc., Faraday Trans. 2, 74, 560 (1978).
    74, 918 (1978).

NEW IDEAS FOR MICROEMULSION STRUCTURE : THE TALMON-PRAGER AND DE GENNES MODELS

C. Taupin
Physique de la Matière Condensée
Collège de France
11 Place Marcelin-Berthelot
75231 Paris Cedex 05
France

ABSTRACT. The droplet model has been very successful to explain the properties of microemulsion, except in the case of (low tension - three phases) systems. Two different models based on a random geometrical repartition of constituents are reviewed.

1. INTRODUCTION

In the preceeding lectures, we have heard about a variety of experimental investigations which are tightly bound with the "droplet model". This model, which appeared in the early studies of microemulsions, has been very successful and illuminating in many monophasic microemulsions but it seems more questioning in the three phase microemulsion systems (Fig. 1).

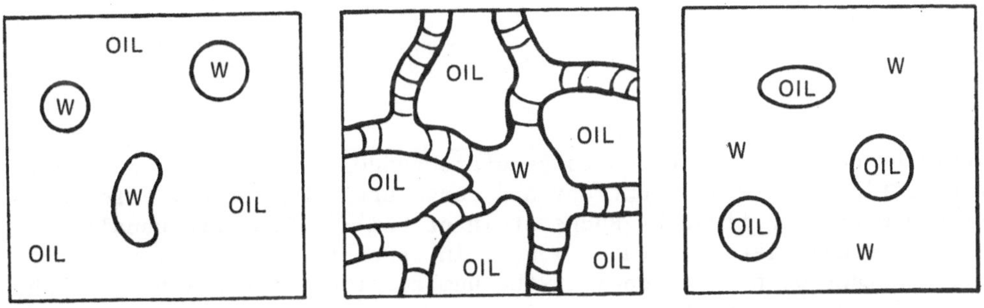

(a) (b) (c)

Figure 1. Structure of microemulsions as a function of the water-to-oil ratio (w/o) : (a and c) swollen micelles (respectively w/o << 1 and w/o >> 1 ; (b) "bicontinuous" structures, first proposed by Scriven. The example shown here can be described as a network of water tubes in an oil matrix (in this figure the individual surfactant molecules are not shown ; the surfactant film is represented as a continuous sheet).

R. Daudel et al. (eds.), Structure and Dynamics of Molecular Systems – II, 195–208.

A different approach is proposed in two recent models[1,2,3] where the statistics of the geometrical repartition in space of the three components (water, oil and amphiphilic film) are studied without focussing on the existence of well-defined structural elements like droplets.

In these models, which are still under experimental investigation, the volume is randomly divided into three different parts ; the interfacial film is supposed to contain all the surfactant molecules and occupy a negligible volume. Then the statistical mechanics of phase equilibria in these pseudoternary systems are developed taking into account the following elements :

- entropy of the geometrical repartition ;
- interfacial energy (i.e. interfacial tension in the film), which is unusually small ;
- contribution of the curvature of the film to the energy ;
- Van der Waals attraction between similar layers.

The original idea for these models resides in a paper by Scriven[4]. This author pointed out that as the oil/water ratio is increased, three types of structure will evolve in succession. Starting from the oil in water emulsion to be expected at low oil content, these should be a smooth transition to a bicontinuous structure, in which neither phase can be said to surround the other, followed by a second transition to a disconnected water in oil emulsion. In this early paper, special attention is given to the variation of the interfacial area which is an important factor in the energy function.

The two models (Talmon-Prager and de Gennes) differ mainly by the way the random geometry is generated. One of them (T.P.) is the Voronoi tesselation for which many properties have been calculated ; the other one is a division of space into cubes of edge ξ_K, ξ_K being determined by the flexibility of the film. Talmon-Prager focus mainly on phase equilibria and de Gennes also considers the competition between microemulsions and lamellar phases.

2. TALMON-PRAGER MODEL

A simple procedure is proposed to generate random two-fluid geometries. Main attention is given to entropy and curvature. The entropy effects have been mentioned first by Ruckenstein et Chi[5]. They are usually difficult to calculate. In their model, in order to determine, at least with a constant of proportionality the number of configurations available to the oil and water regions in a microemulsion specified composition, Talmon and Prager consider the class of geometries resulting from a two step process, consisting of a Voronoi tesselation followed by random segregation into oil and water domains. N points are first distributed completely at random over the volume V to be occupied by the microemulsion ; with each of these Poisson points, is associated the polyhedral region which lies closer to it than to any of the other (N-1) points. In the second step, N_0 of these polyhedra are selected at random to be occupied by oil, and the ramainder filled with water ; the result is a random two-fluid geometry in which the oil occupies the volumic fraction v_0 and the water v_w of the total volume V

The number of distinct positions available to a Poisson point is proportional to the volume V, the proportionality constant 1/w being one of the model parameters. It is the size of w that determines the scale of the microemulsion structure, and it is through w that information about the molecular nature of the fluids enter into the models. Taking into account the number of configurations for the placement of all N points and for each configuration the number of way of assigning the oil and water domains, the configurational entropy is calculated :

$$S = - Vk\ c(\ln(\frac{wc}{e}) + v_0 \ln v_0 + v_w \ln v_w)$$

This equation must be supplemented by a relation between $c \equiv N/V$ the concentration of Poisson points and the true composition variables, namely the surfactant concentration σ.
The mean area S of internal oil/water boundary per unit volume is given by :

$$S = \eta_1\ c^{1/3}\ v_0 v_w \qquad \text{with } \eta_1 = 5.82$$

and the surfactant concentration σ will be taken proportional to S, $\sigma = \alpha S$ where the surfactant capacity α is a second parameter of the system. The phase equilibria of such system which is entirely controlled by the entropy shows that above a certain critical value of surfactant there is only one phase and below that concentration a demixtion of two microemulsions in eqwilibrium with one another. No three phase equilibria exist with only the entropic term.

Talmon and Prager added the effect of curvature in the following way : a curved interface implies a lowering of the surfactant capacity by a factor $(1 - \beta\lambda)$; λ being the edge length per unit area. The curvature is evaluated by the length of the edges of the Voronoi polyhedra which lie on the oil/water interface : $5.83\ c^{2/3} \times 3\ v_0 v_w$. The new expression for σ is :

$$\sigma = 5.82 \quad \sigma c^{1/3}\ v_0 v_w\ (1 - 3\beta c^{1/3})$$

The corresponding expression for the free energy gives rise to interesting new features : a triple tangent plane appears which lies entirely below the free energy surface proving the possibility of three phase equilibria in this model. This is a very new and remarkable results serie ; many ingredients of the model are very simple and reflect the fundamental characteristics of the three phase systems :

- very low interfacial tension ;
- effect of entropy ;
- curvature terms.

It is interesting to note that the curvature energy term in Talmon-Prager model is symmetrical as regards the sense of curvature on the contrary to many other theoretical studies[6]. In spite of the fact that it is clear that such an hypothesis is not true in many systems, it could be realistic in the vicinity of the "optimal composition" where the symmetry of the properties of the surfactant molecule relative to oil and water has been frequently emphasized[7].

Nevertheless some physical phenomena are omitted in this model :

- because in such geometry the interfacial film is wrinkled, a necessary condition is the flexibility of the interfacial film, which is known to be very small in the stiff lamellar phases ;

- Talmon-Prager considered only zero interfacial tension state : phase equilibria could occur close to (but not exactly at) a situation of zero interfacial tension ;

- the Van der Waals interaction, which have been shown to be important in microemulsion systems are neglected.

De Gennes-Taupin[3] model studies the role of these ingredients in a different way of dividing space linking the statistics of this problem to that of the well-known Ising model.

3. THE ROLE OF INTERFACIAL FLEXIBILITY : DE GENNES MODEL

3.1. The existence of a well defined interfacial area per molecule can be correlated with the experimental ultralow interfacial tensions in the following manner (Schulman description)

If one considers a single interface of arbitrary shape separating the oil from the water : the total area of interface is $\mathcal{a}$. It contains a number n_s of surfactant molecules, each of them covering an area $\Sigma = \mathcal{a} / n_s$. In the simple case where : (i) the surfactant is insoluble in bulk oil or water ; (ii) interactions between different portions of the interface are negligible ; (iii) curvature energies are omitted, the free energy may be written :

$$\mathcal{f} = \mathcal{f}_{bulk} + \gamma_{ow} \mathcal{a} + n_s \, G(\Sigma) \tag{1}$$

where the second term corresponds to the bare (i.e. without surfactant) interface (with an interfacial tension γ_{ow}) and $G(\Sigma)$ is a surfactant free energy, depending on the area Σ and containing in particular the effect of surfactant/surfactant repulsions. The Langmuir surface pressure of the film is

$$\Pi(\Sigma) = -\frac{\partial G}{\partial \Sigma} \tag{2}$$

and the actual interfacial tension is

$$\gamma = \gamma_{ow} - \Pi \tag{3}$$

If (1) is minimized, at fixed n_s, with respect to Σ the condition (4) is obtained :

$$0 = \frac{d\mathcal{f}}{d\Sigma} = n_s(\gamma_{ow} - \Pi) = n_s \, . \, \gamma \tag{4}$$

Thus the system will adopt a well defined area per surfactant which is called Σ^* : it is defined by the implicit equation

$$\Pi(\Sigma^*) = \gamma_{ow} \tag{5}$$

The state $\Sigma = \Sigma^*$ will be called the saturated state. It can be reached only if other possible states of the surfactant (such as pure surfactant micelles in water, or in oil) are of higher free energy.

In the saturated state, with a system of zero surface tension, as shown by eq. (4), the area $\mathcal{A}$ is entirely defined by the number of surfactants available

$$\mathcal{A}^* = n_s \Sigma^* \tag{6}$$

It has to be noted that if $\gamma = 0$, several factors become non negligible :

a) entropy of mixing oil and water. Recent calculations were published[8,9],

b) interaction forces (electrostatic and Van der Waals),

c) curvature energy and flexibility.

3.2. Rigidity and spontaneous curvature

The Schulman description ignores all energies associated with the curvature of the interface. Indeed, for many problems involving fluid/fluid interfaces, curvature energies represent only a very minor correction, and the interfacial energy γ dominates the behavior. However, with interfaces where $\gamma \to 0$, curvature effects become relevant.

The two basic ingredients have been defined most clearly in a paper by W. Helfrich[10]. For a curvature $1/R$ one expects an energy contribution per unit area of the form

$$F = \gamma - \frac{K}{R_0 R} + \frac{K}{2R^2} \tag{7}$$

here $1/R_0$ is the spontaneous curvature, and can be of either sign (R_0 is counted as positive when the trend is towards direct micelles). The parameter K has the dimensions of an energy, and may be called the rigidity of the interface. Eq. (7) holds only if R and R_0 are much larger than the interfacial film thickness L. How can one estimates K and R_0 ?

(i) for ionic surfactants the steric considerations of Ninham and Mitchell[6] may give an estimate of the spontaneous curvature (Fig. 2). More detailed computer studies on aliphatic chains anchored at a curved interface have been carried out recently. All these studies assume that the interfacial shell occupied by the surfactant tails is free of oil. This may be correct for the relatively short chains of conventional surfactants.

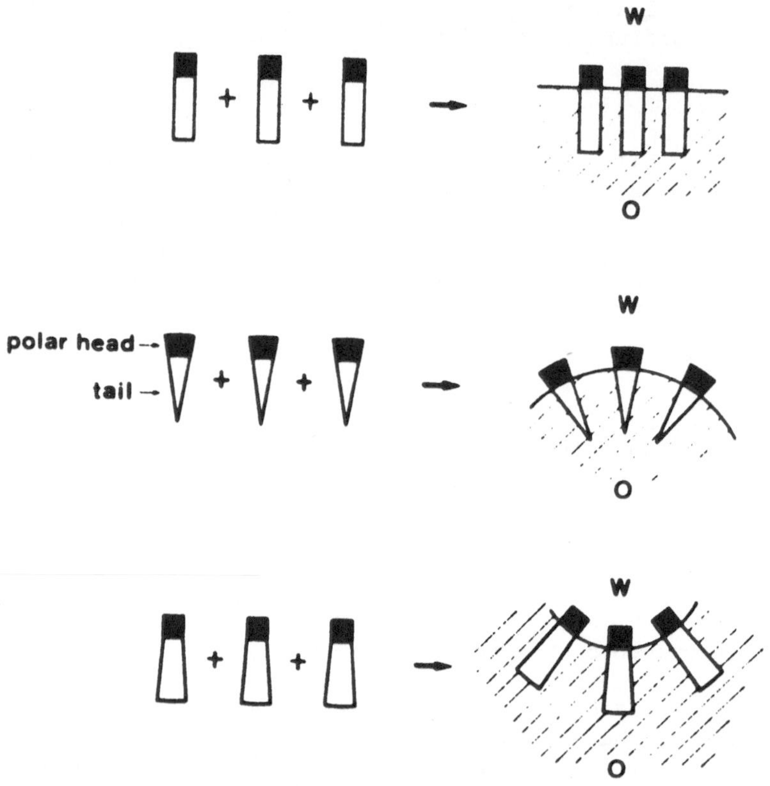

Figure 2. A naive steric model correlating the shape of the amphiphile to the spontaneous curvature of the interface.

(ii) the addition of a cosurfactant may act strongly on $1/R_0$ and also on K. For instance, in the (different but related) soap water systems, Charvolin and Mely showed that a certain mixture of C_{18} and C_{10} soaps could give a cubic phase which is not present with the pure C_{14} soap(11). The cubic phase is believed to be an array of rod portions (with positive curvature) related by branching regions (with negative curvature).

(iii) for films of non ionic surfactants, Robbins has considered curvature effects in some detail(12).

As experiments indicate a small contribution of spontaneous curvature in microemulsions we will focus on the role of the flexibility (K^{-1}).

3.3. Fluctuations of a flexible interface

The general idea is that with $\gamma = 0$ there may be giant fluctuations of the film shape. This approach follow the lines of a previous study of red blood cells scintillation[13].

Let us assume now that the interface has a negligible spontaneous curvature ($1/R_0 \to 0$) and is close to a certain reference plane (xy). The distances between the plane and the interface will be called $\zeta(xy)$. The curvature is then :

$$\frac{1}{R} = \frac{\partial^2 \zeta}{\partial x^2} + \frac{\partial^2 \zeta}{\partial y^2} \equiv \Delta_\perp \zeta \tag{8}$$

The free energy (7) becomes

$$f_K = \frac{1}{2} K(\Delta_\perp \zeta)^2 = \sum_q \frac{1}{2} K q^4 |\zeta_q|^2 \tag{9}$$

where we have gone to two-dimensional Fourier transforms

$$\zeta_q = \int dx \; dy \; \zeta(xy) \exp\left[i(q_x x + q_y y)\right] \tag{10}$$

We shall be mainly interested in the local orientation of the surface, defined by a unit vector $\underset{\sim}{n}$ normal to it

$$n_x = -\frac{\partial \zeta}{\partial x} \qquad n_y = -\frac{\partial \zeta}{\partial y} \qquad n_z \cong 1 \tag{11}$$

For small fluctuations $\partial \underset{\sim}{n} = (n_x, n_y)$ we can write

$$|\partial \underset{\sim}{n}_q|^2 = q^2 |\zeta_q|^2 \tag{12}$$

Applying the equipartition theorem to all modes in eq. (10) we obtain the thermal average of these fluctuations

$$< |\partial n_q|^2 > = \frac{kT}{Kq^2} \tag{13}$$

where T is the temperature and k is Boltzmann's constant. We can now look at the angular correlations between two points (0 and $\underset{\sim}{r}$) on the surface

$$\begin{aligned}\theta^2(r) &= < |\delta \underset{\sim}{n}(0) - \delta \underset{\sim}{n}(r)|^2 > = \sum_q 2 \; 1-\cos(\underset{\sim}{q}.\underset{\sim}{r}) <|\delta n_q|^2> \\ &= \frac{kT}{\pi K} \int_0^{1/a} \left[1 - J_0(qr)\right] \frac{dq}{q}\end{aligned} \tag{14}$$

where $1/a$ is a high q cut off - a microscopic length related to the detergent size. $J_0(x)$ is a Bessel function ; the factor $1 - J_0(x)$ is essentially equal to 1 for $x >> 1$, and to 0 for $x << 1$. Omiting uninteresting constants, the result is thus

$$< \theta^2 > = \frac{kT}{\pi K} \ln (\frac{r}{a}) \qquad (15)$$

For small θ we may also present it in the form

$$< \cos\theta > \simeq < 1 - \frac{\theta^2}{2} > \simeq \exp(- \frac{<\theta^2>}{2}) = (\frac{a}{r})^{\frac{kT}{2\pi K}} \qquad (16)$$

The law (16) shows an exponent $kT/2\pi K$ which is continuously variable with T - a frequent feature of two dimensional fluctuations[14]. We have derived it here only for θ small or not too large r. The question of its validity for larger r values is not solved at present. For our purposes, however, we know enough with eq. (16). In particular we can define a persistence length ξ_K by the following procedure : at distances r smaller than ξ_K the angle θ is small on the average, while at distance $r > \xi_K$ it is large. Choosing for instance $< \cos\theta > = 1/e$ as the cross over value, we obtain from eq. (16).

$$\xi_K = a \exp \left[\frac{2\pi K}{kT}\right] \qquad (17)$$

Thus the persistence length ξ_K is extremely sensitive to the value of the rigidity constant K. If, following Helfrich[15] we assume that a simple monolayer (without any cosurfactant) has a rigidity comparable to that of a thermotropic liquid crystal, we arrive at values $K \sim 10^{-13}$ erg, corresponding to $2\pi K/kT \sim 12$. In such a case ξ_K is exponentially large $\xi_K \sim 10^3 a$, and the interface is stiff. On the other hand, if, by addition of a suitable cosurfactant, we can decrease K by a factor of $5 (K \sim 2.10^{-14}$ erg) then $\xi_K \sim 10a \sim 100$ Å and the interface, observed at scales r larger than 100 Å, is strongly wrinkled. Clearly this distinction must play an important role for the selection of disordered (rather than ordered) structures.

4. SOME APPLICATIONS OF THE CORRELATION LENGTH ξ_K

4.1. The multisurface problem

We now consider a situation where the interface is present in all the sample volume Ω. Numerically, this can be characterized by an amount of surface per unit volume which we call 1/d (since it has the dimensions of an inverse length). If the number of surfactant molecules per cm^3 is n_s, and if they are all located at an interface, we have

$$\frac{1}{d} = n_s \Sigma \qquad (18)$$

Physically d represents a certain average distance between consecutive sheets of the interface : for instance, if we had a lamellar structure, 2d would be the repeat period (each period containing two interfaces).

If our sheets were completely ideal - i.e. if there were no interaction between them - each could show a persistence length ξ_K : it would be rigid at short scales ($r < \xi_K$) and wrinkled at large scales ($r > \xi_K$). However, all sheets interact. We have long range Van der Waals attractions and also repulsive forces : electrostatic forces in the water phase, and steric forces appearing whenever the aliphatic tails from two neighboring sheets begin to overlap.

All these forces tend to restore order in the sheet system : in the following paragraphs, we try to give a qualitative analysis of these very complex effects, steric, interactions and entropic.

4.2. Steric effects

Let us start with non interacting sheets : it is probably correct to visualize any sheet as a system of adjacent platelets, each with a certain typical size ξ_K and area $\xi_K{}^2$. Let us think of them as independent units. Each platelet has a number $\xi_K{}^2/\Sigma$ of detergent molecules, and thus the number of platelets per unit volume is

$$c_p = n_s \frac{\Sigma}{\xi_K{}^2} = \frac{1}{d\xi_K{}^2} \qquad (19)$$

If different platelets cannot intersect each other, they may tend to stack and form a nematic phase of flat objects : this type of liquid has indeed been observed recently with suitable organic molecules, and is currently called a discotic phase(16). Clearly, the discotic phase can exist only if the distance d between platelets is not too large. We can make this statement slightly more quantitative by a transposition of the Onsager argument concerning the nematic alignment of rod like molecules(17). In the present case we may say that each platelet is associated with an interaction volume of order $\xi_K{}^3$ (fig. 4). Whenever two platelets have a finite overlap of their interaction volumes, they are strongly correlated in their orientation. Thus the criterion for nematic order is of the form

$$c_p\, \xi_K{}^3 \gg 1 \qquad (20)$$

Returning now to (19) we see that there are two limits (Fig. 3). These two cases occur depending on the ratio of d, the average distance between sheets, and ξ_K, the persistence length. One should remember that this model is purely steric and does not take into account long-range van der Waals attraction and repulsion forces (electrostatic or structural forces).

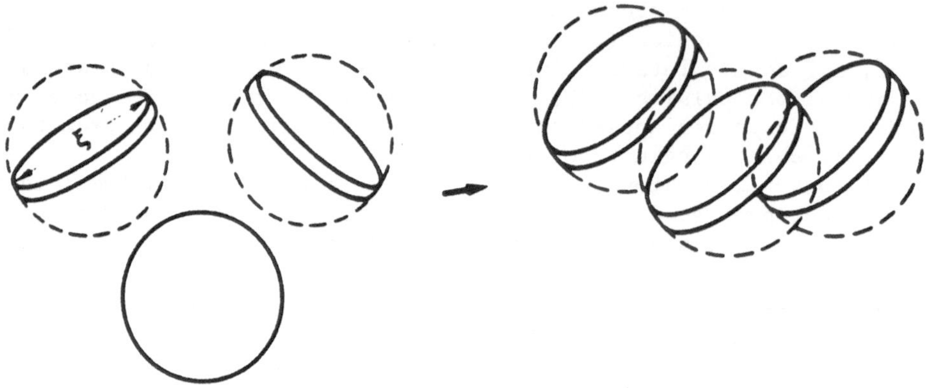

(a) (b)

Figure 3. The stacking of disks ; (a) at low concentration the disks are disordered ; (b) as soon as their "envelope spheres" (dashed) overlap significantly they tend to stack. A "discotic phase" appears.

a) if $\xi_K > d$ the sheets tend to be parallel to each other ;
b) if $\xi_K < d$ the sheets are wrinkled and can build up an isotropic, disordered liquid phase.

Thus for a given surfactant concentration n_s, there is in this case a critical value of the rigidity constant $K = K_c$ corresponding roughly to

$$a \exp\left[\frac{2\pi K_c}{kT}\right] = d = \frac{1}{n_s \Sigma^*} \qquad (21)$$

where we have used eqs. (17) and (18).

4.3. Long-range interaction effects

Our approach here will start from an ordered lamellar phase - adding fluctuations to the average order, and looking for their amplitude. This line of thought has already been followed by Chun Huh[18], but there are important differences. Chun Huh's interfacial energy γ is different from zero (in fact, for the fluctuation calculation, he uses the bare oil/water surface tension). In the present paper we start from a saturated interface with $\gamma = 0$ and thus the fluctuations are automatically much more important.

Namely we look at the motions of one interface in the presence of neighboring layers which are fixed at their equilibrium position. It is an augmented version of eq. (9) where we include a potential energy term ($1/2\ U''\ z^2$) (per unit area). Here U" is the curvature of the potential due to the neighboring layers, and is positive (stable equilibrium). The conclusions of the calculations which was developed elsewhere(19) is that if :

$$K < \frac{(kT)^2}{64\ U''\ d^4}$$

a disordered state is preferred.

4.4. Entropy effects for flexible interfaces

We discuss these effects in a model which is related to, but somewhat different from the original proposal by Talmon and Prager(9). We observe first that the interface must have a certain <u>persistence length</u> ξ_K : (i) it is essentially flat at scales smaller than ξ_K, (ii) consecutive "pieces of interface", with an area $\xi_K{}^2$, have independent orientations. A rough but convenient model is then obtained by dividing all space into consecutive cubes, each of linear size ξ_K. Each cube is either filled with oil (water) is called ϕ_o (ϕ_w). Two adjacent cubes will have no interface, and no energy, if they are of the same type. But if they are different, we must count a free energy contribution $\gamma\ \xi_K{}^2$, γ being the interfacial tension. For the moment, we do not assume $\gamma = 0$. Rather, we say that a given chemical potential μ_s of the surfactant imposes a certain area per surfactant Σ_s, through the thermodynamic condition (obtained from differentiating (1) with respect to n_s and using (2) and $\Sigma_s \equiv \mathcal{a}/n_s$)

$$\mu_s = G(\Sigma_s) + \Pi(\Sigma_s)\Sigma_s \tag{22}$$

The result is then a certain $\gamma(\Sigma_s)$ which depends ultimately on μ_s.

We have now reduced the statistics of the interface to a "lattice gas model". Clearly, the description is very crude : the oil (or water) regions in a microemulsion do not look like an assembly of cubes. But the lattice gas model keeps some essential features of a random surface. Also, the resulting statistical behavior is well known. When the coupling between adjacent cubes ($\gamma\ \xi_K{}^2$) is weaker than kT, or more precisely when

$$\gamma < \gamma_c = \alpha kT/\xi_K{}^2 \tag{23}$$

(where $\alpha = 0.44$ for a simple cubic lattice), we expect a single phase with the oil and water mixed down to the scale ξ_K. But when the inequality (23) is reversed, we may have <u>phase separation</u> (Fig. 4).

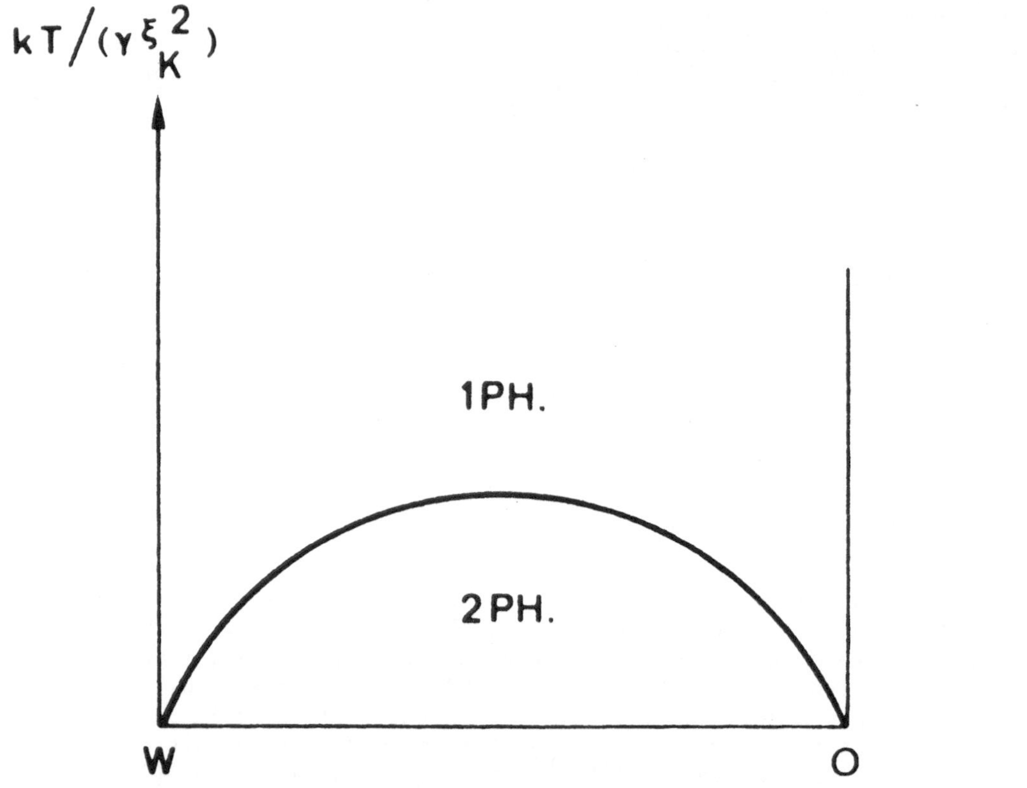

Figure 4. Phase diagram in the modified Talmon-Prager model. At high surfactant contents (low γ) the microemulsion is stable at all water fractions. At lower surfactant contents, a phase separation occurs. The present model does not include the spontaneous curvature of the interface associated with Brancroft's rule. Then (and only then) the plot is symmetrical.

A number of significant properties emerge from the model, and are probably of more general validity :

a) the values of γ involved are weak. As shown by (23) the range of interest is $\gamma \sim kT/\xi_K^2$. The persistence length ξ_K is expected to be rather large, and thus γ should be small : we are not very far from the Schulman criterion.

b) phase separation occurs not because of specific interactions between the droplets, but purely from a balance between interfacial entropy and interfacial energy.

c) in practice one often observes more complex phase diagrams. In particular, certain microemulsions can coexist simultaneously with an oil phase and a water phase. The above lattice gas model generates two-phase equilibria only. However, the model is highly degenerate.

Small perturbations on the structure of the free energy (induced by curvature effects or other corrections) might lead to 3 phase equilibria. A first attempt in this direction is described in ref. (9). Some difficulties are pointed out in ref. (20).

5. CONCLUSIONS

Microemulsions were discovered thirty years ago and described as "extraordinary emulsions". Their structure do not obey to a unique model. Two recent models focussed on a random structure were reviewed. Experimental studies were recently performed in order to measure the flexibility coefficient which was indeed found exceptionnally low[21-23]. A structural study both by X-rays and neutron scattering evidenced the existence of bicontinuous phases[24-25]. The predominant role of the spontaneous curvature was demonstrated in the structure of the interface in agreement with very recent ellipsometric measurements[27].

REFERENCES

(1) Y. Talmon and S. Prager, J. Chem. Phys., 69, 2984 (1978).

(2) Y. Talmon and S. Prager, J. Chem. Phys., 76, 1534 (1982).

(3) P.G. de Gennes and C. Taupin, J. Phys. Chem., 86, 2294 (1982).

(4) L.E. Scriven, Nature, 263, 123 (1976).

(5) E. Ruckenstein and J. Chi, J. Chem. Soc., Faraday Trans. II, 71, 1690 (1975).

(6) D.J. Mitchell and B.W. Ninham, J. Chem. Soc., Faraday Trans. II, 77, 601 (1981).

(7) D.O. Shah and R.S. Schechter, Improved oil recovery by surfactants and polymers flooding, Acad. Press. Inc. (1977).

(8) E. Ruckenstein, J. Chi, J. Chem. Soc., Faraday Trans. II, 71, 1690 (1975).

(9) Y. Talmon, S. Prager, J. Chem. Phys., 69, 2984 (1978).

(10) W. Helfrich, Z. Naturforsch., 28C, 693 (1973).

(11) J. Charvolin, B. Mely, Molecular Crystals Liquid Crystals Letters, 41, 209 (1978).

(12) M.L. Robbins, Micellization, solubilization and microemulsions, (K. Mittal ed., Plenum, NY, 1977), 2, 273.

(13) F. Brochard and J.F. Lennon, J. de Phys., **36**, 1035 (1975).

(14) J.G. Dash, J. Ruvalds, Phase transition in surface films (eds NATO Advances Study Series (Physics), Plenum Press, NY, 1980).

(15) W. Helfrich, Z. Naturforsch., **33a**, 305 (1978).

(16) J. Billard, in Chem. Phys. Ser. Vol. II (ed. Springer, Berlin, 1980), p. 383, 395.

(17) L. Onsager, Ann. N.Y. Acad. Sci., **51**, 627 (1949).

(18) Chun Huh, J. Colloid Interface Sci., **71**, 408 (1979).

(19) P.G. de Gennes and C. Taupin, J. Phys. Chem., **86**, 2294 (1982).

(20) J. Jouffroy, P. Levinson and P.G. de Gennes, J. Physique, **43**, 1241 (1982).

(21) J.M. di Meglio, M. Dvolaitzky, R. Ober and C. Taupin, J. Physique (Lettres), **44**, L-229 (1983).

(22) J.M. di Meglio, L. Paz, M. Dvolaitzky and C. Taupin, J. Phys. Chem., **88**, 6036 (1984).

(23) J.M. di Meglio, M. Dvolaitzky, L. Léger and C. Taupin, Phys. Rev. Lett., **54**, 1686 (1985).

(24) L. Auvray, J.P. Cotton, R. Ober and C. Taupin, J. Physique, **45**, 913 (1984).

(25) L. Auvray, J.P. Cotton, R. Ober and C. Taupin, J. Phys. Chem., **88**, 4586 (1984).

(26) L. Auvray, J. Phys. Lett., **46**, L-163 (1985).

(27) D. Beaglehole, M.T. Clarkson and A. Upton, J. Coll. Inter. Sci., **101**, 330 (1984).

FRACTAL BEHAVIOR AND DYNAMICS ON PERCOLATING CLUSTERS

Panos Argyrakis
Department of Physics
University of Crete
Iraklion, Crete, Greece

ABSTRACT. We calculate very accurately several random walk properties on percolating clusters using various techniques. We derive the associated critical exponents characterizing fractal behavior and compare it to recent conjectures about their exact value. We extend all calculations to cover the whole range from the critical point to the perfect crystal, and thus observe the fractal-to-Euclidean crossover. We find that Euclidean behavior is achieved rather fast above the threshold point. We also investigate correlated motion on fractals, for which we find that it does not belong to same universality class as regular random walk does. Finally we look at long-range interaction clusters, for which we find that random walk is of similar nature as in nearest neighbor clusters.

1. INTRODUCTION

The theory and numerous applications of random walks in a variety of fields ranging from molecular/solid state physics to polymers to biological proteins and many more have attracted a continuous interest over the years, plausibly due to their success in explaining the corresponding phenomena studied [1]. The first pioneering work of Montroll [2], who systematically studied their properties by introducing the generating function method, was improved and refined by Henyey and Seshadri [3] and by Blumen and Zumofen [4]. Because of this we now have available closed form solutions for the number of sites visited in an N-step walk, S_N, the mean-square displacement, R_N^2, the probability for return to the origin, P_O (first passage time), etc. But all this work is concerned with perfect lattices or crystals, while it is recognized that the exact general solution for doped lattices for all concentrations is a formidable task. In this latter but very important problem it has also recently been recognized that a mean-field approach (an

R. Daudel et al. (eds.), Structure and Dynamics of Molecular Systems – II, 209–229.

effective medium approximation) may well provide adequate answers. Still it is expected that such an approach will suffice for small to medium dilutions only, while it will break down close to the critical percolation threshold. The region of the critical point has been a subject of intensive studies in the past, probably because of its connection to classical thermodynamics and phase transitions. But only recently with the advent of the notion of fractals [5,6] did it become apparent that random walk properties at the critical point can be evaluated exactly, something that threw new interest in the general problem discussed here. This is accomplished by introducing the fractal and fracton dimensionalities. These are fractional numbers (dimensions) smaller than the underbedding dimension, but their value gives an indicative measure of the disorder present, both for the structure of the disordered lattice and the dynamics on it. This prompted a surge of publications in a short period of time that discussed conjectures, hypotheses, numerical verifications, corrections to the scaling, etc.

In this paper we discuss our results of calculations that monitor diffusion on disordered lattices via random walks. We find accurate values for the fractal dimensionalities d_s by calculating S_N, the mean number of distinct sites visited in an N-step walk, and for the diffusion exponent D by calculating R^2_N, the mean-square displacement. We investigate in detail the effects of correlated diffusion and compare it to normal random walk. Finally we look into the effects of long-range diffusion. All these properties are calculated using a variety of algorithms (discussed in Section 2) first at the critical point, and then in the whole range above criticality up to a perfect lattice, so that we study in detail the crossover to Euclidean behavior.

It should also be noted that some efforts towards the solution of the general problem of random walks on disordered lattices were first published by us [7] using numerical solutions, and even though the model and methods were rather crude, they were still successful, at least partially, in explaining experimental data of luminescence from organic crystals at low temperatures. But it is only recently that more complete and satisfactory solutions are provided through the ideas of fractals.

2. METHOD OF CALCULATION

2.1. Technique

We use Monte-Carlo simulation methods to monitor several random walk properties. Our algorithm has been considerably improved in recent years, increasing both the speed of operations and the size of the lattice used. Depending on the required application we utilize two techniques for the generation of random lattices. In the first case (a) we use the so called cluster-growth-technique, a method that generates and keeps in memory only the lattice portions used for the random walk. The lattice starts with one site only, and it is built continuously as the random walk proceeds by generating more sites adjacent to the diffusing particle. Once a site is gererated and its identity chosen it remains as such in the memory for the whole run. Using this method only one random walk can be executed on each lattice, which is more time consuming but has considerable advantages from the statistical point of view. The effective lattice size is now usually $4x10^6$ sites i.e. 2000x2000 for the 2-dim square lattice, and 160x160x160 for the 3-dim simple cubic. Consequently, the properties derived using this method pertain to walks that originate on any-size cluster, whether this may be the infinite percolating cluster or a monomer.

In an alternate approach, method (b), the whole lattice is generated and kept in memory before the beginning of the walk. This, although is a slower process, is necessary when it is required to find the critical percolating point exactly. In the previous case (a) the nominal critical occupational probability p_c was assigned, but in the actual numerical computation, due to statistical fluctuations, no exact realization of p_c is attained. After several realizations only its average value is attained. Thus, some realizations are well above p_c while others have not percolated at all. If the exact point is required it may turn out that this average quantity is not good enough, since close to criticality diffusion is not a linear process. Here in case (b) the lattice is initially generated at random well below the critical value, say at p=0.55 for the 2-dim square lattice. Then a certain number of closed sites is changed to open, their exact number and location being recorded. Usually this number is a power of 2, say 2^{10}. If after this change there is still no infinite cluster a new additional set of sites changes identity and the process is repeated until the critical point is surpassed. At this point the last set of sites is removed (i.e. changed back to closed sites), it is cut into two equal pieces and only the first half is now added; the lattice is tested again for criticality, but now with 2^9 sites changing identity. This process continues with the repeated dichotomy of the original number (2^{10}), until it goes down to 2^0. At this point we are assured that we are exactly at the critical point, i.e. one single site has caused the appearance of the percolating cluster. Testing for criticality is done using a new version of the Cluster-Multiple-Labelling-Technique (CMLT). The details of this version are explained in the Appendix. We only need to apply the CMLT as many times as the power of 2, i.e. here we apply it 10 times, something not very time consuming. Using this technique we employ lattices of 2.5×10^5 sites, i.e. 500x500 for the square 2-dim case.

2.2. Computer language

It is commonly agreed that for Monte-Carlo simulations of the type reported here Fortran is not the best available language, since Fortran is best suited for numerical calculations, while the present work involves mainly integer manipulations and conditional statements. Another high level language (and easy to learn) that it better suited is C language. Of course, one would benefit the most by going to a low-level language, such as the assembler that each machine has. However, if it is necessary to use Fortran there are several points worth mentioning which if properly used can provide an added advantage. Since only five alternative pieces of information are needed at each time (the identity of a site, open or closed, and whether a site has been visited before or not, or whether it has not been defined), we can utilize more efficiently the length of each computer word by breaking each 32 bit word in four 8 bit sections. We now store the information for four sites in one word and effectively increase the size of available memory by a factor of 4. In Fortran this is done by use of a subroutine with the main array in Logical*1 and Integer*4 variables declaration, occupying the same memory space and continuously identified via an Equivalence statement. We thus avoid the difficult task of byte manipulation in machine language.

All different topologies in 1-dim, 2-dim, and 3-dim lattices are reduced to a one-dimensional array in the memory, so that it is not necessary to reach into the virtual memory as frequently. Thus the number of page-faults and transfers is decreased, and the overall speed is increased.

All work presented here was performed in a mini-computer, VAX 11/750 by Digital Equipment Corporation, with 4 Mbytes of direct memory and 550 Mbytes of virtual memory, and with the VMS operating system.

3. RESULTS

3.1. Spectral dimension

It is well established by now that the average number of distinct sites visited in an N-step walk behaves as:

$$S_N \sim N^{d_s/2} \tag{1}$$

Here d_s is the scaling exponent called the spectral (fracton) dimension, and it was the initial hypothesis [6] that d_s should have a universal value of d_s=4/3 for all lattices of all dimensionalities. But it was later conjectured [8] that this rule breaks down for d≤2 (where d is the Euclidean underbedding dimensionality), but is still valid for d>2. We will look into these assumptions carefully to check whether they can be verified.

Depending on what clusters does one use for the point of origin the spectral dimension of Eq. 1 will have different values. We use the notation d_s for case (b) of the previous section, i.e. for runs that can originate only on the largest percolating cluster. For case (a), i.e. for runs on any-size cluster we will have an analogous equation:

$$S_N \sim N^{d_s'/2} \tag{2}$$

But it has been shown [9] that d_s and d_s' are related through:

$$d_s' \doteq d_s(2 - \frac{d}{d - \beta/\nu}) \tag{3}$$

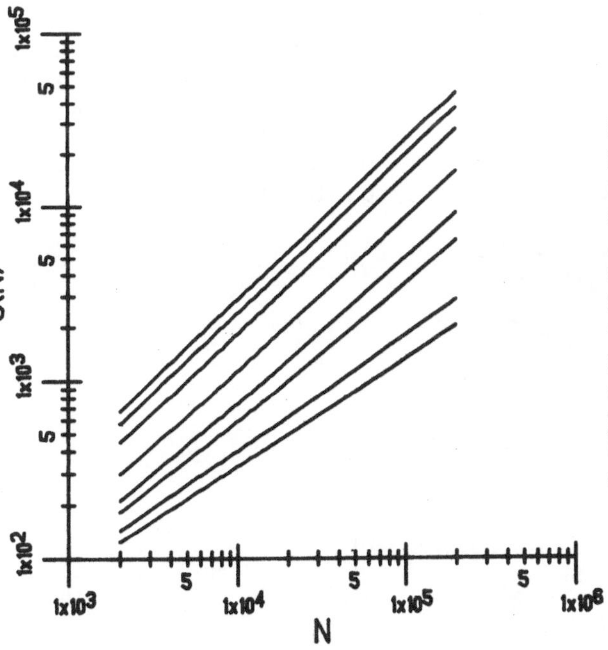

Figure 1: The number of sites visited S_N as a function of the number of steps N for 2-dim lattices. These are averages of 1000 realizations for walks that originate on any-size cluster (method a). Bottom to top: p=0.5931, 0.60, 0.63, 0.65 0.70, 0.80, 0.90, and 1.00.

where β and ν are the static percolation exponents. When one of the d_s or d_s' is calculated the other one can easily be deduced. Figure 1 shows a plot of $\ln S_N$ vs. $\ln N$ for several different occupational probabilities p for the 2-dim lattice, and Figure 2 the same plot for 3-dim lattice. The lowest curve in each case pertains to the critical percolating threshold and from its slope we receive [10]:

$$d_s' = 1.23 \pm 0.02 \quad \text{(2-dim)} \tag{4}$$

$$d_s' = 1.06 \pm 0.02 \quad \text{(3-dim)} \tag{5}$$

Using Eq. 3 and the values of Eq. 4 and 5 we get:

$$d_s = 1.30 \pm 0.02 \quad \text{(2-dim)} \tag{6}$$

$$d_s = 1.33 \pm 0.02 \quad \text{(3-dim)} \tag{7}$$

Using method (b), the method of the exactly percolating clusters, we calculate again S_N, but now for runs on the exactly incipient percolating cluster only. In Figure 3 we plot $\ln S_N$ vs. $\ln N$ for the 2-dim lattice. We observe that the data is almost fit on a straight line, but there are some deviations in the early time part. To avoid any such

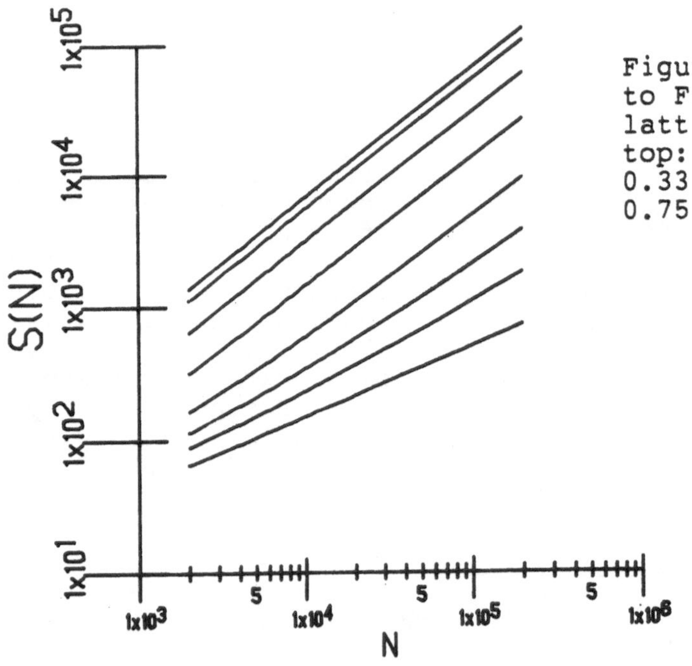

Figure 2: Plot similar to Fig. 1 but for 3-dim lattices. Bottom to top: p=0.3117, 0.32, 0.33 0.35, 0.40, 0.50, 0.75, and 1.00.

complications we notice from Eq. 1 that $S_N/N^{d_s/2}$ should be constant in time. We plot this quantity as a function of N in Fig. 4. Not being sure of the exact value of d_s we treat it as an adjustable parameter, and we plot the range

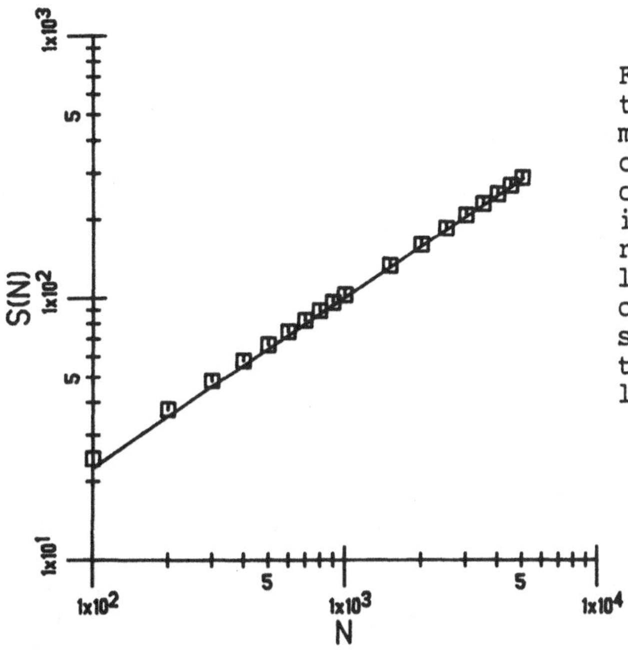

Figure 3: Plot similar to Figure 1 but using method (b). The calculation is at the critical point exactly, i.e. p=0.5931. These results are averages of 10000 realizations. The continuous line is a straight line to show the deviation from linearity.

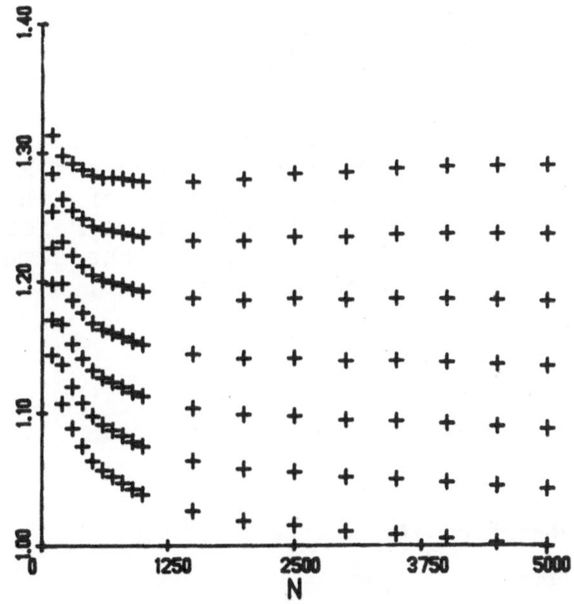

Figure 4: Plot of $S_N/N^{d_s/2}$ vs. N of the same data as in the previous figure. Here: d =1.27, 1.28, 1.29, 1.30, 1.31, 1.32, and 1.33 (top to bottom).

d_s=1.27 - 1.33. We observe that the values of d_s=1.29 or 1.30 are the ones that produce lines independent of time (again, aside from the early time part). Of course, if we use Eq. 3 to calculate d_s' we will receive: d_s'=1.23. We see that the direct calculation of d_s gives a value in agreement with the previous method [11].

Thus, in calculating the spectral dimension we employed two different approaches, methods (a) and (b), and they both produce the same result, i.e. for 2-dim lattices the proper value of the spectral dimension is d_s=1.30±0.02, in agreement with the Aharony-Stauffer [8] prediction, and a deviation of about 2% from the Alexander-Orbach-Rammal-Toulouse [6] theory. In 3-dim lattices d_s=1.33, as originally proposed.

3.2. The diffusion exponent

The mean-square displacement at the fractal limit behaves as:

$$R_N^2 \sim N^{2/D} \tag{8}$$

Depending at the point of origin we again have two exponents from Eq. 8, D and D′ (just like d_s and d_s'). Figure 5 shows $\ln R_N^2$ vs. lnN for several different p values for 2-dim lattices. From the slope of the lowest curve we derive a value for the D′ exponent:

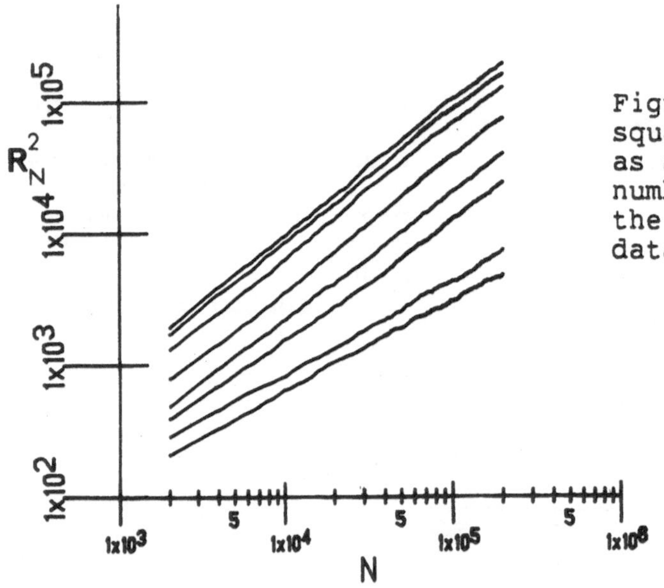

Figure 5: The mean-square displacement R_N^2 as a function of the number of steps N for the same p and the same data as in Fig. 1.

$$D' = 2.89 \pm 0.05 \tag{9}$$

This value of D as calculated for the long time limit is in good agreement with previous work [12], but for small N (N=300 steps).

3.3. Crossover to Euclidean behavior

We can observe the crossover to Euclidean behavior in Fig. 1 by looking at the several curves in this figure. We notice that only the bottom and the top are straight lines. The bottom because it obeys Eq. 1 as it is in the fractal limit, the top because at p=1.00 there is an effectively simple power dependence, in spite of the well known logarithmic correction for the 2-dim walk. But for all intermediate p the slopes are varying and some lines are curved, with the curvature being a function of time, thus showing that each different p has a different effective spectral dimension that is time-dependent. The same behavior is observed at Fig. 2 for the 3-dim lattices. In order to see how fast does this crossover occur we plot the effective d_s, for the long time limit (N=200000 steps) as a function of the occupational probability p. The result is shown in Figure 6, for both the 2-dim and 3-dim lattices. One can see that for both cases the effective d_s sharply increases in the region

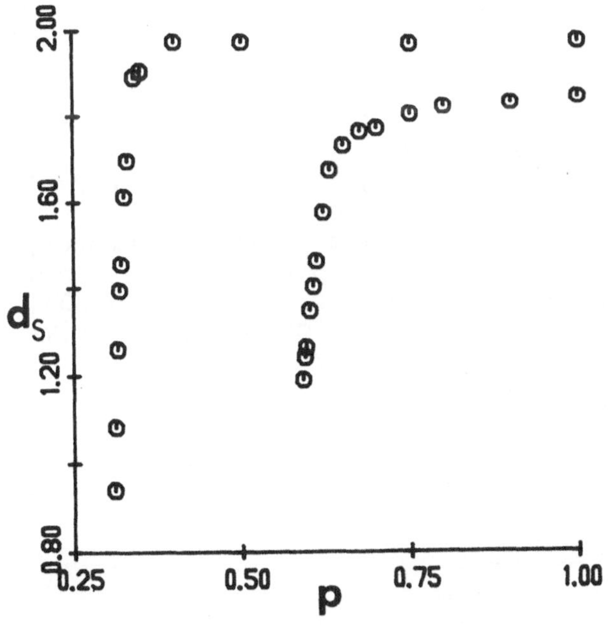

Figure 6: The effective d_s as a function of the occupation probability p, for 2-dim (lower) and 3-dim (upper) lattices, from the long time limit (N=200000 steps).

immediately above the critical point and it is approaching fast the classical value of 2. The limiting values here are: 1.98 for the 3-dim case, and 1.89 for the 2-dim case (a discrepancy from the classical value is expected here due to the logarithmic correction term that the 2-dim formalism contains [2]). From the curvature of the lines in Fig. 1 and 2 we would expect that the shape of the lines in Fig. 6 is time dependent. We found this to be true, however, we also observed that at any time period the main feature of the sharp rise remains intact.

As a consequence of the crossover behavior we expect that scaling will be valid only in a small region close to the threshold point. The scaling relationship is [13]:

$$S_N = N^{d_s/2} \, f\,[(p/p_c - 1)\, N^{1/(2\nu-\beta+\mu)}] \tag{10}$$

where ν, β, and μ are the usual percolation exponents. In Fig. 7 we have a scaling plot for the 2-dim and 3-dim walks. We see that the different p values that fall within the scaling curve are all in the range: 0.31-0.35 (3-dim), and 0.59-0.65 (2-dim). Several time intervals are also included and all fall in the curves shown. Scaling breaks down for $p>0.35$ (3-dim) and for $p>0.65$ (2-dim), since as it can also be seen from Fig. 6 above these limits the Euclidean values are already attained.

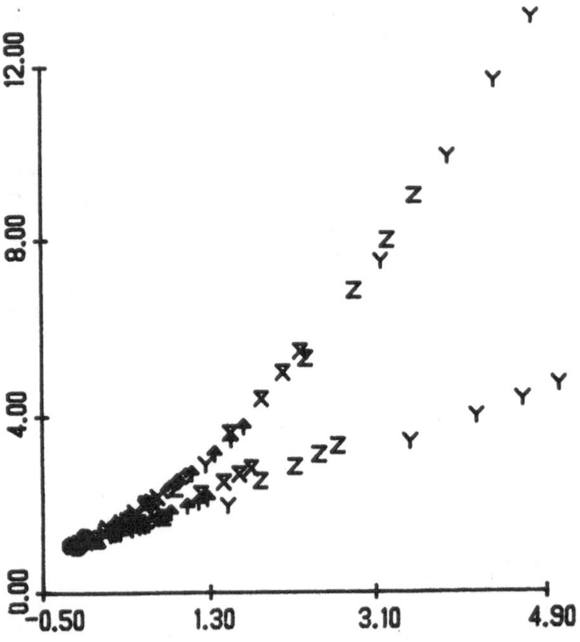

Figure 7: Scaling plot of $S/N^{(d_s/2)}$ vs. $(p/p_c - 1)\ N^{1/(2\nu-\beta+\mu)}$ 2-dim lattices (lower, $0.59<p<0.65$), 3-dim (upper, $0.31<p<0.35$), all for several time intervals.

3.4. Short-time corrections

As it is seen from Fig. 3 and 4 for small values of N, say up to N=1000 steps, Eq. 1 does not hold. In Fig. 3 linearity is not obtained for the entire range tested, while in Fig. 4 we see that the factor $S_N/N^{d_s/2}$ is not constant in time, but it is decreasing up to the value of about N=1000 steps. This leads us to believe that additional corrections to scaling are necessary in order to properly describe the early-time limit. Apparently, in early times the particle samples new lattice areas at a faster pace than after some time has elapsed, and some "state of equilibrium" is reached. This is also in agreement with previous assumptions [14,15]. The additional term is contained in the following:

$$S_N \sim N^{d_s/2} (1 + AN^{\omega}) \tag{11}$$

where ω is the new exponent and is always necessarily negative, so that the contribution of the second term in parenthesis goes to zero for large N. Setting the constants of proportionality (a and b) in Eq. 11 and rearranging we receive:

$$S_N/N^{ds/2} - a = bN^{\omega} \tag{12}$$

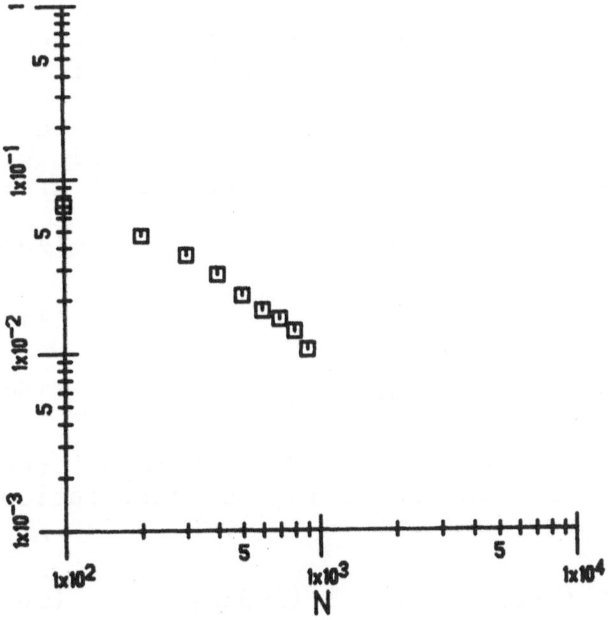

Figure 8: Plot of $S_N/N^{d_s/2}$ vs. N where d_s=1.29 and a=1.184 for the same data as in Fig. 3. The slope (giving the exponent ω) is taken from a straight line drawn at small N.

Here a represents the constant $S_N/N^{d_s/2}$ value for large N, which as seen from Fig. 4 has the value of a=1.18. Therefore, if we plot Eq. 12 directly in logarithmic form we will recover immediately the ω exponent. This is done in Fig. 8 using the same data as in Fig. 4. We consider the slope at short times because this is where the correction term predominates. At the straight line segment in Fig. 8 the slope is: $\omega=-0.48\pm0.08$. We notice that forcing an exponent $d_s=4/3$ gives a=0.99 and $\omega=-0.47$, showing that ω is not as sensitive to d_s as, obviously, a is.

3.5. Correlation effects

Correlation in diffusional motion has been shown in the past to be a necessary idea for the explanation of experimental data ranging from the diffusion of hydrogen in metals [16] and models of diffusion in concentrated lattice gases [17] to the relaxation mechanism of low-lying excited states of organic molecules at low temperatures [7] as studied by the use of random walk hopping models [18]. By correlation here we mean the retention of the directional memory over a certain number of lattice spacings. This is quantitatively described by the fraction p_f, which is the probability of a forward jump, and it is in the range: $1/a<p_f<1.00$, where a is the lattice coordination number. Recently we introduced [19,20] a new model that incorporates the effects of correlation in the usual [4] random walk models, first on perfect lattices [19], and then on mixed binary lattices [20]. Of interest to us here is whether correlated random walks behave the same way as regular walks and if the associated properties belong to the same universality class as the walks of the previous sections. For perfect lattices we extended Montroll's work [2] to include correlated motion. Thus, the formulae for S_N for uncorrelated walk are:

$$S_N = (8N/\pi)^{1/2} + \ldots \quad \text{(1-dim)} \tag{13}$$

$$S_N = \pi N/\ln N + \ldots \quad \text{(2-dim)} \tag{14}$$

$$S_N = cN + \ldots \quad \text{(3-dim)} \tag{15}$$

If we now include correlation by going through the generating function method and incorporating the correlation factor [19] we rederive these formulae as follows:

$$S_N = (8fN/\pi)^{1/2} + (1-f) + \ldots \quad \text{(1-dim)} \tag{16}$$

$$S_N = \frac{\pi f N}{\ln(8fN) + \pi(f-1)/2} + \ldots \qquad \text{(2-dim)} \qquad (17)$$

$$S_N = \frac{(1 - p_r) f N}{1+0.5(f-1) + (1-p_r)} + \ldots \qquad \text{(3-dim)} \qquad (18)$$

where f is the correlation factor (it is a function of p_f, but has different form for each dimensionality), and p_r is the return probability for uncorrelated walk. We see that we arrive at relatively simple modifications as compared to the uncorrelated walk model.

The problem is considerably more complicated in disordered lattices. Our results are given in Fig. 9. We evaluate the spectral dimension d_s' for correlated walks for several different values of the correlation parameter p_f, and for several different occupational probabilities p. The curve marked $p_f = 0.25$ is the limiting case of no correlation at all, since in the square lattice all four directions carry the same 0.25 probability of scattering. It is also the same curve as that in Fig. 6.

We focus attention on the other curves in Fig. 9, which refer to higher p_f ($p_f > 0.25$) values. In the fractal limit (p=0.60) we see that d_s' sharply decreases as p_f increases. For $p_f = 0.95$ (at p=0.60) we see that $d_s' = 0.66$ only. This sharp decrease is a consequence of the fact that at the critical percolation threshold correlated walks have a much

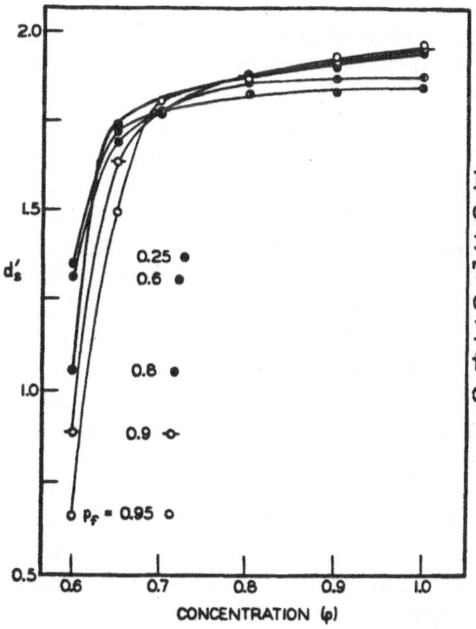

Figure 9: The spectral dimension d_s' vs. p, as a function of the forward correlation parameter p_f. The spectral dimensions are calculated in the long time limit, N=200000 steps, from walks that may originate on any-size cluster.

smaller S_N value than uncorrelated walks because, as was originally shown in the past [18], the particle indulges for long times in revisiting the same row of sites over and over again. We also used this idea [7] to interpret experimental data on mixed naphthalene alloys at 2 K.

As p increases, one observes for each p_f value the corresponding crossover to the classical behavior, since now, above p=0.80, correlated walks are much more efficient than uncorrelated ones. The region 0.70<p<0.80, as seen from Fig. 9, is the "crossover region" between the different p_f values. At p=1.00 our calculations are in excellent agreement with formulae 16-18.

Thus, in this preliminary study for correlated random walk motion on fractal structures we investigated two types of crossover that occur, i.e. the crossover from fractal to Euclidean behavior for any type of walk, and the crossover from uncorrelated to correlated walk at any given p. We found that correlated random walks do not belong to the same universality class as simple walks. We notice that complete analytical solutions for the curves of Fig. 9 exist only for p=1.00, but more work is needed to quantitatively explain the behavior of the other curves.

3.6. Long-range interactions

It has been shown in the past [21] that the static critical percolation exponents β, γ, ν obey the universality hypothesis independent of the interaction range, for finite range cutoffs. It is interesting here to test the same hypothesis for the dynamic exponents examined here, i.e. the spectral dimension and corresponding crossovers. The interest stems from the well known experimental observation that triplet exciton transfer in organic molecules at low temperatures involves such long-range random hops on percolation clusters. The critical occupational probability p_c is a function of the interaction range, and it has been derived for various interaction ranges (R=1-5) by Monte-Carlo simulation [21] and by a position-space renormalization group approach [22]. Some of these p_c values are: p_c=0.5931 (R=1), p_c=0.29 (R=2), p_c=0.16 (R=3), p_c=0.10 (R=4), p_c=0.07 (R=5), p_c=0.05 (R=6), etc. We use a form: $P \sim e^{-\alpha r}$ for the stepping probability, where r is the distance, and α gives the shape of the curve for the distribution of distances.

The random walk process is monitored here in a similar way as in our previous calculations through S_N and R_N^2, with the only difference now that long steps are allowed according to the $P \sim e^{-\alpha r}$ equation. We find a complete analogy with the nearest neighbor case. Figure 10 is a plot of $\ln S_N$

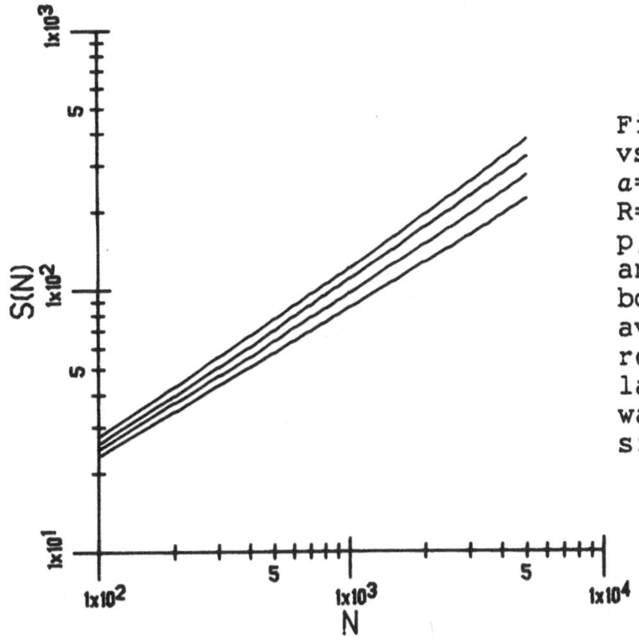

Figure 10: Plot of $\ln S_N$ vs. $\ln N$ for the case $a=0$, interaction range $R=2$, and four different p, $p=0.32$, 0.31, 0.30, and 0.29 (top to bottom). These are averages of 1000 realizations on 500x500 lattices. The random walks originate on any-size cluster.

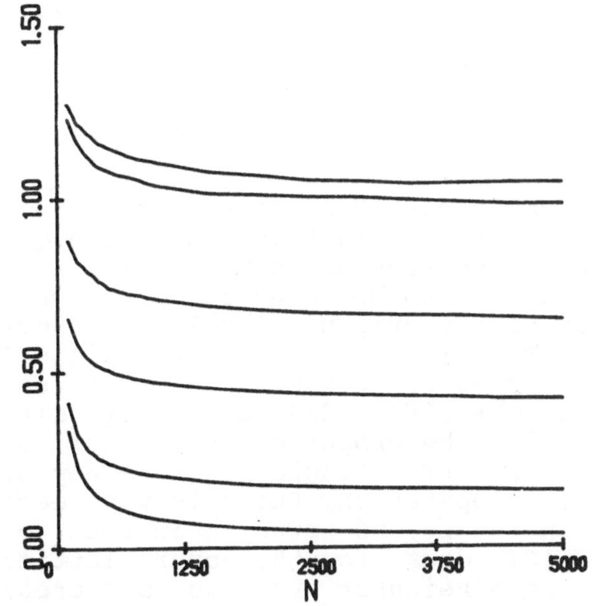

Figure 11: Plot of $S_N/N^{d_s/2}$ vs. N for different values of the parameter a. Here $a=0$, 0.5, 2.0, 3.0, 5.0, and 10.0 (top to bottom). Also, $p=0.29$ (range of interaction $R=2$). We used $d_s=1.23$.

vs. lnN, similar to Fig. 1, but for R=2 and we see that it yields a slope of d_s=1.24, same as Eq. 4. To further investigate the scaling exponent we plot $S_N/N^{d_s/2}$ vs. N in Fig. 11 for different values of the parameter a in the range: $0<a<10.0$, for p_c=0.29 (interaction range R=2). We use d_s=1.23, and we see that for a=0 (all steps equally probable independent of distance) the quantity $S_N/N^{d_s/2}$ above N=2500 steps is constant in time. But for $a>0$, i.e. hopping probabilities that decrease with range, we see it takes longer (more steps) to reach the asymptotic limit. This is expected since it takes, on the average, several nominal time steps before an occurence of a long-range step.

We note that this model is based on defining and using a range (distance) dependent transition probability. But the actual form of this dependence (for example, the $e^{-\alpha r}$ used here) does not enter explicitly in the calculation. Any other proper form, such as for example r^{-n} (n=integer), could have been used with same results. One need only establish the proper correspondence. We conclude that long range random walks on percolating clusters behave similarly to simple random walks, with scaling and universality still intact. But for steeply falling-off step probabilities we find an asymptotic behavior that is approached more gradually in time. It would be interesting to check whether an experimental system (where the cutoff range itself is a function of time) will also exhibit an effective fractal dimension.

ACKNOWLEDGMENT. Parts of this work were performed with K.W.Kehr, A.Keramiotis, and R.Kopelman, whose collaboration is greatly appreciated.

APPENDIX

The cluster distribution is performed here using a new version of the Cluster-Multiple-Labelling-Technique (CMLT). The principal idea is the same as in CMLT, i.e. in cluster coalescence no sites need to be relabelled, and once a site is labelled, it retains its original label throughout the sweep of the lattice. The difference from the original method [23] is a new index processing used here. Instead of applying the routine CLASSIFY [23] at every site labelling in order to determine the proper cluster label of all neighboring sites we perform a single second sweep of the lattice that does the same operation. But this way, each site is checked only once in the second sweep, instead of twice that routine CLASSIFY does, for the square lattice topology, where every site is a neighbor to two different sites. The algorithm is given below:

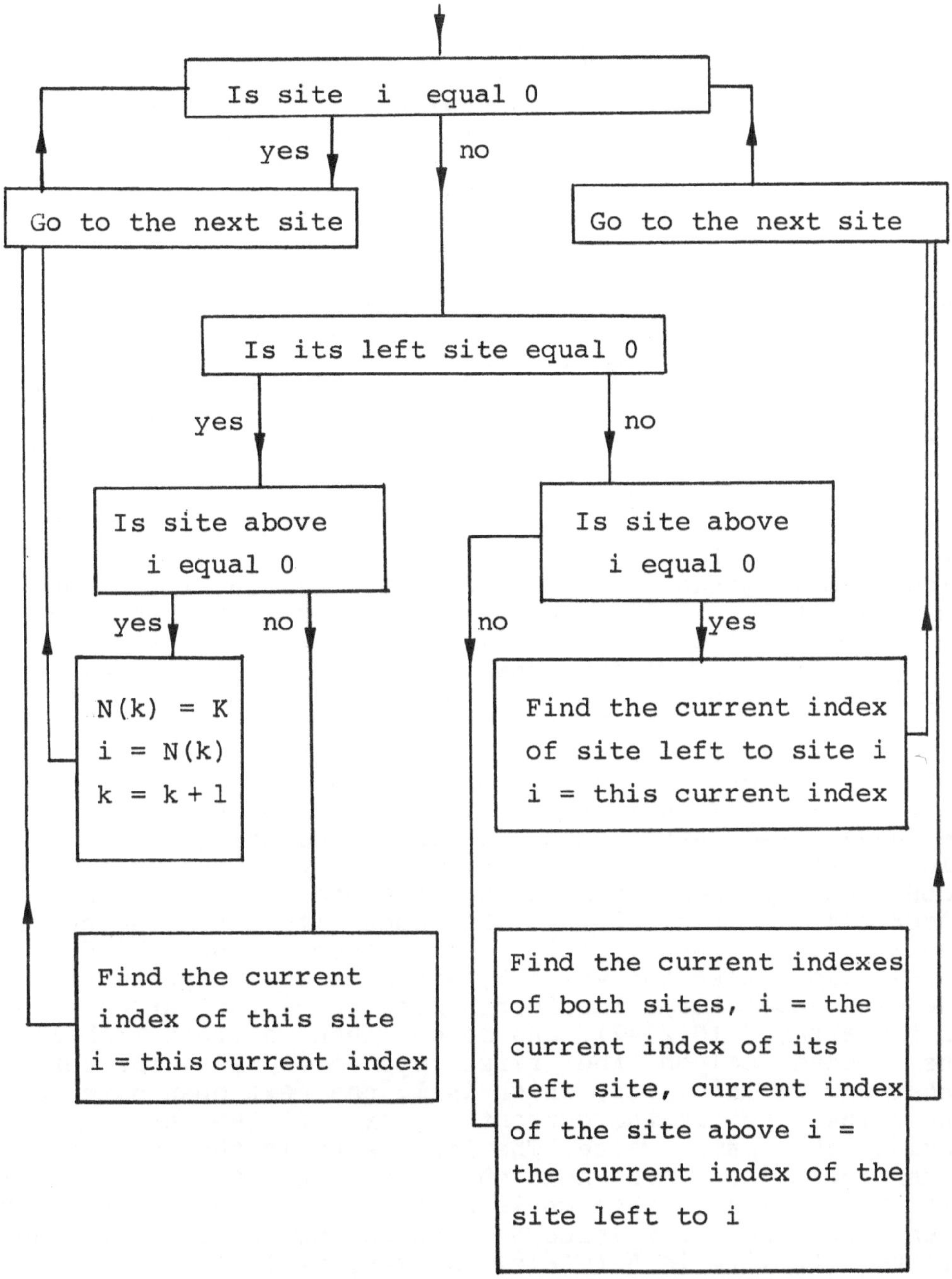

Figure 12: Flow chart for the site labelling assignment, for the first lattice sweep.

Occupied (open) sites are labelled 1 and empty (closed) sites are labelled 0. All labelling is done column by column starting at the top of the left-most column. We label each open site using a one-dimensional array N as follows: If both sites above and to the left of an open site i are closed sites, i takes the current index of the array N, which in the beginning is set equal to 1. If only one of these sites is open then i takes the current index of the open site, which, however, may be different from its original value. If both these sites are open then site i and the current index of the site situated above it simultaneously take the current index of the left site. This is all carried out in the one-dimensional array N, where the current index k (k=1, 2, 3,..., etc.) of site i is the index which satisfies the equation N(k)=k. The value of the index for site i that will be stored in the array N is determined as follows: If the original index of site (i-1) is k_1 and the current index of site (i-1) is k_2, then we set for site i $N(k_1)=k_2$ and $N(k_2)=k_2$ so that k_2 is now the stored index for this site. The flow chart for this algorithm is given in Figure 12. When all index assignment in the lattice is finished we go back for the second sweep to renumber all open sites which do not have the latest assigned index. We then assign all identical indexes to a new one-dimensional array Z, where we store the size and latest index of each cluster.

To make this process better understood we work out an example of index processing for a 20x20 square lattice as shown in Fig. 13. Here the assigned p is p=0.56. In part (a) we have the lattice as it is formed using 0 and 1, for closed and open sites, respectively. In part (b) we begin by labelling the open sites of the first column. Following the algorithm in Fig. 12 the first open site takes the current index of the array N (N(1)=1), the second and third open sites also take the value 1, which is the current index of the site above them. The fourth site takes the current index of the array N (N(2)=2), and so on. When we start labelling the second column the first open site takes the current index of its left site, which is 1, the next open site takes the value 5, while the current index of the array N is taking the same value. The first site in the third column takes the value 6. The next one takes the value 1, while N(6)=N(1). Thus, when we label the fourth column the first open site takes the value 5 because now the current index of its left site is 5 (N(6)=1, N(1)=5, N(5)=5). Using this method we finish labelling all open sites in the lattice. In part (c), in the second sweep we label again all open sites which do not have their latest index. We see that the percolating cluster has as its index the value of 50, and is made of 83 lattice sites, the second largest cluster has a size of 71 and lattice index of 22, and so on.

1	0	1	1	0	0	0	1	1	0	1	0	1	0	0	0	1	1	0	1
1	1	1	1	0	1	0	0	0	1	0	1	1	1	1	1	1	0	1	1
1	0	1	0	1	1	0	0	0	1	1	1	1	1	1	1	1	0	1	0
0	1	1	1	1	0	0	1	1	1	1	0	1	1	1	1	1	0	1	1
0	0	0	0	1	1	1	1	0	1	0	1	1	0	0	0	1	0	1	1
0	0	0	0	0	1	1	1	0	0	0	0	0	0	0	0	1	1	1	0
1	1	1	0	0	1	1	1	1	0	1	1	0	1	1	1	0	0	1	0
1	1	1	0	1	1	1	1	1	0	1	0	1	0	0	0	0	0	1	1
1	1	0	0	1	1	0	0	0	0	1	1	0	1	0	0	0	0	0	1
1	0	1	0	0	1	0	1	1	1	0	0	0	1	1	0	0	0	1	1
1	1	0	1	1	0	1	1	1	1	0	1	0	0	1	1	1	0	0	0
0	1	1	1	1	0	1	0	1	0	1	0	1	0	1	1	0	0	0	1
1	1	1	0	0	1	1	0	0	0	0	0	1	0	0	1	0	1	0	1
1	1	1	1	1	0	0	1	1	0	1	1	1	1	1	0	1	0	0	1
0	1	0	1	1	0	1	1	1	1	1	1	1	1	1	0	0	1	0	1
0	1	0	1	1	1	1	0	1	1	0	1	0	1	0	1	0	1	1	1
1	0	1	1	0	0	0	1	1	1	1	0	0	0	1	1	0	1	1	1
1	1	1	0	0	0	1	0	1	0	1	0	0	1	0	0	1	1	0	1
0	1	1	1	0	0	1	0	0	1	0	0	0	1	0	1	1	0	1	0
0	0	1	0	1	0	0	1	1	0	0	0	1	1	1	0	0	1	1	1

Part (a)

1	0	6	5	0	0	0	18	18	0	26	0	33	0	0	0	43	31	0	52
1	1	1	5	0	13	0	0	0	24	0	30	11	31	31	31	31	0	49	31
1	0	1	0	10	5	0	0	0	24	11	11	11	31	31	31	31	0	49	0
0	5	5	5	5	0	0	19	11	11	11	0	11	31	31	31	31	0	49	31
0	0	0	0	5	5	11	11	0	11	0	31	31	0	0	0	31	0	49	31
0	0	0	0	0	5	11	11	0	0	0	0	0	0	0	0	31	31	31	0
2	2	3	0	0	5	11	11	11	0	27	27	0	37	37	37	0	0	31	0
2	2	3	0	11	11	11	11	11	0	27	0	34	0	0	0	0	0	31	31
2	2	0	0	11	11	0	0	0	0	27	27	0	38	0	0	0	0	0	31
2	0	7	0	0	11	0	20	14	14	0	0	0	38	38	0	0	0	50	50
2	2	0	9	4	0	15	14	14	14	0	32	0	0	38	38	38	0	0	0
0	2	3	3	4	0	15	0	14	0	28	0	35	0	38	38	0	0	0	53
3	3	3	0	0	14	14	0	0	0	0	0	35	0	0	38	0	46	0	53
3	3	3	3	4	0	0	21	4	0	29	22	22	22	22	0	44	0	0	53
0	3	0	3	4	0	16	4	4	22	22	22	22	22	22	0	0	47	0	53
0	3	0	3	4	4	4	0	4	22	0	22	0	22	0	41	0	47	42	42
4	0	8	4	0	0	0	22	22	22	22	0	0	0	40	40	0	47	42	42
4	4	4	0	0	0	17	0	22	0	22	0	0	39	0	0	45	42	0	42
0	4	4	4	0	0	17	0	0	25	0	0	0	39	0	42	42	0	51	0
0	0	4	0	12	0	0	23	23	0	0	0	36	36	36	0	0	48	48	48

Part (b)

50	0	50	50	0	0	0	18	18	0	26	0	50	0	0	0	50	50	0	50
50	50	50	50	0	50	0	0	0	50	0	50	50	50	50	50	50	0	50	50
50	0	50	0	50	50	0	0	0	50	50	50	50	50	50	50	50	0	50	0
0	50	50	50	50	0	0	50	50	50	50	0	50	50	50	50	50	0	50	50
0	0	0	0	50	50	50	50	0	50	0	50	50	0	0	0	50	0	50	50
0	0	0	0	0	50	50	50	0	0	0	0	0	0	0	0	50	50	50	0
22	22	22	0	0	50	50	50	50	0	27	27	0	37	37	37	0	0	50	0
22	22	22	0	50	50	50	50	50	0	27	0	34	0	0	0	0	0	50	50
22	22	0	0	50	50	0	0	0	0	27	27	0	38	0	0	0	0	0	50
22	0	7	0	0	50	0	14	14	14	0	0	0	38	38	0	0	0	50	50
22	22	0	22	22	0	14	14	14	14	0	32	0	0	38	38	38	0	0	0
0	22	22	22	22	0	14	0	14	0	28	0	22	0	38	38	0	0	0	42
22	22	22	0	0	14	14	0	0	0	0	0	22	0	0	38	0	46	0	42
22	22	22	22	22	0	0	22	22	0	22	22	22	22	22	0	44	0	0	42
0	22	0	22	22	0	22	22	22	22	22	22	22	22	22	0	0	42	0	42
0	22	0	22	22	22	22	0	22	22	0	22	0	22	0	40	0	42	42	42
22	0	22	22	0	0	0	22	22	22	22	0	0	0	40	40	0	42	42	42
22	22	22	0	0	0	17	0	22	0	22	0	0	36	0	0	42	42	0	42
0	22	22	22	0	0	17	0	0	25	0	0	0	36	0	42	42	0	48	0
0	0	22	0	12	0	0	23	23	0	0	0	36	36	36	0	0	48	48	48

Part (c)

Figure 13: An example of a cluster distribution for a 20x20 lattice. Part(a) A binary, substitutionally random square lattice with an assigned p=0.56. Part (b)Index assignment after the first sweep. Part (c) Index assignment after the second sweep.

REFERENCES

1. For a collection of papers on random walks see: Proceedings of Conference on random walks, Washington DC, J. Stat. Phys., 36,519-916(1983).

2. E.W.Montroll, Proc. Symp. Appl. Math., 16,193(1964).

3. F.S.Henyey and V.Seshadri, J. Chem. Phys., 76,5530(1982).

4. A.Blumen and G.Zumofen, J. Chem. Phys., 75,892(1982); J. Chem. Phys., 76,3713(1982).

5. B.B.Mandelbrot, The Fractal Geomerty of Nature (Freeman, San Francisco, 1983).

6. S.Alexander and R.Orbach, J. Phys. Lett., 43,L625(1982); R.Rammal and G.Toulouse, J. Phys. Lett., 44,L13(1983).

7. P.Argyrakis and R.Kopelman, Chem. Phys., 57,29(1981); Chem. Phys., 78,251(1983).

8. A.Aharony and D.Stauffer, Phys. Rev. Lett., 52,2368(1984).

9. I.Webman, Phys. Rev. Lett., 52,220(1984).

10. P.Argyrakis and R.Kopelman, J. Chem. Phys., 81,1015(1984); J. Chem. Phys., In press..

11. A.Keramiotis, P.Argyrakis, and R.Kopelman, Phys. Rev. B, 31,4617(1985).

12. D.Ben-Avraham and S.Havlin, J. Phys. A 15,L691(1982); J. Phys. A 15,L311(1982).

13. S.Havlin, D.Ben-Avraham, and H.Sompolinski, Phys. Rev. 27A(1983)1730.

14. R.Rammal, J. Stat. Phys., 36,547(1984).

15. R.B.Pandey, D.Stauffer, A.Margolina, and J.G.Zabolitzky, J. Stat. Phys. 34,427(1984).

16. J.W.Haus and K.W.Kehr, Sol. Stat. Comm., 26,753(1978); V.Lottner, J.W.Haus, A.Heim, and K.W.Kehr, J. Phys. Chem. Sol., 40,557(1979); W.Gissler and H.Rother, Physica, 50,380(1970).

17. K.W.Kehr, J. Stat. Phys., 30,509(1983).

18. P.Argyrakis and R.Kopelman, J. Theo. Bio., 78,205(1978).

19. K.W.Kehr and P.Argyrakis, J. Chem. Phys., Submitted for publication.

20. P.Argyrakis and K.W.Kehr, To be published.

21. J.Hoshen, R.Kopelman, and E.M.Monberg, J. Stat. Phys., 19,219(1978).

22. M.Gouker and F.Family, Phys. Rev. B, 28,1449(1983).

23. J.Hoshen and R.Kopelman, Phys. Rev. B, 14,3438(1976).

NUCLEAR MAGNETIC RELAXATION IN IONIC CONDUCTOR MATERIALS

Hélène THEVENEAU
E.S.P.C.I.
Laboratoire de Physique Thermique
10, rue Vauquelin
75231 Paris Cedex 05
France

ABSTRACT.

We want to present here some results showing how NMR techniques can contribute to the study of superionic conductors. After a presentation of these materials, characterized by a very high ionic conductivity of cationic or anionic type, of their structures and their applications, we make some comments about the NMR interactions and their modulation by the diffusion motion. Discrepancies observed when one tries to interpret the results in the framework of a simple isotropic diffusion model are underlined and some ways of analysis taking into account the dimensionality and boundaries effects in the diffusion process are presented.

1. INTRODUCTION

The use of NMR techniques in materials science has considerably progressed during the last twenty years. The reason is not in a change in the basic principles, which were established forty years ago, but in a huge improvment of the equipments provided to the experimentalists to study solid state systems. The first contribution to this improvment comes from the realization of high magnetic fields by means of superconductor magnets, the second from the realization of spectrometers, often driven by the computer and allowing long accumulations of the signals and sophisticated excitations of the spin system ; as examples one can quote multiple pulses sequences, selective excitation of some transition in the spins energy spectrum, fast Fourier transform of the signal in one or two dimensional modes.

Now one can say that it is possible to look at the resonance of every stable element of the Mandeleev table, provided that it possesses a non zero spin, condition which excludes only four elements [1].

The basic characteristic of the NMR techniques is that in a fixed magnetic field Ho, the resonance frequency of a particular nucleus is determined and given by the expression :

$$\omega_o = \gamma H_o \qquad (1)$$

R. Daudel et al. (eds.), Structure and Dynamics of Molecular Systems – II, 231–254.

where γ is the gyromagnetic ratio of the nucleus. This is the case in absence of any interaction. But, in matter, where the nucleus feels the numerous interactions due to its neighborhood, its resonance spectrum reflects the whole information about these interactions in their structural and dynamical aspects. All the interactions which can be experimented by a spin system are known, nevertheless going up again to the microscopic details of a material once one has identified some particular interactions is not yet so easy. It requires very important theoretical efforts associated with a good knowledge of the basic structure of the material, obtained, for instance, through X-Ray and Neutron diffraction experiments.

In these conditions, NMR can be a good tool to study the localisation and the motions of ions in materials. It is the reason why it has been applied to the study of superionic conductors since thirty years.

Superionic conductors (or solid electrolytes) are ionic materials possessing in the solid phase a ionic conductivity comparable to that of liquid electrolytes.

The use of NMR in these materials is based on the hypothesis that the mass diffusion process causes the decrease of the correlation functions coming into the expressions of the relaxation interactions of the spin system considered in the material, i.e. one expects that these correlation functions relaxe in a time τ, comparable to the jump time, υ^{-1}, of the ions in diffusing motion. In other words, one expects that the υ frequency lies in the frequency range of the NMR characteristic spectral densities.

Then one hopes that combining measurements of transport macroscopic parameters, like the diffusion coefficient and the electrical conductivity, and NMR results, one can characterize the diffusion process in terms of mobile particles density and nature of the motion and elementary paths.

In part 2, I shall indicate some properties and applications of the superionic conductors. Then in part 3, devoted to NMR, I shall briefly present the interactions in the spin system and the motional correlation functions the spectral density of which are related to the measured parameters. In part 4, I shall give some results obtained in various materials. In the conclusion, I shall emphasize the need of more realistic diffusion models for the calculation of correlation functions and finally I shall present a new NMR method of 2-dimensional Fourier transform which may be useful in the field of superionic conductors.

2. SUPERIONIC CONDUCTORS

2.1 Definition and typical parameters

Superionic conductors are solid materials with a ionic conductivity much larger than that of ordinary ionic crystals. This high conductivity is reached at temperatures much lower than that of fusion. As an example we have reported on figure 1 the conductivity of some famous superionic conductors versus the ratio $\frac{Tm}{T}$, where T is the temperature and Tm the fusion temperature [2].

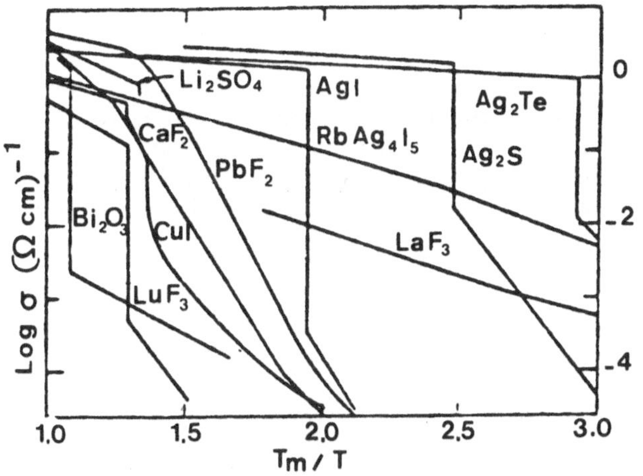

Figure 1. Ionic conductivity versus Tm/T, where Tm is the fusion temperature, for some famous superionic conductors. From [2].

The phenomenon was first observed in 1839 by Faraday [3] who pointed out the two types of conductivity : cationic, like in silver sulfide Ag_2S, where the mobile ions are the silver ions and anionic, like in lead fluoride PbF_2, where the mobile ions are the fluoride. This mass transport can be or not accompagnied by electronic conduction.

The materials can be decribed as ionic crystals where one of the ion sublattices has the translation freedom of a fluid and where the rest of the ions vibrate around the equilibrium positions like in a solid. This exceptional behaviour has raised up important studies oriented towards the origin of the ions mobility and the characterization of their motions.

If one applies a field on an ionic material a current occurs, due to the motion of the ions and generally the conductivity is expressed by the Nernst-Einstein relation :

$$\sigma = \frac{c(Ze/^2D)}{kT} \qquad (2)$$

where c and D are respectively the concentration and the diffusion coefficients of the mobile ions, Z the charge number, e the electron charge, k the Boltzmann constant and T the temperature.

If the ion is supposed to make non correlated jumps of length l, at a frequency υ, the diffusion coefficient can be expressed as :

$$D = \frac{\upsilon l^2}{2d} \qquad (3)$$

where d represents the system dimensionality (d = 1,2, or 3). If there are departures from this simple model υ and l may be redefined as effective quantities [4].

The jump time υ^{-1}, is often supposed to depend upon temperature according to an Arrhenius law :

$$\upsilon^{-1} \equiv \tau = \tau o \exp\left(\frac{E}{kT}\right) \tag{4}$$

where τo is said to be the preexponential factor ($\upsilon o \equiv \tau o^{-1}$ is the attempt frequency) and E is the potential barrier height.

These three equations supply the basis for the discussion of the factors affecting the conductivity, which is found to follow the experimental law :

$$\sigma = \frac{\sigma o}{T} \exp\left(- \frac{E}{kT}\right) \tag{5}$$

Typical values of the conduction parameters are the following for T = 500 K :

$c = 10^{22} cm^{-3}$
$\sigma = 10^{-1} \Omega^{-1} cm^{-1}$
$D = 10^{-6} cm^{-2} s^{-1}$
$E = 0,1$ eV
$\upsilon_o = 10^{12}$ to 10^{14} Hz

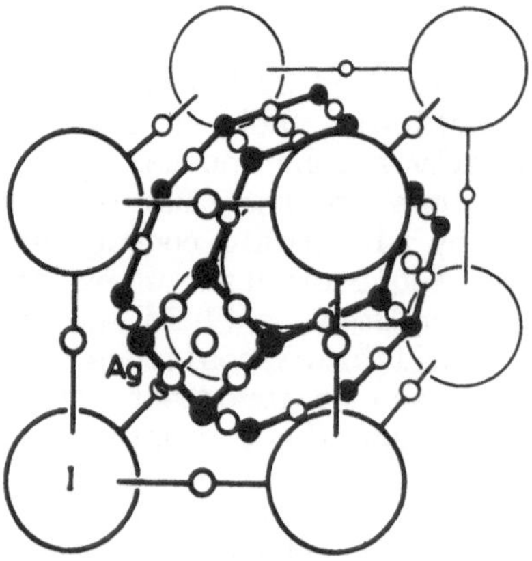

Figure 2. Crystal structure of α-AgI. The I ions form a bcc lattice. The silver distribution is shown by the small circles. From [5].

In summary, in a good solid electrolyte the number of mobile ions and the number of vaccant lattice sites of nearly equivalent energy that the mobile ions can occupy must be great. The barrier energy to the motion must be low. A good example is silver iodide AgI [5] represented figure 2.

Finally the compound must be stable in the temperature range required for applications.

2.2 Classification and structures

The number of superionic conductors is considerable, growing up every day. Detailed lists of materials are given in [6] [7] [8]. I shall present here the materials according to the nature of their mobile ions, cationic

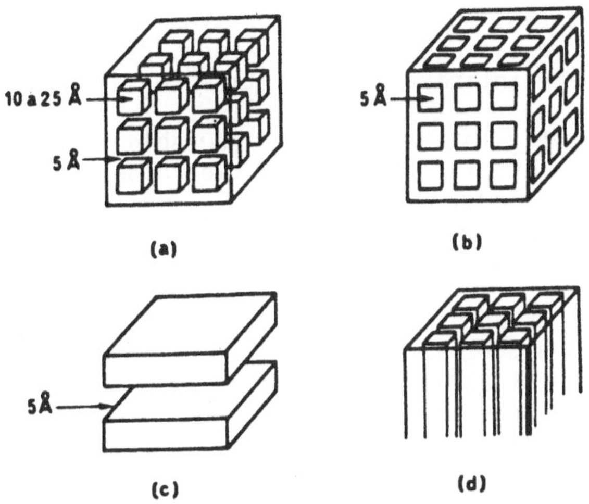

Figure 3. Schematical structures of ionic conductors : (a) globular ; (b) with interconnected channels (3d) ; (c) layered (2d) ; (d) 1d with channels. From [13].

and anionic, and show some structures favourable for ionic conduction.

2.2.1. Cationic conductors. A high ionic conductivity is obtained in materials where the anions constitute the rigid atomic lattice in which the cations, which are smaller, diffuse. One observes that the more mobile ions are those of group IA and IB. Groug IB cations like Cu^+ and Ag^+ lead to the materials with the best conductivity. As exemples one can quote CuBr, CuI, CuS, Cu_2Se, AgI, Ag_2S, Ag_3SI, $RbAg_4I_5$. The barrier height is low : 0.05 to 0.3 eV and the structures are formed by chains of octahedra or tetrahedra shearing faces and forming tunnels where ions can diffuse in 3-dimensions. The disagreements of these conductors are their price, their instability in the proper temperature range ($RbAg_4I_5$ is unstable in air atmosphere at 27°C) and the weakness of the e.m.f. obtained in battery application (< 0.7 eV).

The IA group provides numerous materials for making electrodes and electrolytes in batteries developping important e.m.f.. Examples of these materials are β-alumina doped with sodium or potassium (general formula : $M_2O_{11}Al_2O_3$) Nafion, Nasicon, LiI, Li_3N, Li_2SO_4, Lithium intercalated Tungsten oxide, hollandites (K^+).

Some other cationic conductor are investigated which may present a conductivity comparable to that of β-alumina. For instance, Thallium compounds like $TlNb_2O_5F$ [9] or $TlZrF_5$ are very good candidates because of the high polarisability of Tl^+ ion.

The protonic conductors constitute a very important class of cationic conductors [10],[11],[12]. In these compounds the role of water is not yet elucidated, and the nature of the mobile ion has to be determined: H^+ ion, H_3O^+ oxonium cation, $H_2O_5^+$ dioxonium cation or the OH^- anion.

One can find some hydrates, acids like $H_2Sb_4O_{11}nH_2O$, hydrated-oxides like $SnO_2,2H_2O$, some bronzes $HxWO_3$, $HNbO_3$, uranyl dihydro-genophosphate (HUP). They present very interesting properties for electrochromism.

2.2.2. Anionic conductors. The two anions favourable for conduction are oxygen O^{2-} and fluorine F^-. The ionic conduction appears in some oxides of IVB group metals like ZrO_2, HfO_2, CaO_2, ThO_2 and in the fluorites like CaF_2, SrF_2, LaF_3. These materials are used in gas sensors and work at high temperature, the energy barrier being relatively high (1 eV).

2.2.3. Structures. The structures favourable for the ionic conduction have been summarized schematically by A. Potier [13] and are presented on figure 3. 3-dimensional structures are encountered in ReO_3, pyrochlore or perovskite type compounds as illustrated on figure 4(a) : pyrochlore $TlNb_2O_5F$ [14], and figure 4(b) : niobic acid $HNbO_3$ [15].

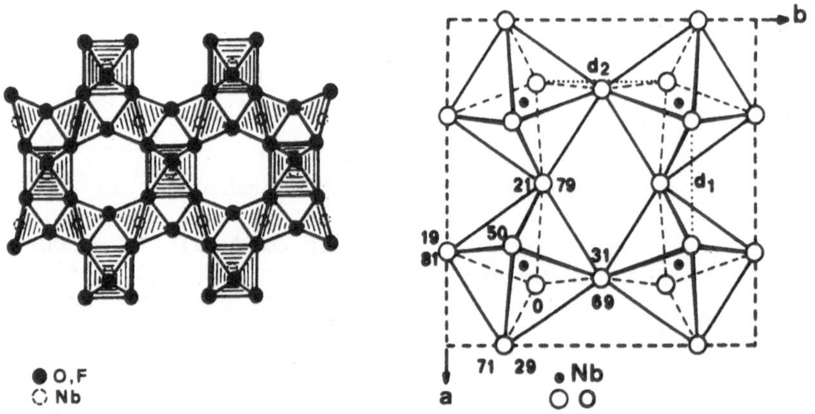

(a) (b)

Figure 4. Examples of 3d structures : (a) the pyrochlore $TlNb_2O_5F$ (after [14]) and (b) the niobic acid $HNbO_3$. After [15].

An example of a 2-dimensional system is found in Li_3N for which two values of the conductivity were measured depending on the direction [16], [17], (See figure 5). Another example is that of Na-β-alumina $Na_2O_{11}Al_2O_3$ represented figure 6 [5], [18].

Finally, on figure 7 is shown the structure of β-eucryptite, $LiAlSiO_4$, as an example of 1-dimensional system [19].

2.3 Applications

The applications are those of a high exclusively ionic conductivity or of a mixed ionic and electronic conductivity. These two features lead to four types of realization :

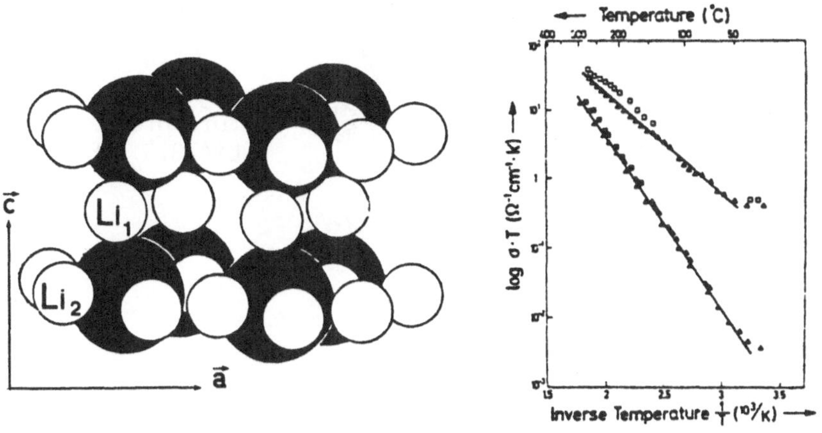

(a) (b)

Figure 5. Example of 2d structure : Li_3N.
(a) The two types of Li nuclei. From [16].
(b) The conductivity versus temperature showing the anisotropy. From [17].

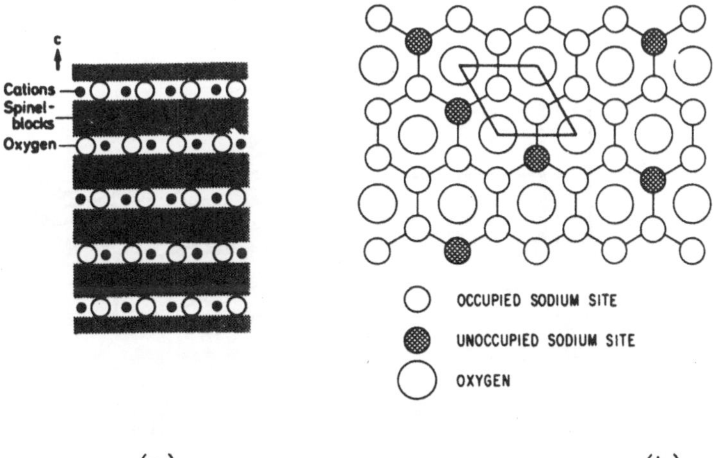

(a) (b)

Figure 6. Crystal structure of Na-β-alumina.
(a) Schematic view parallel to the c-axis. The conductivity layers are formed by bridging oxygens and the conducting ions. The conducting layers are separated from each other by spinel $-(Al_2O_3)-$ blocks. From [5].
(b) Section through the conducting layer. From [18].

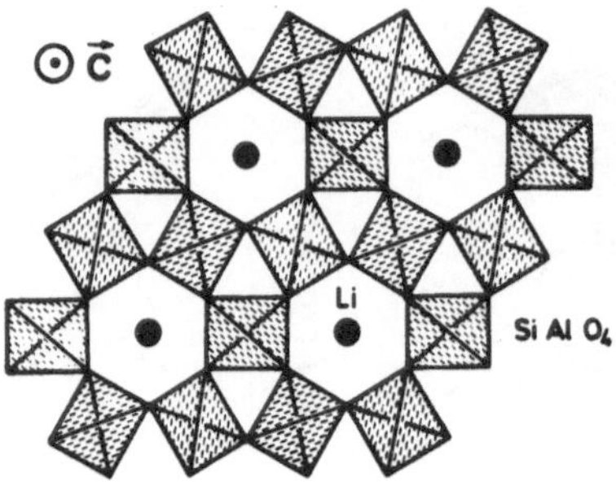

Figure 7. The 1d structure of β-eucryptite $LiAlSiO_4$. From [19].

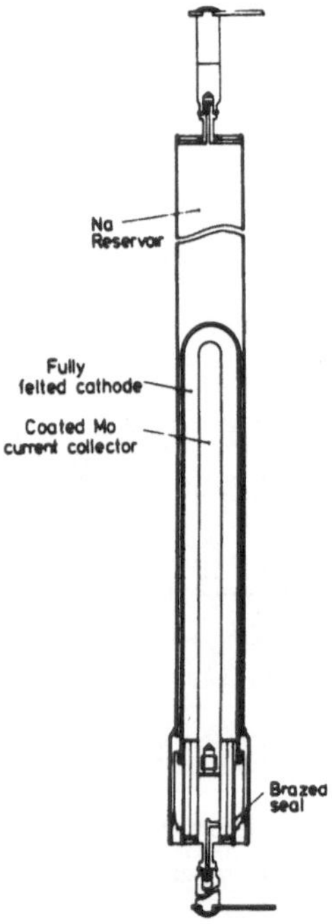

(a)

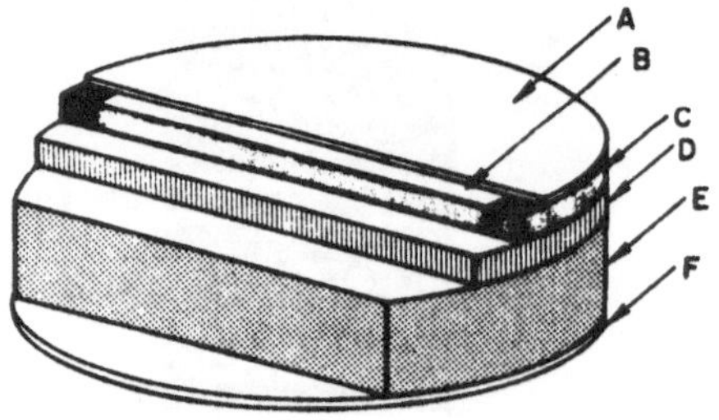

(b)

Figure 8. Examples of batteries.
(a) The sodium/sulfur battery of British Railway. It is a central sulfur cell with Na-β alumina as the electrolyte. Overall length = 1 m. From [20].
(b) Sectioned view of the solid electrolyte cells $Li/LiI(Al_2O_3)/PbI_2$, PbS,Pb. A, anode current collector : steel ; B, anode : Li ; C, anode retaining ring ; D, solid electrolyte : powder mixture of LiI and Al_2O_3 ; E, cathode : well blended mixture of PbI_2,PbS and Pb ; F, cathode current collector : Pb. From [21].

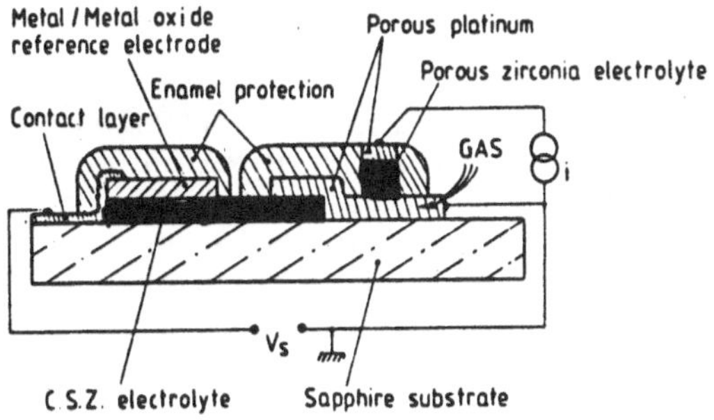

Figure 9. Oxygen gas sensor - after [22]. The solid electrolyte is zirconia.

- electro chemical generators all solid or not, in which pure ionic conductors or mixed conductivity materials are used as electrolyte and electrodes respectively ;
- gas or ions sensors ;
- vizualization devices based on the electrochromism phenomenon ;
- supercondensators.

Examples of batteries are presented figure 8 ; (a) : the sodium-sulfur battery of British Railway [20] ; (b) the all solid-state battery system based on lithium [21]. Many informations about lithium cells can be found in [7]. This device can provide e.m.f. of 1.9 eV at 25°C. Lithium batteries have extremely long storage life and are used as pacemakers.

A gas sensor is presented figure 9. It is an oxygen sensor developed by Thomson CSF [22].

Finally, we present on figure 10 the scheme of a vizualization device [23]. In electrochromic material a change of color, persistent but reversible, can be induced by a field or a current.

As one can see, for most of the applications, the materials must be obtained and characterized in thin layers. It is a new field of solid-state science which has been called by G. Velasco, Microionics [24].

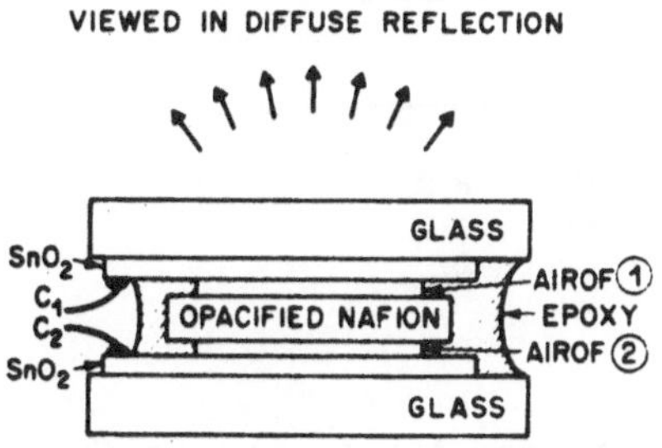

Figure 10. Schematic diagram of an all solid state electrochromic cell based on Anodic Iridium Oxide Films (AIROFs) and opacified Nafion (perfluorosulfonic acid resin). From [23].

3. NMR IN THE STUDY OF SUPERIONIC CONDUCTORS

Many reviews have been devoted to the subject [25], [26], [27], [28].

The aim of the study is to characterize the diffusion motion which causes the high conductivity and modulates the interactions of the spin systems present in the material. But one must keep in mind that all the other motions, even the non diffusive ones, also contribute to the NMR relaxation. Therefore the role of the spectroscopist is to distinguish among the measured parameters which are affected by the diffusion process and to analyze them on the basis of relevant correlation functions corresponding to the particular material investigated. For this work it will be useful to investigate the resonance of the different nuclei species of the sample, i.e. the mobile ones and the others because they can experience different interactions modulated by different motions.

3.1 The interactions in the spin system

The first step of the inverstigation is to identify, for a given type of nucleus, the interaction responsible for the relaxation process. All the interactions are known and the spin Hamiltonian may be written as : [29], [30].

$$H = H_Z + H_{RF} + H_\sigma + H_D + H_J + H_{SR} + H_Q \qquad (6)$$

In this expression, the two first terms represent the interactions with the static magnetic field H_0 (Zeeman term) and the radiofrequency field H_1 respectively. The others represent the interactions inside the material ; H_σ : chemical shift, H_J:indirect, H_{SR} : spin-rotation, H_D : dipolar and H_Q : quadrupolar interactions.

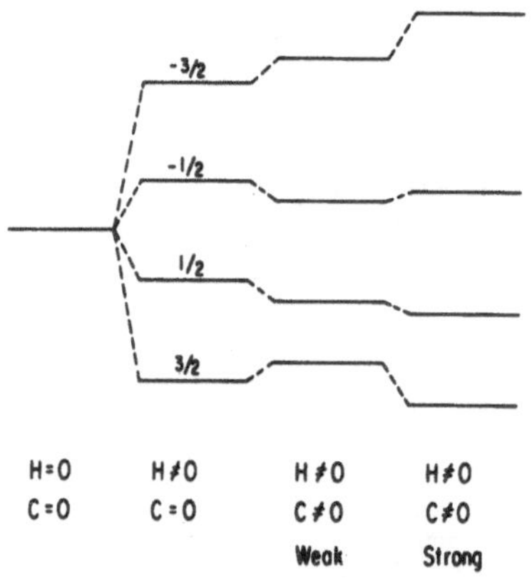

Figure 11. Schematic energy level diagram for a spin I=3/2. H represents the external magnetic field. C is the quadrupole coupling strength. The energy level modification by C is small compared with the Zeeman splitting (C=0) and depends on the direction of the magnetic field relative to the principal axis system. From [31].

These Hamiltonians are dependent on the orientation of the spin system in the static field H_0 and on the resonance frequency. So a study versus frequency may be useful to identify an interaction.

Generally a type of interaction is greater than the others. In solids, the most important are the chemical shift, the dipolar and the quadrupolar interactions. This last interaction exists only for nuclei with spin $I > 1/2$ and submitted to an electric field gradient. In the superionic conductors it is a very informative interaction as one will see further. On figure 11 is represented the energy spectrum of a $I = 3/2$ spin system in presence of quadrupolar interaction [31] ; the transitions between the levels are no longer equivalent ; they must be excited selectively, and one can observe for instance only the central transition or one satellite.

3.2 Spectral densities and correlation functions

The interactions in the spin system are motionaly averaged and one can define for each type of interaction correlation functions the decay of which is characterized by a correlation time τ. Their Fourier transforms are the spectral densities in terms of which are expressed the observed NMR parameters, like the relaxation times T_1 or T_2 and the linewidth. So, retrieval of the desired information in the case of superionic conductors requires proper models for the diffusion and calculation of the motional correlation function [25],[32],[33].

The simplest model is the one of Bloembergen, Purcell and Pound (BPP) [34], developped for isotropic 3-d translation. In this model, the correlation functions are supposed to decay exponentially with a time constant τ :

$$G(t) \simeq \exp(-\frac{t}{\tau}) \tag{7}$$

which gives a lorentzian spectral density function :

$$J(\omega) \sim \frac{\tau}{1 + \omega^2\tau^2} \tag{8}$$

The spin-lattice relaxation time is expressed as :

$$T_1^{-1} = A\left[a\,\tau + \frac{b\tau}{1 + (\omega_0\tau)^2} + \frac{c\tau}{1 + (2\omega_0\tau)^2}\right] \tag{9}$$

and similar expression for T_2, the spin-spin relaxation time and $T_{1\rho}$, the spin-lattice relaxation time in the rotating frame (with $\omega_I = \gamma H_1$ in place of ω_0 .

A is characteristic of the interaction. If A if frequency independent, like in dipolar or quadrupolar interaction, one has for the limiting cases (see figure 12) :

$$\begin{array}{ll} \text{high temperature} : \omega\tau \ll 1 & J(\omega) \sim \tau_1 \\ \text{low temperature} : \omega\tau \gg 1 & J(\omega) \sim \dfrac{1}{\omega^2\tau} \end{array} \tag{10}$$

If τ fits the Arrhenius law of (4) one can deduce from the data the

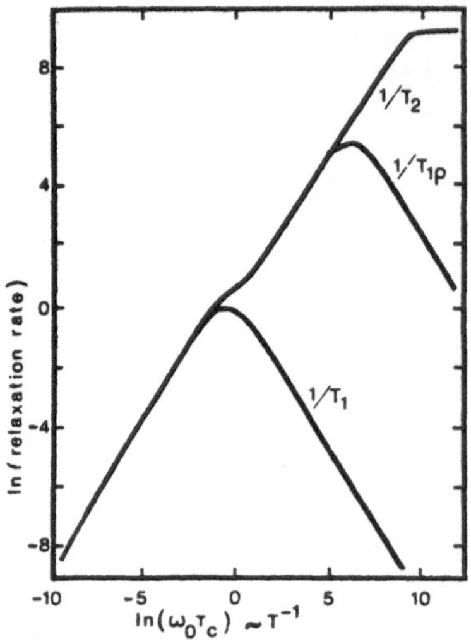

Figure 12. BPP behaviour for the relaxation rates $T_1^{-1}, T_{1\rho}^{-1}$ and T_2^{-1} ($\sim\Delta Hpp^{-1}$). Here $\omega_0/\omega_1 = 10^3$, $\omega_0^2/\langle\omega^2\rangle = 10^8$ and $\langle\Delta\omega^2\rangle^{1/2} = 10^8 sec^{-1}$. From [27].

activation energy τ and the preexponential factor τ_o.

If A varies with frequency like in the chemical shift interaction, the relaxation time behaviour versus temperature and frequency is reverse (see figure 13).

The BPP model serves as a reference for analyzing the data as a first step but many discrepancies arise due to the use of improper correlation functions or/and non realistic diffusion model. Generally it is found that the T_1 over T_2 ration is not equal to 1.6 ; the curves repre-

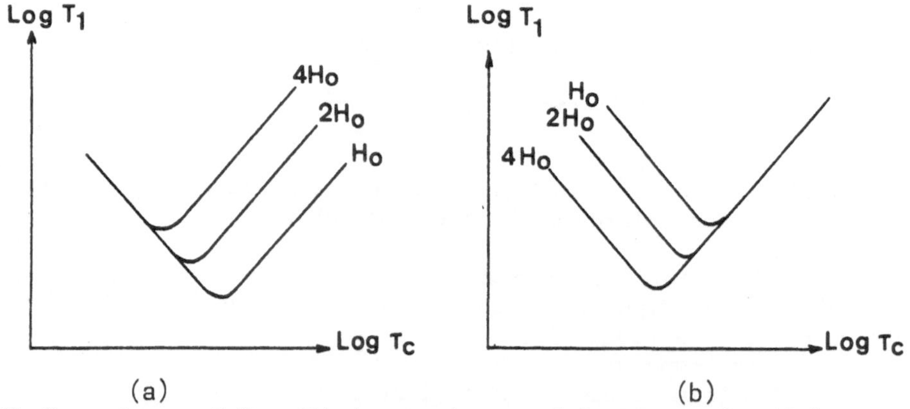

Figure 13. Dependence of T_1 with temperature and frequency for the case of a dipolar interaction (a) and a chemical shift interaction(b).

Conditions		T_1 proportional to	Slope d ln T_1/d(1/kT)
$\omega_0\tau >> 1$	n = 1,2,3	$\omega_0^2\tau$	E
$\omega_0\tau << 1$	n = 3	$\frac{1}{\tau}$	- E
	n = 2	$\frac{1}{\tau \ln \frac{1}{\omega_0\tau}}$	$-E[1 + \frac{1}{\frac{E}{kT} + \ln\omega_0\tau}]$
	n = 1	$(\frac{\omega_0}{\tau})^{1/2}$	$-\frac{E}{2}$
	n = 1 $\omega_0\tau_\perp >> 1$	$(\frac{\omega_0}{\tau})^{1/2}$	$-\frac{E}{2}$
	n = 1 $\omega_0\tau_\perp << 1$	$\frac{1}{(\tau\tau_\perp)^{1/2}}$	$-\frac{E + E_\perp}{2}$

Table I. Limiting behaviour of T_1 for independent particle diffusion as a function of hop time τ, frequency ω_0 and dimensionality d. Apparent activities energy $d(\ln T_1)/d(1/kT)$ is derived by assuming an activated $\tau = \tau_0 \exp E/kT$. In quasi 1d.case the interchannel hop time is $\tau_\perp$ and the activation energy $E_\perp$. From [4].

sentating the thermal behaviour of the relaxation times are asymetric and the BPP ω^2 law is not fulfilled ; experimentally a ω^β law is found with $\beta < 2$; the slopes of the thermal dependence curves for the different relaxation times are not the same and the activation energies deduced are not equal to that obtained in conductivity measurements. Finally anomalous values of τ_0 are found ; normally the values of τ_0^{-1} should be in the phonon frequency range, i.e. 10^{11}- 10^{14} Hz.

Numerous ways of interpretation have been proposed and are reviewed in [4], [25], [32], [33].

Concerning the correlation functions, one obtains for instance in the dipolar case particle-particle correlation functions for both the case when the resonnant spin is mobile and that when only the interacting particle diffuses. But quadrupolar interaction gives different correlation functions for mobile nuclei and non mobile ones ; in this last case one obtains the site occupancy correlation function.

The aspect of dimensionality has been introduced to obtain more realistic description of the diffusion. Richards has proposed expressions for the apparent parameters determined in NMR versus the 'true' activation energy E and prefactor τ_0. See for instance the T_1 results on table I [4].

Bjorkstam and Villa reported calculations of spectral density functions adequate for 2D and 1D systems [25]. On figure 14 are shown the

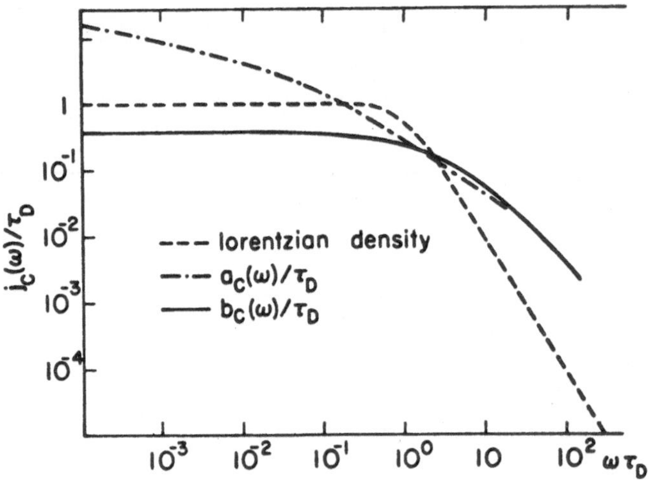

Figure 14. Dipolar spectral densities, $a_c(\omega)$ and $b_c(\omega)$, for 2D unlimited continuum diffusion compared with the Lorentzian density. (in τ_D units versus $\omega\tau_D$). See [25].

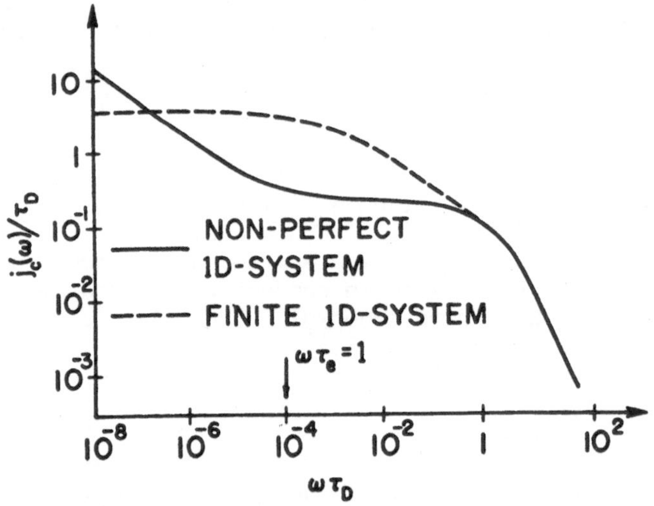

Figure 15. Spectral densities for continuum diffusion in a non-perfect and in a finite 1D system. From [25].

BPP Lorentzian density and two particular densities, calculated for a 2D unlimited diffusion, which present a broadening of the maximum at $\omega\tau \sim 1$, a weaker frequency dependence than ω^{-2} and, for one of them, a divergence at $\omega = 0$.

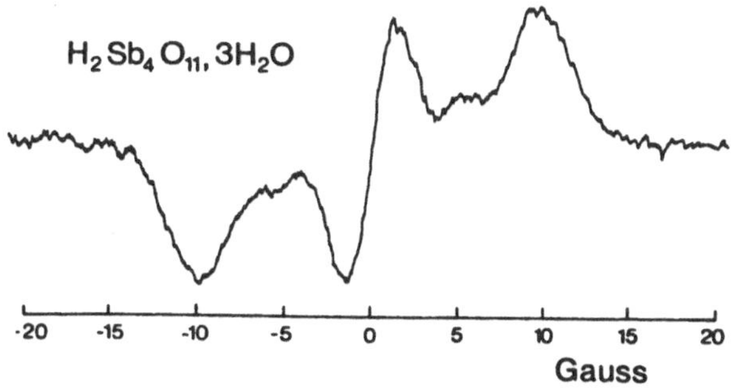

Figure 16. Derivative of the proton spectrum of antimonic acid, $H_2Sb_4O_{11}3H_2O$ at 4K. From [38].

But the notion of low-dimensional diffusion is related to the fact that the particle is subject to some constraining boundaries. On figure 15 are compared the densities obtained in 1D diffusion for the infinite case (non-perfect 1D system) and for a case with reflections (limited system). This case leads to a significant increase of the low-frequency spectral density.

Some 'ad hoc' models have been also proposed where the measured spectral densities are fit by non lorentzian expressions [35].

Finally another way proposed essentially for the interpretation of NMR data in protonic conductors, is to consider correlation time distribution and not a unique correlation time for describing the modulation in the spin system [36].

4. SOME RESULTS

A good analysis must involve the determination of several NMR parameters like the spectra and the different relaxation times. An example of such a complete analysis can be found in [37].

4.1 Spectra

The spectra are useful to identify the mobile ionic species and may lead to structural informationslike the occupation of some particular sites. I shall give three examples.

The first one concerns antimonic acid $H_2Sb_4O_{11}3H_2O$ [38] . Looking at the proton spectrum at low temperature (figure 16) one observes three resolved lines corresponding to three kinds of protons : those of water molecule , oxonium ions H_3O^+ and hydroxyl ions OH^- : when the temperature is raised the authors observed the narrowing of the linewidth characteristic of the occurence of protonic conduction.

The second example concerns the identification of the mobile ions in the pyrochlore $TlNb_2O_5F$. This problem can be resolved by observing the linewidth thermal behaviour of the different nuclei of the structure.

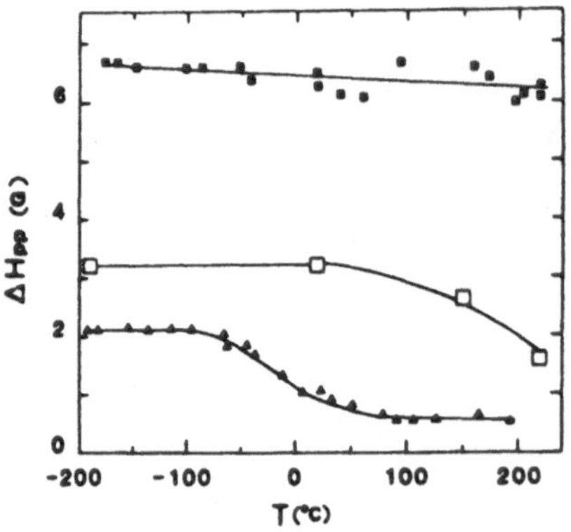

Figure 17. Thermal behaviour of the thallium isotopes ^{203}Tl, □ , and ^{205}Tl, ▲ , and of the fluorine ^{19}F, ■ , linewidths in the pyrochlore $TlNb_2O_5F$ [14].

On figure 17 are represented the results obtained for the three nuclei ^{203}Tl, ^{205}Tl and ^{19}F, all with spin I = 1/2 [14]. One can see the narrowing of the thallium linewidths while the fluorine linewidth is rather constant with temperature. The difference between the two thallium isotopes linewidths comes from the scalar interaction.

In the protonic conductor, niobic acid $HNbO_3$, the structure of which is represented figure 18 numerous sites are possible for the proton and the question was to determine if two protons could be linked to the same oxygen in the same time. A study of the second moment of the protonic line led us to the conclusion it was the case [15].

4.2 Relaxation times

I have represented figure 19 the 'cage' of the thallium ion in $TlNb_2O_5F$. From X-ray and neutron scattering experiments we knew that four types of sites were accessible for the thallium ions : 8b at the center of the cage, $32e_1$ and $32e_2$ inside the cage and 16d on the faces of the cage, at the entrance of the 3D conductivity channels. The figure 20 represents

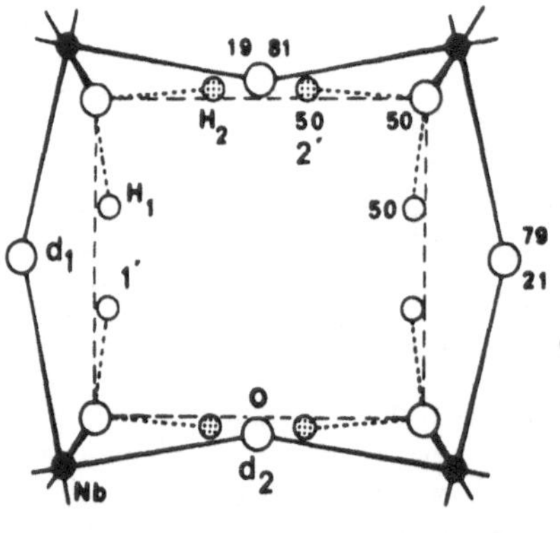

Figure 18. $HNbO_3$: positions of oxygen and hydrogen atoms in the xy 1/2 plane. From [15].

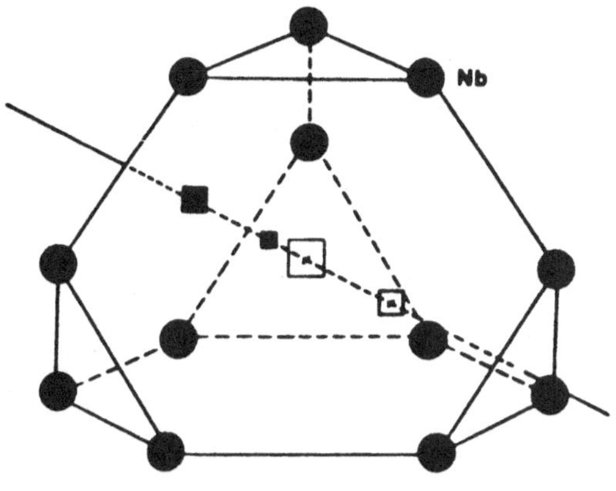

Figure 19. A cavity in the pyrochlore structure. The anions are not presented. 8b site : □, $32e_2$ site : □, $32e_1$ site : ■, 16d site : ■ From [14].

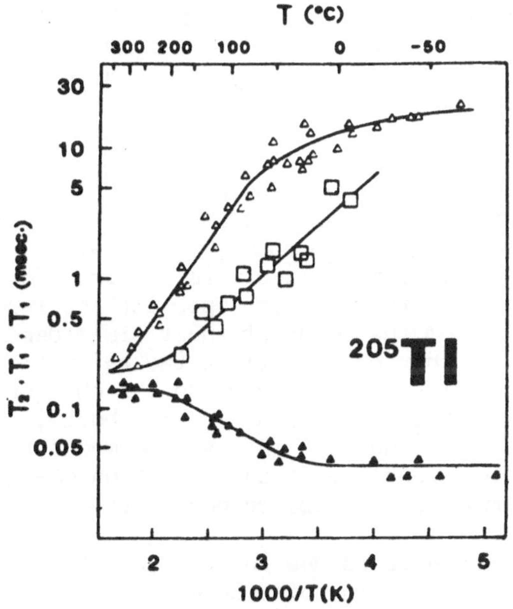

Figure 20

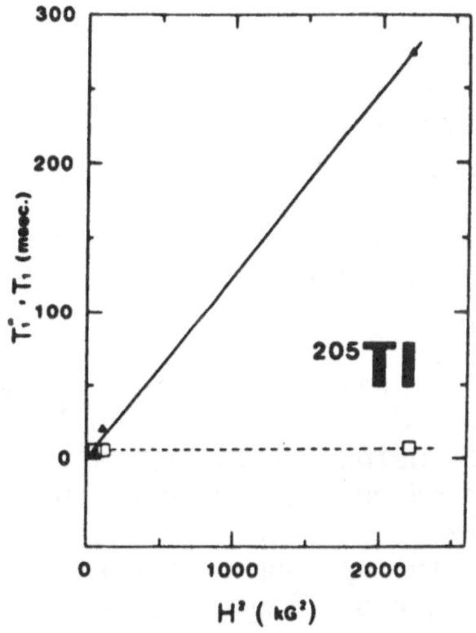

Figure 21

Figure 20. Thermal behaviour of the ^{205}Tl relaxation times in $TlNb_2O_5F$, T_2 : ▲ , T_1'short': □ , T_1'long': △ , showing the extreme narrowing [14].

Figure 21. Field dependence of the ^{205}Tl two components of the spin-lattice relaxation time : T_1'short' : □ , T_1'long' : ▲ , at room temperature [14].

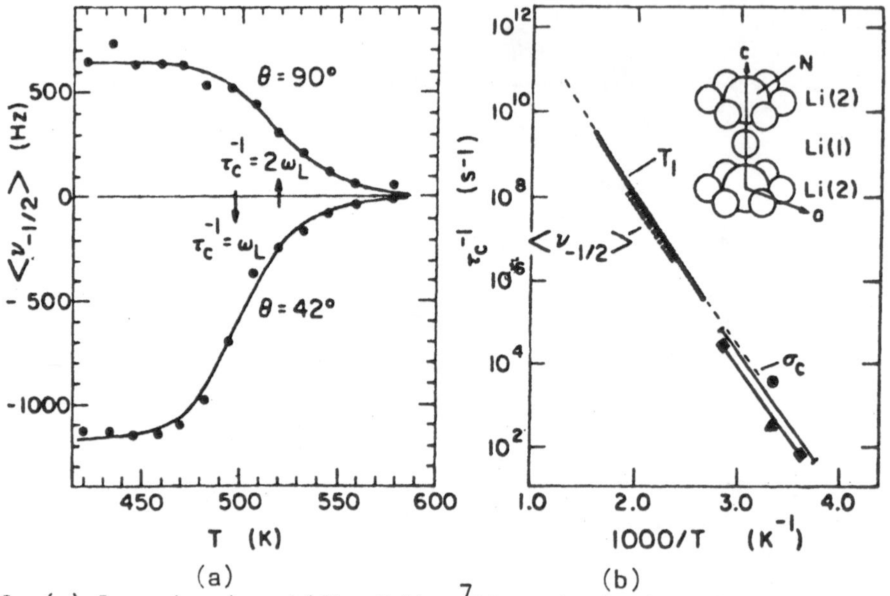

Figure 22. (a) Second-order shift of the ^7Li central line of Li_3N at two orientations [39] ; (b) temperature dependence of the correlation time for interlayer motion as obtained from second order line shift (xxx), T_1(●●●) and conductivity [17]. From [28].

the thermal behaviour of the spin-spin T_2 and spin-lattice T_1 'short' and T_1 'long' relaxation times of the ^{205}Tl nucleus and shows the extreme narrowing characteristic of conduction obtained near the highest temperature realizable in this experiment. Same results were obtained for the ^{203}Tl isotope. Two components were observed for the spin-lattice relaxation in the range : -20°C →300°C. From the study of their field dependence (figure 21) we concluded that they characterized the relaxation of two types of thallium ions feeling different interaction : a chemical shift interaction modulated by an intracage motion for T_1 'short', with an activation energy of 0.14 eV and an indirect interaction modulated by a diffusing intercage motion for T_1 'long', with an activation energy of 0.25 eV corresponding to the value obtained from conductivity measurements.

The spin-spin relaxation T_2 seems to be due to a third intercage motion requiring a very low activation energy of 0.08 eV [14].

Now I want to present results concerning nuclei with spin I > 1/2 i.e. experimenting quadrupolar interactions [28].

Brinkman et al. [39] made an extensive study of Li_3N through the resonance of the two lithium isotopes ^6Li (I = 1) and ^7Li (I = 3/2). The crystal is made up of layers of pure Li(1) and layers of Li(2)N and two types of lithium motion are present : an intralayer one and an interlayer.

They studied the interlayer one using two NMR parameters : the ^7Li relaxation time T_1, which is dominated by quadrupolar fluctuations due to Li(1)-Li(2) exchange along the c axis for temperature higher than

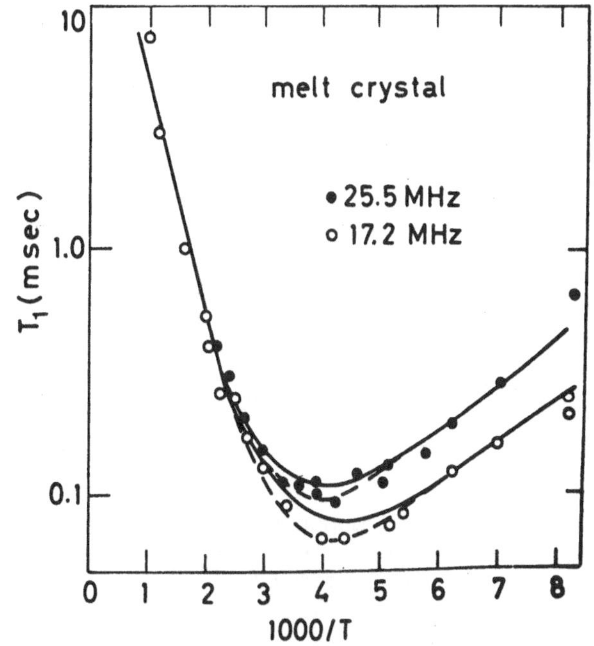

Figure 23. ^{23}Na T_1 versus 1/T in Na-β alumina. From [40].

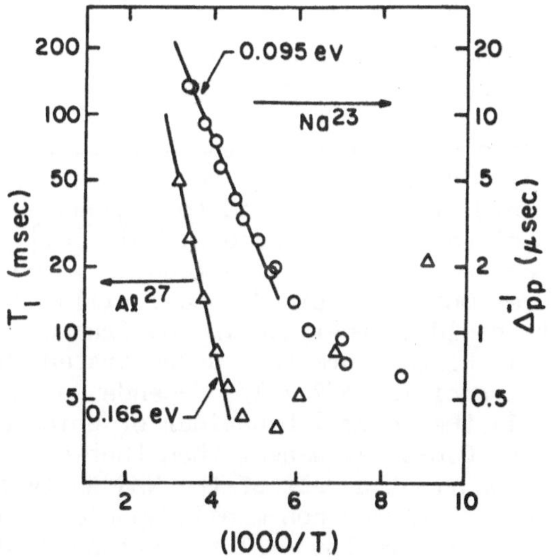

Figure 24. ^{27}Al T_1 (Δ) and ^{23}Na satellite linewidth (o) in β-alumina. From [25].

room temperature, and the ^7Li quadrupolar second order shift of the central line (see figure 11). As seen on figure 22a, the shift can be measured for different orientations of the single crystal in the magnetic

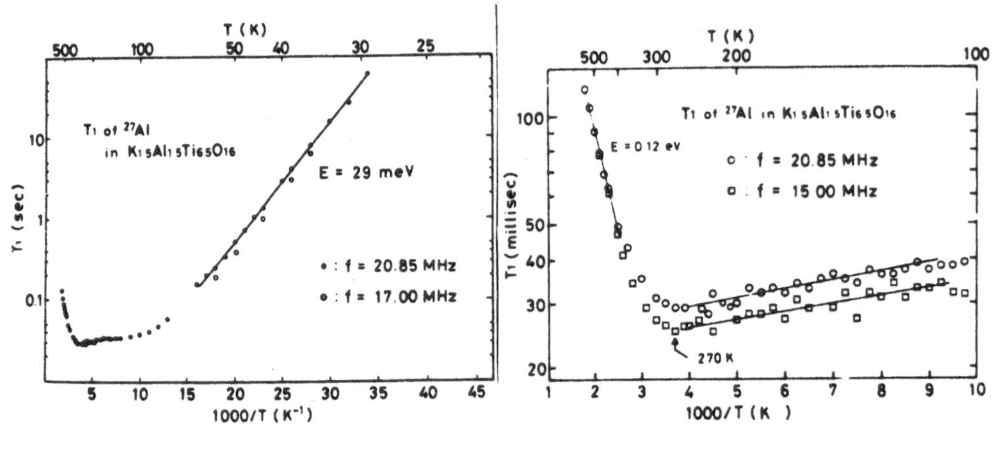

(a) (b)

Figure 25. Temperature and frequency dependence of T_1 of ^{27}Al in priderite : $K_{1.5}Al_{1.5}Ti_{6.5}O_{16}$; (a) the whole temperature range ; (b) the high — temperature region above 100K. From [43].

field.And from the thermal behaviour of the shift $\Delta\nu_{1/2}$ and of T_1 they obtained for the interlayer motion the same correlation time, which agrees perfectly with the one of conductivity measurements (figure 22b) [39], [17].

Compounds of the β-alumina family have been studied extensively, particular Na-β-alumina. On figure 23 is shown the thermal behaviour of ^{23}Na (I = 3/2) T_1 [40]. One can observe the asymetry of the curve and the non adequacy of the simple BPP theory. So the authors fit their data with a BPP model calling for a distribution of activation energies. (On figure 23 the solid and dashed lines correspond to a gaussian and a lorentzian distribution respectively). Nevertheless, Wolf [41] pointed out that it might be inconsistent with conductivity results which give a single activation energy over more than six decades. Villa and Bjorkstam have shown that the results could be analyzed in the framework of the 2D continuum diffusion model. Furthermore they demonstrated that the linewidth of the satellite transition 3/2 → 1/2 depends on the $a(\omega)$ spectral density of figure 14 the thermal behaviour of which gives an apparent activation energy of 0.095 eV, weaker than that of diffusion. Finally, they underline an important aspect of the NMR analysis of superionic conductors : the resonance of the non mobile species can reveal directly the diffusion process : here the ^{27}Al relaxation is driven by the sodium diffusion and the thermal behaviour of T_1 leads to the activation energy of this process (see figure 24).

In the same way the motion of K^+ ion in the 1D ionic conductor Al-priderite was investigated through the ^{27}Al NMR [43]. On figure 25 are reported the T_1 results versus temperature. The low-dimensional aspect of the diffusion is reflected in the asymetry of the curve which can be fitted by an $\omega^{1,5}$ law at low temperature and an $\omega^{0,4}$ law for higher temperature, leading to a very weak apparent activation energy.

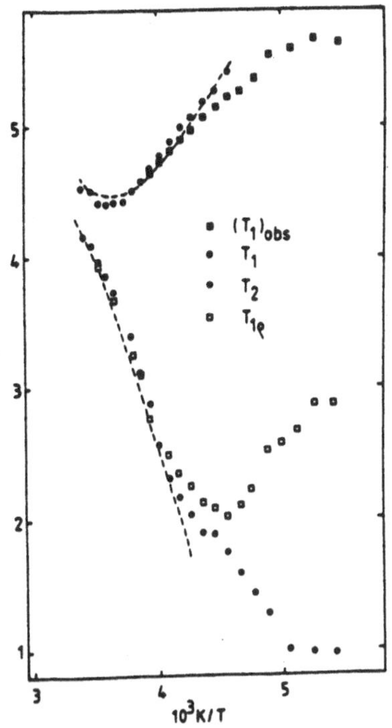

Figure 26. ^{1}H NMR relaxation times (60 MHz) versus 1/T for 12-Tungstophosphoric acid (TPA)$H_3PW_{12}O_{40}21H_2O$ The dashed lines correspond to the calculated fit to T1 and T2. See [44].

The last result concern a family of protonic conductors like the 12-tungstophosphoric acid [44] . On figure 26 are reported the temperature dependence of the ^{1}H relaxation times. It has been analyzed through a distribution of translational correlation times, an approach which has been used with some success in study of sorbed species. Figure 27 shows the similitude of the correlation time distributions obtained here and in the case of water adsorbed on charcoal or zeolites.

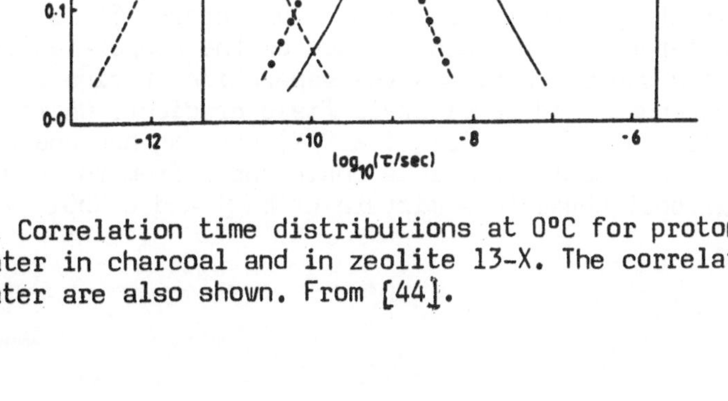

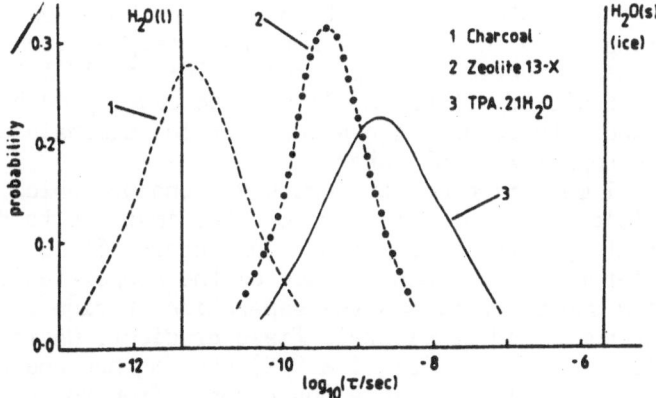

Figure 27. Correlation time distributions at 0°C for protons in TPA. 21 H_2O, water in charcoal and in zeolite 13-X. The correlation times for ice and water are also shown. From [44].

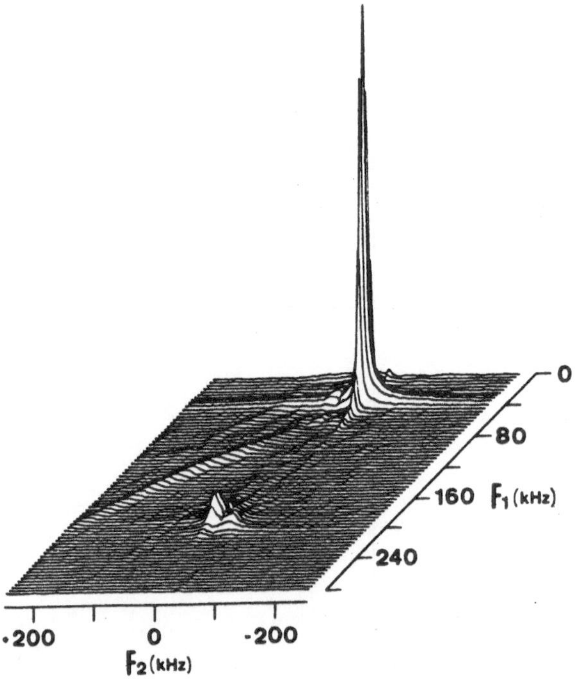

Figure 28. ^{93}Nb (I=9/2) 2-dimensional Fourier Transform NMR spectrum for a mixture of NbO_2F (large peak) and $LiNbO_3$ (small peak). From [48].

5. CONCLUSION

As said in the introduction, the whole information about the structure and the dynamics of a material is contained in the NMR response of a spin system belonging to the material but the interpretation of the response is not always obvious. To progress in this field I think one needs ab initio models of diffusion, like those developped by Korb [45], [46], [47], taking into account the aspects of low-dimensionality and boundaries which appear in real materials.

Finally I want to present a new NMR technique which will be useful in the field of solid state physics, for instance to distinguish sites of different symetry in superionic conductors. It is a 2-dimensional NMR Fourier transform method based on the properties of quadrupolar interaction and which allows the separation of signals of nuclei submitted or not submitted to electric field gradient. On figure 2 are presented results for ^{93}Nb nuclei (I = 9/2) showing how one can separate the signals of the niobium in a cubic phase from the signal of niobium in an hexagonal phase in a mixture of NbO_2F and $LiNbO_3$ [48].

REFERENCES

1. NMR and the periodic table edited by R.K. Harris and B.E. Mann, Academic Press (1978).
2. J.B. Boyce and B.A. Huberman, Physics Reports 5 1, 189 (1979).
3. M. Faraday, Experimental researches in electricity, Taylor and Francis, London (1839).
4. P.M. Richards,'Magnetic Resonance in superionic conductors' in Physics of superionic conductors, Topics in current physics, edited by M.B. Salamon, Springer Verlag, chapter 6, 141 (1979).
5. H. Schulz,'Relation between crystal structures and ionic conductivity', in The physics of superionic conductors and electrode materials, édited by J.W. Perram, NATO ASI Series, B 9 2, 5 (1983).
6. P. Mac Geehin and A. Hooper, 'Review fast ion conduction materials', J. of Mat. Science, 12, 1 (1977).
7. K. Shahi, J.B. Wagner and B.B. Owens, 'Solid electrolyte lithium cells' in Lithium batteries, edited by J.P. Gabano, Academic Press, Chapter 15, 407 (1983).
8. Proceedings of the 5th International conference on solid state ionics, Lake Tahoe, USA, August 18-24 (1985).
9. J.L. Fourquet, M. Rousseau and R. de Pape, Mat. Res. Bull.14, 937 (1979).
10. Solid state protonic conductors for fuel cells and sensors - I. edited by J. Jensen and M. Kleitz, Odense University Press (1982).
11. Solid state protonic conductors for fuel cells and sensors - III. edited by J.B. Goodenough, I. Jensen and A. Potier, Odense University Press (1985).
12. P. Colomban and A. Novack, 'Hydrogen containing β and β" alumina' in 10 page 153.
13. A. Potier, 'Quels supraconducteurs protoniques pour une microionique' in De la microélectronique à la microionique, édited by P. Colomban Colloque, 3-4 March 1983, Palaiseau, p. 149.
14. P.P. Man, H. Theveneau, P. Papon and J.L. Fourquet, 'NMR study of the thallium ions mobility in the pyrochlore $TlNb_2O_5F$', J. of Chem. Phys. 80 (3), 1272 (1984).
15. J.L. Fourquet, M.F. Renou, R. de Pape, H. Théveneau, P.P. Man, O. Lucas and J. Pannetier, '$HNbO_3$: structure and NMR study', Solid state ionics 9 and 10, 1011 (1983).
16. R.A. Huggins, Electrochim. Acta 22, 773 (1977).
17. U. Valpen, A. Rabneau and G.H. Talat, Appl. Phys. Lett. 30, 621 (1977).
18. W.C. Bailey, H.S. Story, A.R. Ochadlick Jr, G.C. Farrington, 'NMR study of sodium ion motion in β" alumina' in Fast ion transport in solids electrodes and electrolytes, edited by P. Vashishta, J.N. Mundy and G.K. Shenoy, Elsevier North Holland (1979), p. 281.
19. U.V. Alpen and M.F. Bell, 'Electrochemical properties of some solid lithium electrolytes" in the book of 18 p. 463.
20. R.M. Dell and R.J. Bones, 'Sodium/sulphur traction batteries : a review of progress and problems in their development', in the book of 18, p. 29.
21. C.C. Liang and L.H. Barnette, J. Electrochem. Soc. 123,453 (1976).
22. C. Velasco, J.P. Schnell, N. Nouailles and M. Croset in 10, page 27.

23. W.C. Dautremont-Smith, G. Beni, L.M. Schiavone and J.L. Shay, in the book of 18, p. 99.
24. P. Colomban and G. Velasco, La Recherche, 148, 1292 (1983) and P. Colomban in the book of 13 p. 3.
25. J.L. Bjorkstam and M. Villa, 'NMR and EPR studies of superionic conductors', Magnetic Resonance Review, 6 (1), 1 (1980).
26. D.C. Ailion, 'NMR techniques for studying ionic diffusion in solids' in Nuclear and electron spectroscopies applied to Material science, edited by Kaufmann and Shensy, Elsevier North Holland (1981) p. 55.
27. H. Arribart, 'NMR in solids showing fast proton diffusion' in 10 p. 111.
28. J.L. Bjorkstam,'First and second order effects in NMR spectral averaging', J. of Molecular Structure 111, 135 (1983).
29. R.K. Harris, Nuclear magnetic resonance spectroscopy, Pitman Book limited, London (1983).
30. U. Haeberlen, High resolution NMR in solids - Selective averaging, Advances in Magnetic resonance, Suppl. 1, Academic Press (1976).
31. H.S. Story, W.C. Bailey, I. Chung and W.L. Roth, 'NMR studies of superionic compounds, primarily beta-alumina' in Superionic conductors, edited by G.D. Mahan and W.L. Roth, Plenum Press (1976) p. 317.
32. D. Wolf, Spin temperature and nuclear spin relaxation in matter, International series of monographs in physics, Clarendon Press, Oxford (1979), chapter 12.
33. M. Villa, 'NMR studies of fast ion conductors', Proceedings of NATO Summer School, Erice (1982).
34. N. Bloembergen, E.M. Purcell and R.V. Pound, Phys. Rev. 73, 679 (1948).
35. M. Villa and J.L. Bjorkstam, 'Prefactor anomalies', Solid State Ionics, 9 an 10, 1421 (1983).
36. R.C.T. Slade, Solid State Comm. 54, 12, 1035 (1985).
37. T.C. Jones and T.K. Halstead, J. of less-common metals, 73,209 (1980).
38. Y. Piffard, H. Arribart and C. Doremieux-Morin, in ref.9 page 247.
39. D. Brinkmann, M. Mali, J. Ross, R. Messer and H. Birli, Phys. Rev.B, 26,9,4810(1982).
40. R.E. Walstedt, R. Dupree, J.P. Remeika and A. Rodriguez, Phys. Rev.B, 15,3442 (1977).
41. D. Wolf, Bull. Am. Phys. Soc. 24, 351 (1979).
42. M. Villa, J.L. Bjorkstam and S. Manzini, in the book of 18, p. 289, 293.
43. Y. Onoda, Y. Fujiki, S. Yoshikado, T. Okachi and I. Taniguchi, Solid State Ionics, 9 and 10, 1311 (1983).
44. A. Hardwick, P.G. Dickens and R.C.T. Slade, Solid State Ionics, 13, 345 (1984).
45. J.P. Korb, D.C. Torney and H.M. Mac Connell, J. Chem. Phys., 78, 9, 5782 (1983).
46. J.P. Korb, M. Winterhalter and H.M. Mac Connell, J. Chem. Phys. 80, 3, 1059 (1984).
47. J.P. Korb, 'Theory of spin relaxation by translational diffusion in two-dimensional systems' in Structure and Dynamics of molecular systems, edited by R. Daudel, J.P. Korb, J.P. Lemaistre and J. Maruani, Reidel, Dordrecht, volume I, April (1985) p. 245.
48. P.P. Man, H. Théveneau and P. Papon, J. Mag. Res., 64 (1985).

THEORETICAL INVESTIGATION OF METAL HYDRIDES : ELECTRONIC PROPERTIES AND SUPERCONDUCTIVITY

M. GUPTA
Centre de Mécanique Ondulatoire Appliquée du C.N.R.S.
23, rue du Maroc, 75940 PARIS CEDEX 19 — FRANCE
and Bât. 506, 91405 ORSAY-FRANCE

ABSTRACT. The electronic structure of stoichiometric metal hydrides has been investigated by means of the augmented plane wave band structure method and detailed determinations of the densities of states, wave-function analysis and studies of the properties at the Fermi level. The hydrides under study are on one hand binary compounds of transition and rare earth metals and on the other hand ternary intermetallic hydrides which have potential technological applications. Several experimental data such as electronic specific heat and magnetic susceptibility measurements, metal-insulator transitions, densities of occupied states observed by photoemission or X-ray emission etc... are interpreted in light of the present results. A discussion of the nature of the bonding properties and of the role of the factors which control the stability of metal hydrides is also given. We also tried to understand why some metal hydrides are superconductors while hydrogen destroys superconductivity in some other metals. Our analysis is based on a theoretical evaluation of the electronic contribution of the electron-phonon coupling constant in conjunction with an estimation of the phonon contribution from available experimental data.

1. INTRODUCTION

Theoretical and experimental studies of hydrogen in metal have been for several decades an active field of research for chemists and physicists. Fundamental aspects as well as various technological applications have been thoroughly investigated.(1-4)

For theorists a broad spectrum of problems can be studied since hydride forming metals exist over a wide range of hydrogen concentrations from the very dilute limit up to concentrated stoichiometric compounds, the phase diagrams being often quite complex. We shall leave aside here the problem of hydrogen <u>on</u> metal surfaces which is of course of crucial importance for the understanding of the mechanism of hydride formation and restrict this paper to the review of some theoretical aspects of hydrogen <u>in</u> metals. The first theoretical investigations started in the early 1950's with the pionneering work of Friedel (5) in which he studied the

R. Daudel et al. (eds.), Structure and Dynamics of Molecular Systems – II, 255–288.

screening of a hydrogen impurity and the heats of solution of H in noble metals. More recently, with the development of the density functional formalism[6] and the availability of computers, self-consistent studies of the non-linear screening of a proton in an electron gas including exchange and correlations have been developped in several groups[7-12]. The proper treatment of the screening of the proton together with an adequate description of the metal lattice ions play an important role in the study of the heat of solution of H in metals, in the determination of the prefered site occupancy of the H atoms, in the characterization of the diffusion paths and in the estimation of the activation energies for the overbarrier jump processes involved in the classical diffusion mechanisms, in the calculation of the frequencies of the local mode vibrations of H etc... The problem of the strain field around the H impurity and the subsequent metal lattice relaxation, the questions of possible vacancy trapping of H atoms in simple metals and the H-H or H-impurity interactions have also been considered.[7-12] However, even in simple metals, due to the approximate treatment of the metal ion cores, no general consensus has yet been reached among theorists to give quantitative answers to the problems mentioned above. We wish to point out here that, besides their fundamental importance, these theoretical approaches are of great relevance for their applications in the field of metallurgy. Theoretical investigations of H as an impurity in transition metals remain a challenge for theorists due to the difficulty in the treatment of the d electrons. Semi empirical calculations using the tight-binding method[13] and ab-initio supercell[14,15], molecular cluster [16-18] and Korringa-Kohn-Rostoker (KKR) scattering theory calculations [19-20] have however been performed.

At the other end of the spectrum, in the high hydrogen concentration range theoretical investigations of the electronic structure of stoichiometric binary hydrides were initiated in the early 1970's by Switendick [21] and later pursued in several groups.[22-27] This theoretical effort has led in most cases to a satisfactory interpretation of a large number of experimental data [3] on the physical properties of binary hydrides such as resistivity, magnetic susceptibility, specific heat, Fermi surfaces, superconductivity measurements etc... as well as to a successful interpretation of various photoemission, X ray emission and absorption spectra. Up to recently, however, theoretical investigation of ternary hydrides have been lacking due to the complexity of their crystal structures Since these hydrides have numerous potential applications [3,4] such as in chemical heat pumps, hydrogen burning engines, electrodes in fuel cells, hydrogen storage etc... it appears important to broaden our fundamental knowledge of their electronic properties.

In the remainder of this paper I shall present some aspects of the electronic properties of stoichiometric binary and ternary hydrides of transition and rare earth metals. Section 2 is devoted to a brief description of the theoretical method which I used to study the band structure, wavefunctions and densities of states of metal hydrides as well as the theoretical approach used to evaluate the electron phonon coupling constant which plays a crucial role in the occurrence of superconductivity. In Section 3, I summarize the most important results obtained for binary transition metals (TM) and rare earth (RE) metals. The problem of the ocurrence of superconductivity in some metal hydrides is discussed in section 4 ; our approach is based on a theoretical evaluation of the electron-phonon coupling in these materials and on an estimation of the phonon properties from available experimental work. Section 5 is devoted to a presentation of our most recent results obtained for some technologically important ternary hydrides. A summary and concluding remarks are given in Section 6.

2. FORMALISMS INVOLVED IN THE THEORETICAL STUDIES

2.1. Band Structure

The method which has been mostly used for studying the band structure of stoichiometric hydrides [21-27] is the augmented plane wave [29] (APW) method in which the Bloch functions are expanded in a set of energy dependent partial waves. The crystal potential is assumed to have a muffin-tin (MT) shape : it has a spherical symmetry inside touching spheres and is constant in the remaining interstitial region. In the present work on hydrides we introduced however the so-called warping corrections to account for the deviations of the interstitial potential from a constant value. Inside the MT spheres where the potential is assumed to be spherically symmetric the wavefunctions are expanded in terms of energy dependent partial waves χ :

$$\Psi(E,\vec{r}) = \sum_{i,\ell,m} c_{i,\ell,m} \; \chi_{i\ell m} \; (E,\vec{r}_i) \; , \qquad (1)$$

$$\chi_{i\ell m} \; (E,\vec{r}_i) = \phi_{i\ell}(E,r_i) \; y_{\ell m}(\hat{r}_i) \; , \qquad (2)$$

where $y_{\ell m}(\hat{r}_i)$ are spherical harmonics and the radial function $\phi_{i\ell}$ is the solution of the radial Schrödinger equation

$$[r \; \phi_{i\ell} \; (E,r)]'' - [\ell(\ell+1) \; / \; r^2 + V_i(r) - E] r\phi_{i\ell}(E,r) = 0 \qquad (3)$$

The one electron energies in Eq. (3) are the values of E for which expansion coefficients c can be found so that the matching condition of the partial wave expansion at the frontier between the regions is fulfilled. These coefficients are obtained by solving a set of linear homogeneous equations

$$\underset{\sim}{M} \; (E) \; \underset{\sim}{c} = 0 \qquad (4)$$

Here, unlike in the usual Rayleigh-Ritz eigenvalue eigenvector problem, the matrix elements are energy dependent. This results in a diagonalization procedure which is not straithforward In the APW method, all the informations concerning the crystal potential enter the matrix elements of Eq. (4) through the energy dependent logarithmic derivatives of the radial functions evaluated at the MT radius R_i

$$D_{i\ell}(E) = R_i \; \phi'_{i\ell}(E,R_i) \; / \; \phi_{i\ell}(E,R_i) \qquad (5)$$

For hydrides with fcc lattices, I calculated the energy eigenvalues $E(\vec{k})$ and wavefuntions at 89 $\vec{k}$ points in the 1/48th wedge of the irreducible Brillouin Zone (BZ). The energy eigenvalues have then been interpolated using symmetrized plane wave expansions. For the calculation of the densities of states (DOS), the BZ integration was performed by an accurate analytical method the linear-energy tetrahedron method [30] in which the BZ is divided into joining microtetrahedra, typically of the order of 6000 tetrahedra in the f.c.c.BZ; a linear interpolation of the energies was performed inside each microzone. Accuracies of the order of a few n Ry are typically obtained. Decompositions of the DOS into site and angular momentum characters inside the MT metal and hydrogen spheres have also been calculated.

2.2. Electron-phonon coupling

Since some metal hydrides are superconductors we have evaluated the electron phonon coupling constant λ which essentially determines the superconducting critical temperature T_c. Following the work of McMillan [31] for transition metals, an approximate expression of λ has been proposed for compounds with a large mass ratio between the constituent atoms

$$\lambda \sim \lambda_{metal} + \lambda_H = \frac{\eta_{metal}}{M_{metal} <\omega^2>_{acoustic}} + \frac{\eta_H}{M_H <\omega^2>_{optic}} \qquad (6)$$

where η is the ' electronic contribution', M the atomic mass, and $<\omega^2>$ the second moment of the renormalized phonon frequencies defined by McMillan. Using the rigid ion approximation in which the renormalization effects accompanying the lattice vibrations are ignored, Gaspari and Gyorffy [32] have shown that the mean square of the electron phonon matrix element can be conveniently expressed in terms of quantities obtained from ab-initio band structure calculation. They have shown that

$$\eta \sim \sum_k \frac{E_F}{N\uparrow(E_F)\pi^2} \sum_\ell 2(\ell+1)\sin^2(\delta^K_{\ell+1} - \delta^K_\ell) \frac{n^K_\ell(E_F)n^K_{\ell+1}(E_F)}{n^{K(1)}_\ell(E_F)n^{K(1)}_{\ell+1}(E_F)} \qquad (7)$$

where the summation on k runs on all the atoms in the unit cell, δ^K_ℓ is the single-site scatterer phase shift at the Fermi energy E_F, n^K_ℓ the partial DOS of angular character ℓ at site K, and $n^{K(1)}_\ell$ the corresponding partial DOS of a free scatterer.

In the present work we have thus calculated the electronic contributions η_{metal} and η_H from our band structure results while the phonon contribution appearing in the denominators of Eq. 6 have been estimated from available experimental data.

3. ELECTRONIC PROPERTIES OF BINARY HYDRIDES

As shown in Table I, concentrated hydrides form with the TM of the end of the periodic Table such as Pd and Ni which form monohydrides and

ScH_2	TiH_2	VH VH_2	CrH	(Mn) --	(Fe) --	(Co) --	NiH
YH_2 YH_3	ZrH_2	NbH NbH_2	(Mo) --	(Tc) --	(Ru) --	(Rh) --	PdH
See Rare Earth Series	HfH_2	TaH	(W) -	(Re) --	(Os) --	(Ir) --	(Pt) --

LaH_{2-3}	CeH_{2-3}	PrH_{2-3}	NdH_{2-3}	Pm ?	SmH_2 SmH_3	EuH_2	GdH_2 GdH_3	TbH_2 TbH_3	DyH_2 DyH_3	HoH_2 HoH_3	ErH_2 ErH_3	TmH_2 TmH_3	YbH_2 YbH_3(?)	LuH_2 LuH_3

Table 1. The occurence of binary hydrides from Ref.21.

with the early members of the TM series which form dihydrides. The lanthanides form also dihydrides ; trivalent RE metals and yttrium can absorb even larger amounts of hydrogen and form trihydrides.

3.1. Monohydrides PdH and NiH

The monohydrides PdH and NiH crystallize within the NaCℓ structure. The hydrogen uptake is not accompanied by a structural change in the metal lattice which remains fcc but only by an increase in the volume ; the relative expansion of the metal lattice parameter $\frac{\Delta a}{a}$ is of the order of 5% in the case of Pd. The H atoms occupy the octahedral interstices of the metal fcc lattice.

The energy bands znd DOS of pure Pd or Ni are characterized by narrow metal d states overlapped and hybridized with a wider metal s-p band. As shown in Figs 1 and 2 for PdH (22b) and NiH (22d), the electronic structure of a monohydride (21a,b 22b,d 23a,d) can be characterized essentially i) by the presence of a low-lying band below the metal d states, ii) by an important difference in the Fermi level position which in the pure metals falls just below the top of the bands while in the hydrides, the metal d bands are filled and the Fermi level cuts only one band, the metal s-p band as in Cu or Ag.

The partial density of states analysis inside the Muffin-Tin spheres into their s,p,d components at the metal and at the H site plotted in Fig. 3 shows that the low lying states are formed of a bonding combination of H-S and metal-d and s-p states this feature is characteristic of the formation of the compound. Spectroscopic techniques such as photoemission (33) and X-ray emission data (34) reveal clearly the presence of these states upon formation of a monohydride and are in good agreement with the theoretical predictions.

A detailed analysis shows that in the monohydrides the bonding structure observed in the occupied DOS at low energy is formed mostly out of states already filled in the metal ; in the case of PdH, one estimates

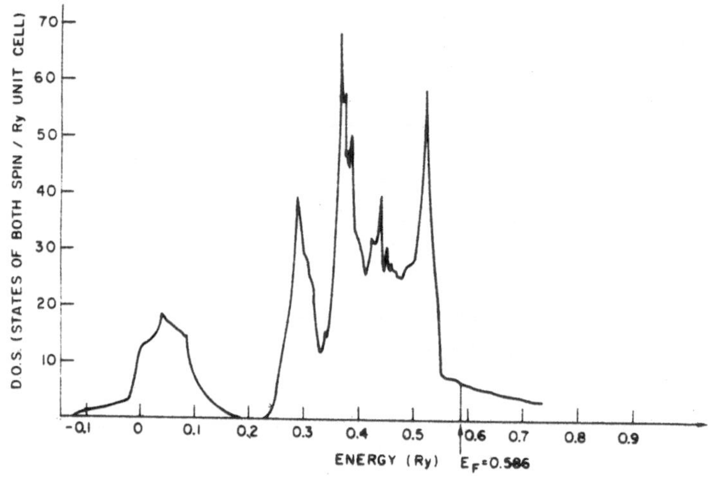

Figure 1. Total DOS of PdH from ref. 22-b.

to only 0.25 the number of additional electrons in this structure. This explains that the Fermi level of the hydride is necessarily higher than in the pure metal since the bands need to accomodate the additional electron brought in the unit cell by the H atom. The value of the DOS at E_F of the monohydrides PdH and NiH is much lower than in the pure metals. For PdH $N(E_F) = 6{,}81$ states / Ry unit cell compared to 28.9 states/ Ry unit cell in pure Pd for which E_F falls close to a peak of the DOS just below the top of the d bands. This feature explains the drastic reduction of the electronic specific heat coefficient from $\gamma = 9.48$ in J. Mole^{-1} K^{-2} in pure Pd [35] to $\gamma = 1{,}54 \pm 0.13$ in J.mole^{-1} K^{-2} in PdH_x [36] for x = 0.876 as well as the decrease of the Pauli susceptibility X_p [37].

The shape of the Fermi surface of PdH and NiH bears similarities with that of the noble metals ; however the Fermi level of the monohydrides is closer to the top of the d bands than that of the noble metals. The main difference between the Fermi level properties of PdH and NiH stem from the fact that pure metal Ni has more holes in its d bands than pure Pd and consequently the Fermi level of NiH is closer to the top of the d bands ; the states at E_F have thus a larger metal d and a smaller H-s character in NiH than in PdH. This difference plays a role in the value of the matrix elements of the electron-phonon interaction and thus on the superconducting properties as explained in section 4.

3.2. Transition-metal and Rare-earth dihydrides and trihydrides

The early TM and RE dihydrides cristallize within the CaF_2 structure. The metal atoms form an fcc lattice while the H atoms are located in tetrahedral interstices of the metal lattice ; the H atoms form a cubic array and are located at the 1/4 (111) positions. Besides the lattice

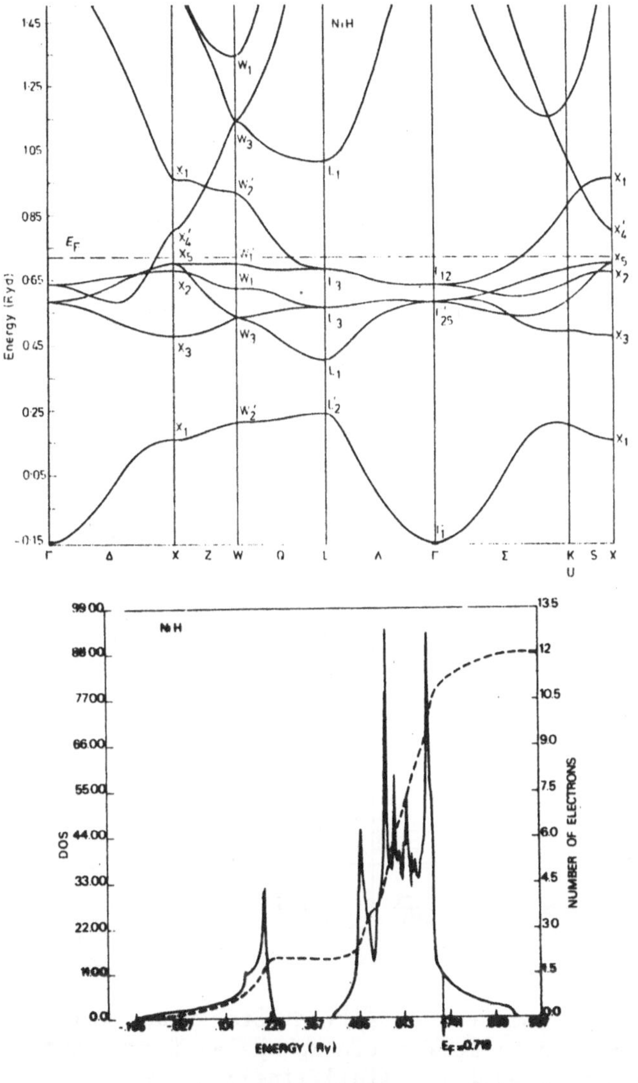

Figure 2. The energy bands of NiH (upper graph) along several high symmetry directions. The Fermi energy E_F is represented by a broken line. In the lower graph : the total DOS of NiH (full line, left hand side scale) units are states of both spin per Rydberg unit cell ; the number of electrons (dashed line right hand side scale). From Ref. 22-d.

expansion the formation of the dihydrides of the early TM is thus accompanied by a structural change in the metal lattice from h.c.p. (for Sc, Ti,.) or b.c.c. (V,Nb) to f.c.c.
As shown in Figs. 4 and 5, the electronic structure of the dihydrides is characterized by the presence of two low-lying bands below the metal states ; in addition to the lowest band which is also observed in the

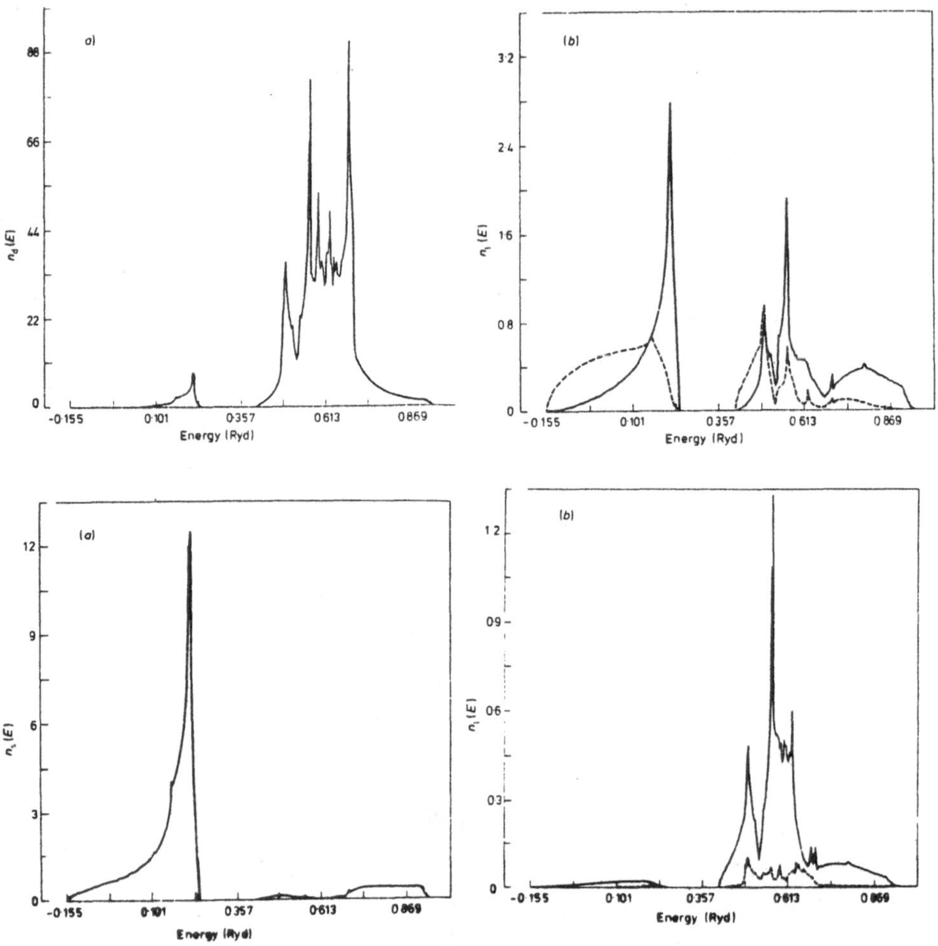

Figure 3. The angular momentum density of states analysis n_ℓ of NiH: upper graphs inside the Ni muffin-tin sphere a) $\ell = 2$ contribution b) $\ell = 0$ (broken line) and $\ell = 1$ (full line); lower graphs inside the H muffin-tin sphere a) $\ell = 0$ contribution b) $\ell = 1$ (full curve) and $\ell = 2$ (broken curve). Units are states of both spin per Rydberg unit cell.

case of the monohydrides, a second band appears which is formed at the BZ center from an antibonding combination of the two H-s states. The presence of this second additional Γ_2' state at low energy is an important factor for the stability of the compound. Since Γ_2' is an antibonding combination of the two H-1s wavefunctions, its position in energy depends sensitively upon the H-H distance. Thus, as seen in Figs 4 and 5, for the TM hydrides, there is an overlap between the antibonding

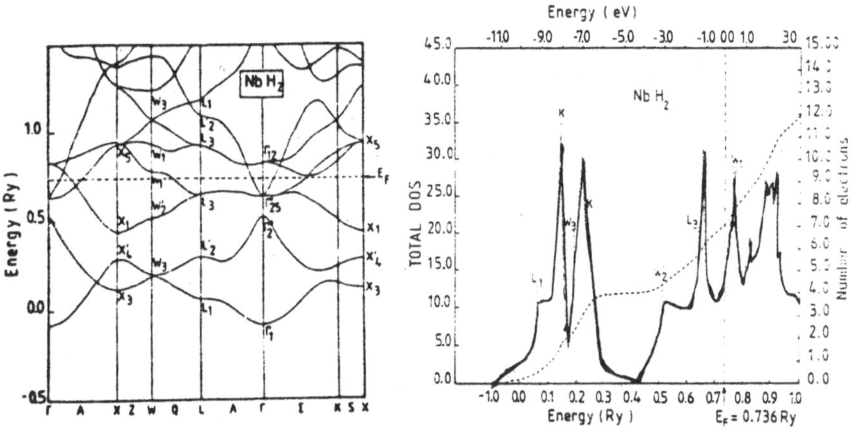

Figure 4. The energy bands of NbH_2 (left graph). Right graph : the total DOS of NbH_2 (full line, left-hand side scale) from Ref. 22-h. Units are states of both spin per Rydberg unit cell. The number of electrons (dashed line, right-hand side scale).

H-H band and the metal d bands located at higher energies whereas in LaH_2 which has a much larger lattice constant the antibonding band is more stable and is separated by a gap from the metal d bands. This feature is certainly connected to the larger stability of the RE dihydrides and can also explain why the dihydrides do not form with the TM on the right of column V. As we go from the left to the right of the TM series in the Periodic table, the lattice parameter of the pure metals contracts consequently, the H-H distances decrease, the antibonding band is destabilized and the formation of the dihydride becomes energetically unfavorable. We want to point out however that the position of the antibonding H-H band is related not only to the H-H distance which appears to be in the hydrides always larger than about 2Å, but also to the relative position of the metal s-p band and the d bands in the pure metal which is related to the position of the metal atom in the Periodic Table. Thus, for example, a calculation of fictitious PdH_2 even when assuming a large lattice constant does not lead to a stabilization of the antibonding H-H band below the metal d states [38].

For dihydrides, the structure observed in the DOS below the metal d bands which, as explained above results from metal-hydrogen and H-H bonding and from H-H antibonding interactions has been observed in photoemission experiments [39,40]. As shown in Fig. 6 in the case of LaH_2, the width of the La d bands as well as the two peak structure at low energy observed experimentally are in general good agreement with our theoretical work.

For dihydrides since two bands are formed below the metal d states they accomodate two electrons, thus if Z is the valency of the metal atom, only Z-2 electrons occupy the bottom of the metal d bands. It thus

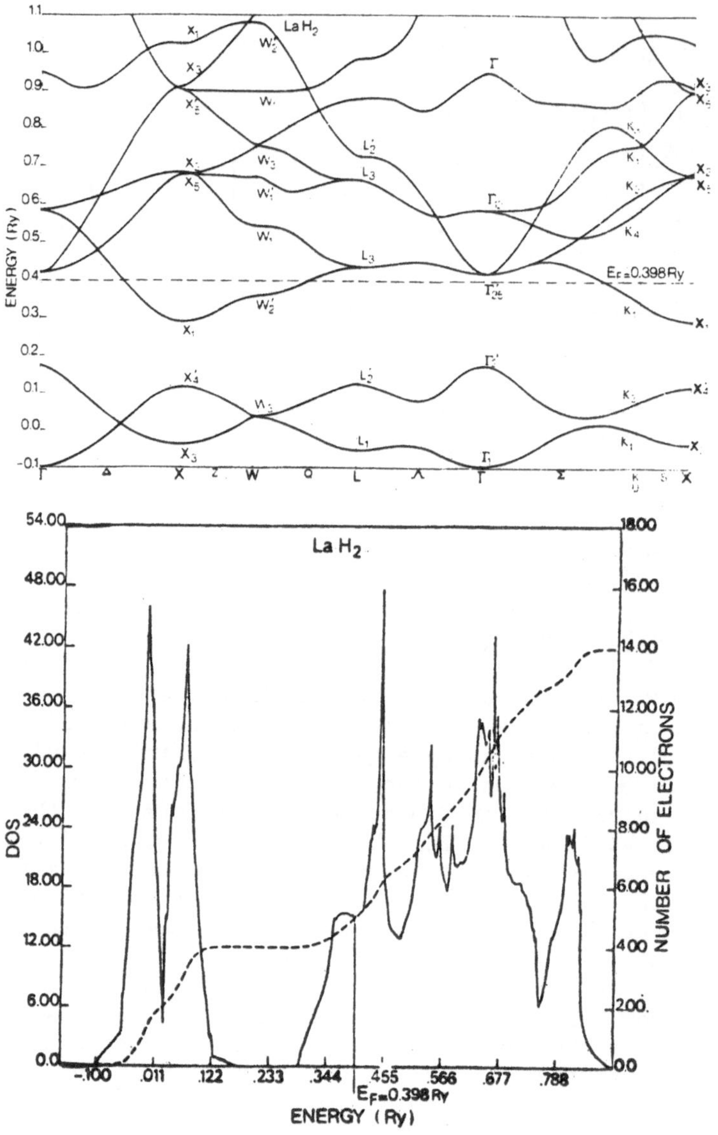

Figure 5. Energy bands of LaH_2 (upper graph). Lower graph : the total DOS of LaH_2 (full line, left-hand side scale) from Ref. 22-i. Units are states of both spin per Rydberg unit cell. The total number of electrons (dashed line, right-hand side scale).

appears, at first sight that the metal d bands have been depopulated in favour of the low-lying bands. This statement has however to be taken with caution since, as shown by the partial DOS analysis of Fig. 7, a substantial portion of the metal d bands has been deformed and shifted to lower energies by the M-H interaction and forms an important part of the low-lying bands. In view of this important deformation of

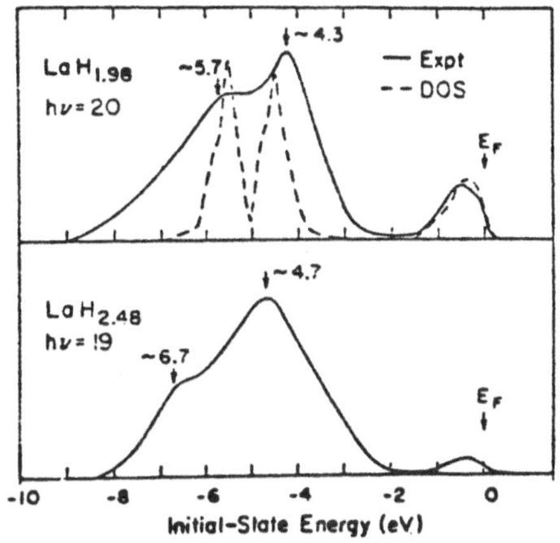

Figure 6 . Photoemission spectra of Weaver et al. from Ref. 39 for two hydrogen concentrations in La. The dashed curve is the theoretical DOS of Ref. 22-d.

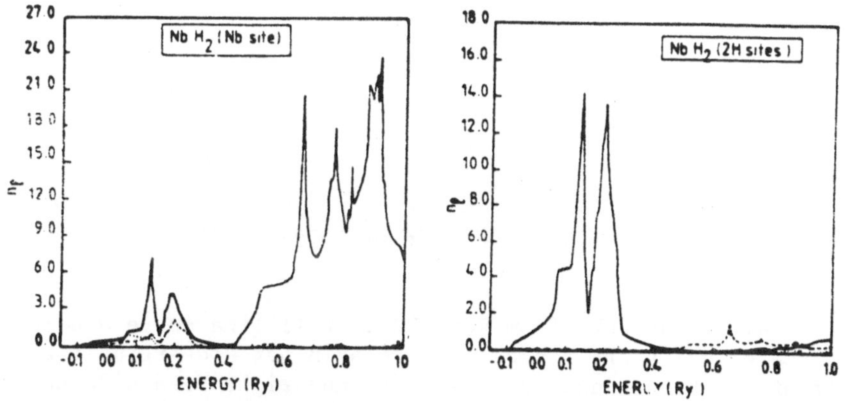

Figure 7. The angular momentum DOS analysis of NbH_2, n_ℓ. Left graph : inside the muffin-tin Nb sphere ; ℓ = 0 (dotted curve) ; ℓ = 1 (dashed curve), and ℓ = 2 (full line) contributions.
Right graph : inside the two H muffin-tin spheres ; ℓ = 0 (full line) ; ℓ = 1 (dashed line), and ℓ = 2 (dotted line) contributions.
Units are states of both spin per Rydberg unit cell.

the metal d bands in the hydride, the appearent depopulation of the metal d states should not thus be understood in a rigid band model sense. In the dihydrides of the group III TM and of the trivalent RE, only one electron is accommodated at the bottom of the metal d bands, while there are two and three electrons respectively in the case of the dihydrides of the group IV and V TM. The divalent RE dihydrides are semiconductors since there is a gap between the filled two low-lying bands and the empty metal d states.

The band structure of rare earth trihydrides which cristallize within the BiF_3 structure is characterized as shown in Fig. 8 for LaH_3 by the presence of three bands at low energy which can thus accomodate the 6

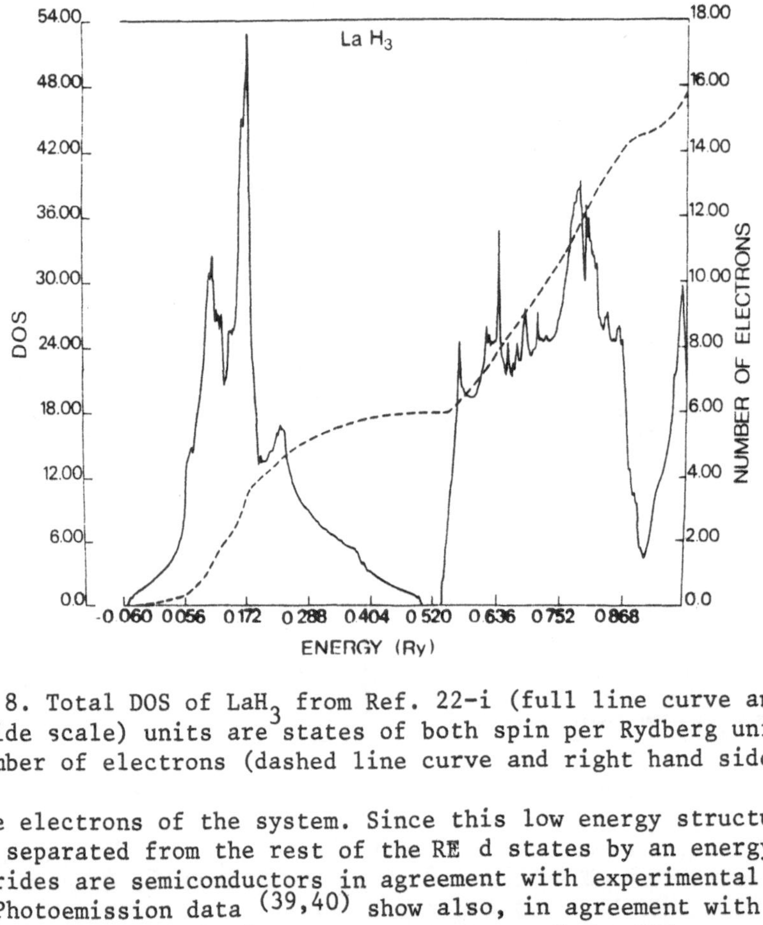

Figure 8. Total DOS of LaH_3 from Ref. 22-i (full line curve and left hand side scale) units are states of both spin per Rydberg unit cell ; the number of electrons (dashed line curve and right hand side scale).

valence electrons of the system. Since this low energy structure of the DOS is separated from the rest of the RE d states by an energy gap, the trihydrides are semiconductors in agreement with experimental observation. Photoemission data (39,40) show also, in agreement with the theoretical result that the low energy structure of the DOS grows from LaH_2 to LaH_3 and that the metal d bands disappear progressively as the H/La ratio increases above the value of 2.

The change in the metal lattice crystal structure accompanying generally the formation of di and trihydrides and the lattice expansion together with the strong modification of the lowest portion of the metal d bands due to the H-H interaction, lead to strong modification of the Fermi level position and of the value of the DOS at E_F from the pure metal to the hydrides. The DOS at E_F is much lower for the dihydrides of RE than for the pure RE metals. For example we obtain $N(E_F) = 15.05$ states of both spin per Ry-cell for LaH_2 compared to $N(E_F) = 27.48$ in the same units for pure fcc La (41). This drastic decrease in the value of $N(E_F)$ explains why both the electronic specific heat (42) and the Pauli susceptibility (43) of La decrease upon hydrogenation. It is to be noted in connection with the decrease of γ, however, that the electron-phonon mass enhancement decreases also from pure La to LaH_2 as we shall see in section 3.

As emphasized previously, because of the lowering of some metal states due to the M-H interaction and because of the presence of additional H-H antibonding states at low energy, the net result appears as a depopulation of the d bands which in the trivalent RE dihydrides accomodate only one electron. This decrease in the number of conduction electrons and also the large expansion of the lattice upon hydrogenation should lead to a net decrease in the strength of the Ruderman-Kittel-Kasuya-Yoshida (RKKY) indirect exchange interaction of the RE local moments mediated by the conduction electrons. This can explain why the magnetic transition temperatures are far lower for the dihydrides than for the pure RE metals. For example, the Neel temperature is 229K for pure Tb while TbH_2 shows an antiferromagnetic order below 17K only (44).

Another interesting property of the trivalent RE dihydrides is that they are all much better conductors than the corresponding RE (the dihydrides of divalent RE, as explained previously, are semiconductors) ; the resitivity of the dihydrides (45) is reduced from that of the corresponding RE by a factor which varies from 2 to 7. The change in the electronic properties of the RE upon hydrogenation which is characterized by a depolulation and deformation of the metal d bands and by a reduction of the DOS at E_F leads certainly to a different value of the $s \to d$ scattering mechanism of the electrons at the Fermi surface and also, as we shall see in Section VI to a change in the electronic term of the matrix elements of the electron-phonon coupling. Moreover, the hydrogenation process which is accompanied by a large lattice expansion and a change on the screening of the phonon field, results in drastic changes in the phonon properties. This is indicated by the variation of the value of the Debye temperature which is $\theta_D = 140K$ for pure La (46) while it becomes $\theta_D = 243K$ for $LaH_{2.03}$ (47). This increase of θ_D leads also to the reduction of the phonon resistivity from the pure RE to the dihydrides.

3.3. Bonding properties and stability of binary metal hydrides

Numerous experimental data have been prior the results of ab-initio calculations interpreted in terms of either the protons H^+ or the anionic H^- model.

The protonic model was used mostly for TM hydrides on the right of the periodic Table such as PdH or NiH. It was assumed that the additional electron brought in the unit cell by the H atom was shifting rigidly the Fermi level of the pure metal to accomodate one electron. As we explained in paragraph 3.1, although the Fermi level of the pure metals Pd and Ni is shifted towards higher energies upon formation of the corresponding hydrides, the protonic model is not quantitatively correct. We have emphasized in this connection the importance of the M-H bonding and the deformation of some portions of the metal d bands in the bonding process. Moreover, a charge analysis reveals the existence of a small charge transfer from the metal to the H site. The anionic H^- model has been often used on the other hand to interpret experimental data on early TM and RE dihydrides ; in particular the crystal field spliting pattern of the RE ions [48]. The existence of a charge transfer from the metal to the hydrogen atom is also indicated by the study of the core level shifts obtained from photoemission experiments [39,49]. The results indicate a shift of the metal core levels towards higher binding energies, upon hydrogenation. This in turn points out, at least qualitatively, to the existence of some charge transfer from the metal to the H atoms. For example, in the case of $ZrH_{1.86}$ it has been found that the 4p levels of Zr are shifted by about 1 eV towards higher binding energies [49].

The detailed theoretical results of the electronic structure indicate, as emphasized previously, that inspite of an apparent depopulation of the metal d bands in the dihydride, the study of the bonding effects and the deformation of the d bands upon formation of a compound show that the picture of H^- ions is oversimplified. The proton is well screened in the dihydrides however, part of the charge found in the MT hydrogen spheres is due to a charge overlap rather than to a charge transfer. Moreover, the metal-hydrogen interaction cannot be considered as fully ionic ; covalent and metallic interactions are also important. Theoretical analysis [25] indicate charge transfers from the metal to the H site in agreement with experimental trends however the magnitude of the charge transfer is never higher than 0.15 electrons,it is too small to justify quantitatively the anionic model.

In our discussion of the electronic properties of TM and RE metal hydrides we have shown that the stability results from the interplay of several competing factors. The factors which positively affect the stability are (i) the general lowering of the metal d bands by the attractive H potential ; (ii) the lowering of the metal states which have an s symmetry at the H sites, by the M-H interaction. This effect is particularly important if the metal states lowered were previously empty in the pure metal ; (iii) the H-H interactions which lead to the presence of additional states below the metal d bands such as the antibonding states observed in the di- and trihydrides of TM and RE. The destabilization of these low-lying states due to the increase of the H-H distances and the difference in the relative positions of the metal d and s bands as Z increases in a series of TM, leads to the non-existence of dihydrides beyond column VI. This feature also explains why the dihydrides of lanthanides which have large lattice constants and thus have low antibonding H-H states are more stable than the TM dihydrides (iiii) the nature of the M-H bonds which are partially covalent and partially ionic with of course also a metallic contribution affect also favorably the total energy of the metal hydrides.

We have also indicated the factors which adversely affect the stability of the hydrides in particular the upward shift of the Fermi level of the pure metal. In fact, as shown previously in this section, there is never an exact balance between the number of additional empty states brought below E_F by the M-H interaction and the number of additional electrons due to the presence of H atoms in the unit cell. Besides the effects of the deformation of metal d bands, the Fermi level of the metal necessarily shifts upon formation of a hydride. The most unfavorable case for the hydride formation corresponds to a large upward shift of E_F ; this occurs in particular when the DOS of the pure metal is low, as in the noble metals, and when few empty metal states are lowered below E_F by the M-H interaction. The experimental data[1-3] of the enthalpies of formation of hydrides show a general upward trend as one goes from left to right in the periodic table, the most stable hydrides form on the left. In addition to this general trend in the curve of the heats of formation, an anomalous dip is observed for Pd and Ni. Theoretical studies have confirmed this trend [24].

4. OCCURENCE OF SUPERCONDUCTIVITY IN METAL HYDRIDES

Superconductivity of hydrides was first discovered for Th_4H_{15}($T_c \sim$ 8K)[50] and for PdH ($T_c \sim$ 9K) and PdD ($T_c \sim$ 11K) [51]. Other systems such as alloys of Pd with noble metals [52] have fairly high values of $T_c \sim$ 16.6K. The ion implantation technique [52,53] has been widely used to load metals and alloys with Hydrogen and the search for superconductivity in these systems has given in some cases positive results. However if hydrogen makes Pd a superconductor, on the other hand hydrogen affects negatively the superconductivity of the group V TM and of fcc La. For example, TaH, VH_2, NbH_2 [54] and LaH_x [55] for $1.8 < x < 2.36$ are not superconducting down to 1K. NiH which is diamagnetic and isoelectronic to PdH does not show any superconducting transition [56] above 1K.

Using the formalism briefly summarized in Section 2.2 we have evaluated the electron-phonon coupling constant λ in order to explain the existing data on stoichiometric hydrides such as PdH, NiH, TiH_2, ZrH_2, NbH_2, LaH_2. From the trends derived from our calculations we also try to make some speculations about the possible occurrence of high T_c superconductors amongst metal hydrides.

4.1. Electron optical phonon coupling

For the PdH_x system it has been established that in contrast to most of the superconducting TM compounds, the electron-optical phonon coupling plays an essantial role. The importance of the low energy optical phonons ($\hbar\omega \sim$ 56 in eV for PdH) [57] in the electron phonon coupling mechanism has been revealed by superconducting tunneling experiments [58]. Theoretical estimates of the electron-phonon coupling constant (23,22-g) have shown, in agreement with the prediction of Ganguly [59], that the largest contribution to λ is provided by the electron-optical phonon coupling term λ_H It is thus interesting to find out wether this term is also responsible for the high value of T_c or for the lack of superconductivity in other hydrides.

The results listed in Table 2 show that the hydrogen potential scatters strongly the s waves ; the s(ℓ = 0) scattering phase shift δ_0^H

Table 2. Values of the various parameters entering the calculation of λ from Ref. 22-g. Symbols are defined in Eqs. (6) and (7). The angular-momentum-dependent phase shifts δ_ℓ are given in radians. (a) Theoretical values obtained by Butler (Ref. 62) for the corresponding metals.

		δ_0	δ_1	δ_2	δ_3	η (eV/Å^2)	$M<\omega^2>$ (eV/Å^2)	λ
LaH_2	La	-1.1006	-0.4556	0.5326	0.0006	0.753	7.35	0.103
	1xH	1.5290	0.0400	0.009	0.0	0.043	3.35	0.013
TiH_2	Ti	-0.7913	-0.2262	0.3156	0.0016	3.898		
	1xH	1.2502	0.0374	0.0009	0.0	0.067		
ZrH_2	Zr	-0.9504	-0.3662	0.9042	0.0049	2.352 (3.87)[a]	6.24	0.377
	1xH	1.2462	0.0373	0.0009	0.0	0.088	4.92	0.018
NbH_2	Nb	-1.0219	-0.4075	1.3749	0.0067	2.975 (7.39)[a]		
	1xH	1.1593	0.0408	0.0011	0.0	0.102		
BbH	Nb	-0.8077	-0.2852	1.3732	0.0054	2.855		
	H	1.2413	0.0405	0.0012	0.0	0.0898		
NbH_0	Nb	-0.5105	-0.1402	1.3396	0.0032	3.6848		
PdH	Pd	-0.5115	-0.1094	2.8066	0.0030	0.886	5.971	0.15
	H	1.1931	0.0280	0.0006	0.0	0.641	1.062	0.60
NiH	Ni	-0.3857	-0.0235	2.7916	0.0025	0.810	10.0	0.08
	H	1.0604	0.0318	0.0008	0.0	0.275	3.44	0.08

is large at the Fermi energy for all the metal hydrides, δ_0^H is always close to a resonance $\delta_0{}^H \sim \pi/2$. The phase shifts of higher angular momentum of η_H is dominated by the s-p scattering mechanism and its variation across a series of metal hydrides is controlled by the magnitude of the partial s and p DOS at E_F, relative to the value of the total DOS. As shown in Table III, the value of n_s at the H site is found to be

Table 3. The partial wave analysis n_ℓ of the DOS inside the MT metal and hydrogen spheres at the Fermi energy E_F for several stoichiometric metal hydrides (from Ref. 22-g) $N_\uparrow(E_F)$ is the total DOS per spin at E_F.

		n_s	n_p	n_d	n_f	$N_\uparrow(E_F)$
LaH_2	La	0.017	0.168	3.250	0.014	7.550
	1xH	0.017	0.154	0.0168	0.000_7	
TiH_2	Ti	0.0015	0.0590	18.7405	0.035	23.519
	1xH	0.015	0.8755	0.0535	0.0015	
ZrH_2	Zr	0.004	0.073	10.805	0.035	16.460
	1xH	0.008	0.432	0.021	0.000_8	
NbH_2	Nb	0.004	0.051	4.369	0.034	6.440
	1xH	0.016	0.119	0.010	0.001_2	
NbH	Nb	0.0925	0.3555	9.3800	0.066	12.425
	H	0.0202	0.3049	0.0283	0.004	
NbH_0	Nb	0.337	1.464	11.405	0.1652	15.41
PdH	Pd	0.058	0.163	2.618	0.012	3.405
	1xH	0.255	0.029	0.001_6	0.001	
NiH	Ni	0.048	0.133	4.741	0.009	5.390
	H	0.161	0.035	0.002_5	0.001	

sizeable for the TM hydrides at the end of the series, especially for PdH. Since for the early transition and rare earth metal hydrides the Fermi level falls at the bottom of the metal d bands, it is not surprising to find small values of n_s^H in this energy range. For PdH and NiH, the d bands are filled and the values of n_s^H at E_F become sizeable ; the Fermi level is not high enough to fall in the antibonding metal hydrogen band ; nevertheless, in this energy range some metal states of s symmetry at the H site have been lowered by the H potential, but not enough to fall below the d bands. The difference in the values of n_s^H between PdH and NiH is due to the fact for NiH, the Fermi energy is closer to the top of the d bands than in PdH. The increase of n_s^H as a function of energy, if the rigid band model is applied to the Fermi level of PdH has been invoked (23) to explain (in part) the increase of T_c in the Pd-noble-metal-H_x systems. We remind the reader that this is not the only factor which explains the increase of T_c upon alloying since there is also experimental evidence of a softening of the acoustic phonons. The DOS of p type ($\ell = 1$) at the H site is vanishingly small for hydrides of the end of the TM series such as PdH and NiH while hydrides of the beginning and the middle of the series, for which n_s^H is very small, have larger values of n_s^H. This contribution to the partial DOS arises from the metal d states having a p symmetry at the H site, which have not been perturbed by the H potential and remain almost unperturbed from the pure metal to the hydride. Thus the s-p scattering mechanism which essentially determines the electronic contribution η_H in the angular momentum representation presently used should not be viewed as an intra atomic effect since the partial DOS of p type at the H site arises from the tails of the metal d states. We can conclude from the results listed in Table VIII that, except for the simple metal hydrides which have large values of η_H and for the hydrides of the end of the TM series such as PdH, the values of η_H per site remain small for most of the metal hydrides studied here.

To date, ab-initio prediction of the position of the optic modes in metal hydrides have not yet been made ; nevertheless some neutron scattering data are available for these compounds (60). A compilation of the experimental results shows that the occupation of the octahedral sites by the H atoms such as in the fcc Pd metal or in the fcc β $VH_{0.4}$ seems to lead to lower optic modes ($\hbar\omega_{opt} \sim 50$ meV) than the occupation of tetrahedral sites. In the cubic CaF_2 structure or in the bcc metal hydrides, the energies of the optic modes is of the order of $\hbar\,\omega \sim 120$ meV. In the case of NiH, although H occupies the octahedral interstices, there is some experimental evidence that the optic modes have a much higher energy than in PdH. A study of the temperature dependence of the electrical resistivity of NiH [110] shows that the optical phonon contribution occurs at higher temperature than in PdH. The Einstein temperature of the optic modes of NiH has been estimated to be at least a factor of 1.8 larger than that of PdH. The values given in Table II for the phonon contribution $M_H \langle\omega^2\rangle_{optic}$ are taken from realistic neutron scattering data in the case of PdH. In view of the lack of data for other dihydrides, we used a scaling factor derived from the average position of the optic modes when the corresponding data are available (60).

From the results obtained in the present work, we find that the values of the electronic parameter η_H are sizeable for the TM hydrides with filled d bands, especially when E_F is not too close to the top of the d bands. λ_H in NiH is smaller than PdH because η_H is smaller and the phonon contribution is larger, thus the two isoelectronic compounds PdH and NiH have different superconducting properties. For the dihydrides of the beginning of the TM series and for LaH_2 we obtained small values of η_H^T. Moreover, for these compounds the energies of the optic phonons are high and this leads to small values of the electron-optical phonon coupling. From our study of the electronic properties of the TM dihydrides we could however expect to obtain large values of η_H for the unstable dihydrides of the middle of the TM series, since in this case the antibonding H-H band cuts E_F and gives a large density of H-s states at the Fermi level. For these compounds, which could eventually be obtained by the ion implantation technique, one could expect sizeable values of T_c due to the electron-optical phonon coupling provided that the optic-mode frequencies are not too high.

4.2. Electron acoustic phonon coupling

From the results listed in Table 2 we can see that for the TM hydrides, the s and p phase shifts at the metal site are negative ; this indicates a repulsive character of the metal potential for the s and p waves, due to the orthogonalization conditions to the corresponding core states. The d wave phase shifts at the Fermi energy are positive and increase with the filling of the metal d bands thus δ_2^{Metal} is small for hydrides whose Fermi energy falls at the bottom of the metal d bands like LaH_2 ; it increases and reaches a resonance $\delta_2^{Metal} \sim \pi/2$ in the middle of the d bands, the sharpness of the resonance being related to the width of the bands. When the d bands are filled like for PdH of NiH, the d wave phase shift is large and close to the value of π. A study of the partial wave decomposition of the DOS at E_F shows that for all the TM hydrides, and LaH_2 the 'd' character is dominant at the metal site as it can be seen from the results listed in Table 3. From the trend obtained in the values of the phase shifts and of the partial DOS at the metal site we found that the value of η_{Metal} for the TM hydrides of the middle and of the end of the series is dominated by the d-f scattering mechanism while the p-d mechanism is important also for the early TM and LaH_2. As an example, 60% of the contribution of η_{La} in LaH_2 arises from the p-d scattering term and 37% from the d-f contribution while for TM hydrides of the end of the series such as NiH and PdH more than 80% of the value of η_{Metal} is provided by the d-f scattering mechanism. We wish at this point to remind the reader that in the angular momentum representation used here, the d-f scattering should not be considered as an intra-atomic effect since the partial DOS of f type at one metal site is provided by the tails of the metal d functions of the neighboring sites ; thus the physical origin of this term should rather be understood in terms of a metal d-d interaction, in a tight-binding picture [61].

A comparison of the values of η_{Metal} in the hydrides with the values of the corresponding pure metals [62] shows, in most of the cases under study here, an important decrease of η_{Metal} upon formation of the hydride ; the theoretical values for the pure TM obtained using similar method [62] are listed in parenthesis in Table 2. It is interesting to focuss on the change of η_{Metal} observed for the high T_c metals such as Nb and fcc La, upon dihydride formation . In order to understand the origin of the disappearance of superconductivity between bcc Nb and the dihydride of Nb which has an fcc metal lattice, we calculated η_{Nb} assuming an fcc phase which we will call here after NbH_0 since it corresponds to the

NbH_2 lattice where the H atoms have been removed. We obtained a strong reduction of η_{Nb} from 7.39 eV/Å^2 in bcc Nb[62] to η_{Nb} = 3.68 eV/Å^2 in fcc NbH_0 in spite of the fact that the total DOS at E_F increases from 11.41 states/Ry spin unit cell in bcc Nb to 15.41 states/Ry spin unit cell in NbH_0. This reduction in η_{Nb} can be ascribed to the lattice expansion effect (we assumed the same lattice constant for fcc NbH_0 and NbH_2) which leads to an increase of the Nb-Nb nearest neighbor distance from 5.369 a.u. in the bcc phase to 6.088 a.u. in the fcc phase and thus reduces the strength of the d-d interaction. As shown in Table 2 the value of η_{La} decreases from 2.62 eV/Å^2 in pure fcc La[58] to 0.753 eV/Å^2 in LaH_2. It is also smaller than η_{Metal} in trivalent pure metals like Y and Sc[62]. This decrease is due (i) to the 6.7% increase in the lattice constant which leads to a decrease of the width of the d bands and thus to a change in the metal d-d interaction. A similar effect is observed in the pure metal since η_{La} decreases under pressure (ii) to the reduction of the DOS at E_F from 13.74 states/Ry spin unit cell [41] to 7.55 states/Ry spin unit cell. We have previously emphasized the electronic differences between pure fcc La and LaH_2 which are characterized by the deformation of the lower portion of the metal d band and its apparent depopulation in favor of the low-lying states in the dihydride.

Of course, even when the electronic contribution η_{Metal} decreases upon hydrogenation an enhancement of γ metal could still be obtained if the acoustic modes becomes soft enough in the hydrides. Unfortunately experimental data on the acoustic phonon frequencies of all the hydrides investigated here are not always available [60]. For PdH, the value of $M_{Pd} \langle\omega^2\rangle_{acoustic}$ listed in Table 2 has been obtained from the experimental phonon DOS of Rowe et al[63]. Since similar data are not available for other hydrides, we have thus used the approximation

$$M_{Metal} \langle\omega^2\rangle_{acoustic} \sim \frac{1}{2} M_{Metal} \theta_D^2$$

for the hydrides whose Debye temperature θ_D is known.

For LaH_2 for example, the increase of θ_D from the pure metal (θ_D = 140K) [46] to the dihydride (θ_D = 243K for $LaH_{2.03}$)[47] leads to a reduction of λ_{Metal} ; since we have seen in the study of λ_H that the electron-optical phonon coupling provides a negligible contribution in LaH_2, we find a drastic decrease in the total value of λ upon formation of the dihydride, which leads to a vanishingly small value of T_c. Thus, according to our theoretical estimate, LaH_2 should not be a superconductor ; this result is in agreement with the experimental investigation of Merriam and Schreiber [55].

For the bcc metals of group V such as Nb, there exists some experimental evidence of a hardening of the acoustic phonons with hydrogenation [60] ; it has been observed in this connection that the anomalous dips in the acoustic branches of bcc Nb disappear upon absorption of hydrogen ; this factor will further reduce the decrease of λ_{Metal} caused by the reduction of η_{Metal}. For NiH we have used the value [56] of θ_D =366K which is larger than that of PdH and leads to a reduction of λ_{Metal}. Our theoretical values of the electronic contribution η together with the experimental estimates of the phonon contribution lead to values of the electron-phonon coupling λ which explain the experimental trends.

Thus we found that substantial values of λ are obtained for the TM with filled d bands such as PdH. We have explained why NiH is not superconducting above 1K, unlike PdH. Our results explain also why the dihydrides of the early TM and LaH_2 are not superconducting down to 1K, in agreement with the experimental observations (55). From our calculations we can speculate that large values of the electron-optical phonon coupling could be obtained for the TM dihydrides of the middle of the series if they could be stabilized and prepared by techniques such as the ion implantation. Moreover, our simulations of the effect of alloying on the metallic matrix show that sizeable values of η_{Metal} could be reached ; thus, nothing prevents in principle the occurence of large values of the electron acoustic phonon mechanism in metal hydrides as the Fermi energy sweeps through the metal d bands. These speculations will hopefully stimulate further experimental investigations.

5. ELECTRONIC PROPERTIES OF TERNARY HYDRIDES

Amongst the hydrides of intermetallic compounds which are important for their potential technological applications(3,4) La Ni_5 H_x, Fe Ti H_x, and the compounds of the Mg_2 Ni H_x family have been the most extensively studied from the experimental point of view. The choice of these materials is guided by several factors such as the maximum hydrogen absorption capacity per volume and also per weight unit the easyness of the activation process, the kinetics of absorption and desorption, the sensitivity of the metallic matrix to impurities, the temperature and pressure conditions of absorption and desorption of hydrogen and of course the cost of the material used for storage of H...

I shall discuss in what follows the essential features of the electronic structure of hydrides of FeTi and of the Mg_2 NiH_4 family and give an interpretation of several experimental data.

5.1. Fe TiH and Fe TiH_2

As first shown by Reilly and Wiswall (64), FeTi absorbs hydrogen reversibly and forms, besides the dilute α phase observed up to $x \sim 0.1$, two concentrated β and γ phases 5 based respectively upon the composition $x \sim 1.0$ and $x \sim 2.0$. Elastic neutron scattering data (65) on β-FeTiH have been interpreted in terms of an orthorhombic structure which can be viewed as resulting essentially from the doubling of the cesium chloride unit cell of pure FeTi, followed by a large tetragonal distortion due to an expansion along the (110) directions of the cubic CsCl lattice, and a further orthorhombic distortion. The unit cell of β -FeTiH as well as that of γ -$FeTiH_2$ contain two formula units. γ-$FeTiH_2$ has a monoclinic structure (65) which can be regarded as resulting from a distortion of a tetragonal cell; the H-atoms are located at octahedral interstices. Among the four H octahedral interstices in the monoclinic cell, three are located in the basal plane at positions similar to those observed for β-FeTiH while the fourth octahedral H atom is located in the a/2 plane where it is surrounded by two Fe and four Ti atoms. In the present work, the monoclinic distortion with an angle of about

97° in $FeTiH_2$ was ignored for simplicity and the latter was treated as tetragonal.

The energy bands [(28a)] of β1-FeTiH are plotted in Fig. 9. Keeping in mind the remarks made above concerning the crystal structure, our results can be understood (i) by a folding of the bands of FeTi due to the doubling of the size of the CsCl cell (ii) the expansion of the cell in the (110) direction of the CsCl structure and the lifting of the degeneracies due to the orthorhombic distortion and (iii) by the role played by the H atoms in the lattice. Since the electronic structure of FeTi in

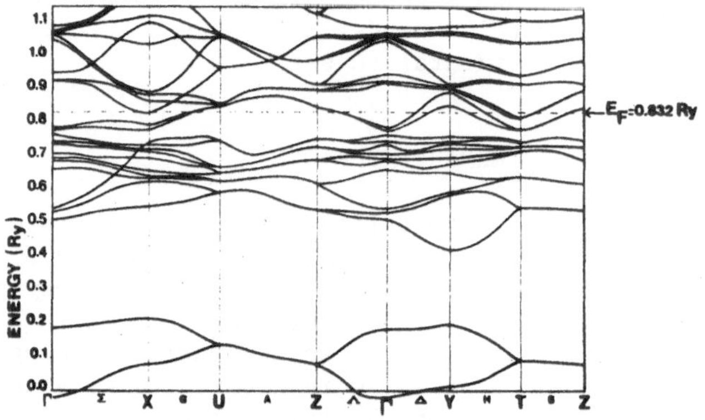

Figure 9. The energy band of β -FeTiH from Ref. 28-b along several high symmetry directions of the orthorhombic Brillouin Zone. Energies are in Rydbergs. The dashed lines indicates the Fermi level position.

the CsCℓ simple cubic BZ is characterized by 10 metal d bands overlapped and hydridized with a wider metal s-p band, these will lead to 22 bands in β-FeTiH. The presence of the H atoms results in a drastic lowering of two bands by the metal-hydrogen interaction. The lowering of metal states having an s symmetry at the H interstitial site by the H potential which strongly scatters s waves, is a feature common to all the metal hydrides previously studied it is an important factor for the stability of the hydride. The two low-lying bands give rise to the structure of the DOS observed in Fig. 10 ; the second band in Fig. 9 should be understood as resulting from the folding of the low-lying metal-hydrogen band due to the doubling of the size of the unit cell. These metal-hydrogen low-lying bands are formed out of states already filled in the pure intermetallic FeTi ; however the M-H interaction is strong enough to bring inside the metal d bands, below the Fermi level of the hydride a full branch of metal p states located above E_F in pure FeTi. These additional states do not participate in the two low-lying bands, however they bring some additional electrons, about 0.15 electrons, below E_F. As a consequence, less than one electron brought by the H atom has to be accomodated at the top of the d bands and thus, although the Fermi level of the pure intermetallic is shifted towards higher energies in the hydride, the protonic rigid band model is not quantitatively

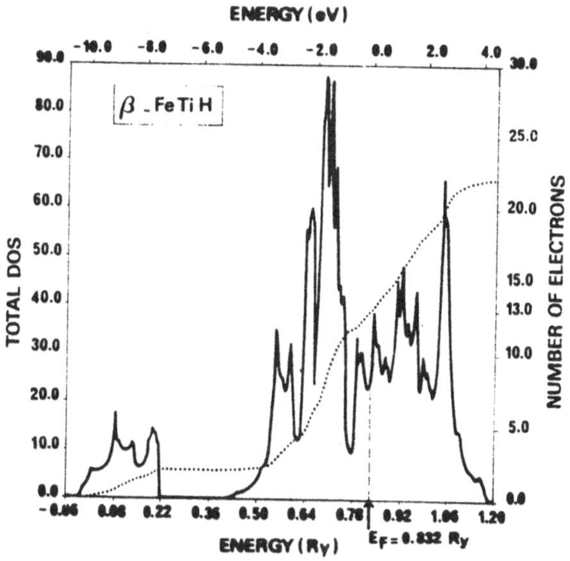

Figure 10. The total DOS of β -FeTiH (full curve, left-hand side scale) from Ref. 28-b . Units are states of both spin per Rydberg FeTiH. The total number of electrons (dotted curve, right-hand side scale).

correct. Further, the metal d bands are deformed in going from the intermetallic to its hydride. The DOS of β-FeTiH plotted in Fig. 10 can be characterized (i) by a structure centered at 9 eV below E_F which is due to the M-H interaction, and (ii) the metal d bands at higher energies in which we identify, as pure FeTi [66], the two peaks structure characteristic of the bonding and antibonding metal states in the bcc metals. The width of the lowest metal d states measured up to the energy of the valley of the DOS is slightly smaller in FeTiH than in FeTi due to the lowering of the lowest portion of the Fe-d states by the H potential ; the increase in the lattice parameter plays also a role in the slight narrowing of the d bands. The increase by 2.3% of the FeTi bond length from FeTi to β -FeTiH results also in a weakening of the bond and a larger overlap between the bonding and antibonding metal d states in FeTiH, the valley of the DOS being narrower in the hydride. Pure FeTi is isoelectronic to Cr and the Fermi energy falls in the valley of the DOS [66], the Fermi level of FeTiH is located at higher energies in the antibonding states but, as discussed above less than one electron brought by the H atom is added at the top of the d bands. The number of states at E_F increases substantially from FeTi to FeTiH. We find $N(E_F)$ = 23.93 states of both spin/Ry-FeTiH which corresponds to an electronic specific heat coefficient of 2.02 $mJ.mole^{-1}K^{-2}$ in the noninteracting electron model and without electron-phonon enhancement factor. This is to be compared with the values of $1 mJ.mole^{-1}K^{-2}$ and 0.53 $mJ.mole^{-1}K^{2}$ given respectively by the authors of Ref. 66 for pure FeTi.

A partial DOS analysis into its angular momentum components inside the H and metal MT spheres is plotted in Fig. 11 and reveals that the low-lying energy states have essentially a H-s and also a Fe-d character ; the Ti-d contribution and the metal s and p components are substantially smaller. Similar partial DOS analysis shows that the lowest

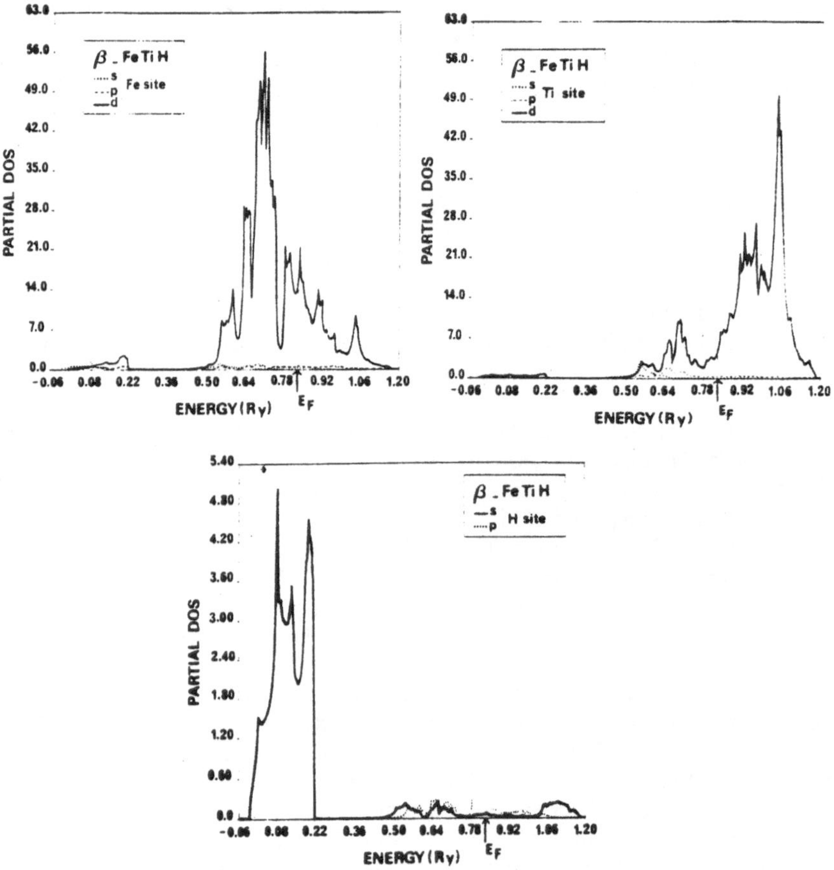

Figure 11. The angular momemtum DOS analysis n_ℓ of β-FeTiH
inside the Fe muffin-tin sphere
inside the Ti muffin-tin sphere
inside the H muffin-tin sphere
Units are states of both spin per Rydberg FeTiH.

portion of the metal d bands is essentially dominated by Fe-d states while Ti-d states have their most important contribution for energies larger than that of the valley in the DOS ; this feature reminiscent of the relative position of the atomic d levels, is also observed in the pure intermetallic FeTi. It would be very instructive to compare the features of the band structure described above with photoemission and X-ray emission spectra.

A comparison of the DOS of $FeTiH_2$ (28-b) in Fig. 12 to the DOS of FeTiH reveals that upon hydrogen uptake, the M-H related structure at low energy grows in width and in intensity. Indeed, this structure which

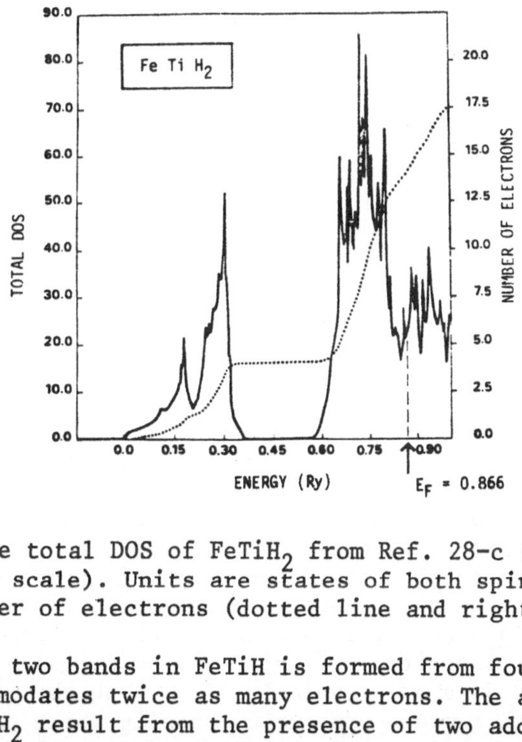

Figure 12. The total DOS of $FeTiH_2$ from Ref. 28-c (full line curve, left-hand side scale). Units are states of both spin per Rydberg $FeTiH_2$. The total number of electrons (dotted line and right-hand side scale).

corresponds to two bands in FeTiH is formed from four bands in $FeTiH_2$ and thus accommodates twice as many electrons. The additional two bands found for $FeTiH_2$ result from the presence of two additional H atoms in the unit cell ; the H states forming these low-lying bands are also hybridized with the metal-d and to a lesser extend with the metal s and p states. The Fermi energy of $FeTiH_2$ falls, as for FeTiH, above the valley of the DOS however, it is closer to the bottom of the valley of the DOS in the case of $FeTiH_2$ due to the presence of a large number of additional states in the low-lying bands. For $FeTiH_2$ we obtain $N(E_F)$ = 22.82 states of both spin/Ry - $FeTiH_2$ which corresponds to an unenhanced value of $\gamma = 1.93$ $mJ.mole^{-1}K^{-2}$. The results of γ obtained for FeTiH and $FeTiH_2$ are in agreement with the large increase observed experimentally in the electronic specific heat coefficient (67) after hydrogenation of FeTi (68). The large increase in the DOS from FeTi to its hydrides results, in agreement with experimental data, in an increase in the Pauli susceptibility χ_p (69). It is to be noted that magnetization data are not trivial to analyse since in order to obtain χ_p, the important contribution of superparamagnetic Fe particles has to be subtracted.

In the absence of photoemission and X-ray emission data, isomer shift (IS) studies provide useful informations on the change in the electronic properties of a metal upon absorption of hydrogen. Since the compounds under study contain already a Mössbauer nucleus, iron, IS studies are not hindered by the introduction of an impurity atom which always affect the electronic properties of the matrix. A systematic decrease of the contact -s electron density, ρ_s, at the metal site with hydrogen uptake has been observed for the intermetallic compounds $FeTiH_x$ (70) ; a decrease of ρ_s occurs also in binary hydrides such as TaH_x, PdH_x, and NiH_x (71). In the case of $FeTiH_x$, unlike for PdH_x, simple renormalization of the 4s wave-functions due to the lattice expansion accompanying H uptake did not appear to be sufficient to explain the observed decrease of ρ_s. Our ab-initio calculation of ρ_s at the Fe nucleus in FeTiH and $FeTiH_2$ shows indeed (28-c), in agreement with the experimental observation (70) that the depletion of the 4s metal states at the Fe site observed upon hydrogen uptake can be in large part ascribed to M-H bonding effects.

5.2. Hydrides of the Mg_2NiH_4 'family'

Mg_2Ni absorbs hydrogen reversibly (72) and leads to high hydrogen capacity compounds since the Mg_2Ni-H system forms an α phase up to H/M = 0.3 and a β phase higher hydrogen concentrations, which is based upon the stoichiometric Mg_2NiH_4 composition. This technologically important material (3,4) belongs to a family of ternary hydrides $M_2M'H_x$ where M is a divalent simple or rare earth metal (M≡ Mg, Ca, Sr, Eu, Yb) and M' is a transition metal (M'≡Fe, Ni, Rh, Ir, Ru etc.). In this family the maximum hydrogen content corresponds to x = 6. It is interesting to note that the maximum value of x which can be achieved for given elements M and M' is such that the total number of valence electrons is 18. For example, the hydrides with the highest hydrogen capacity are Sr_2RuH_6, Sr_2IrH_5 and Mg_2NiH_4. This observation suggests an electronic origin for the hydrogen absorption capacity, and this will be discussed below. Various crystallographic data are available (73-77) which all indicate the existence of a high temperature cubic phase in which the transition metal elements M' form an fcc array. The divalent metal elements M are located at the $(\frac{1}{4}, \frac{1}{4}, \frac{1}{4})$ positions in the cubic cell ; they form a cubic array, and every other cube of M atoms has a transition element M' at its center. The metal atoms thus have an antifluorite-type arrangement in which the positions of the transition metal atom M' are those of the cations and the positions of the divalent metal M are those of the anions in the fluorite CaF_2. Allotropic transformations have been investigated experimentally for Mg_2NiH_4, but we shall be concerned only with the cubic structure here. The localization of the hydrogen atoms has been studied by elastic neutron diffraction on deuterated samples. The experimental data indicate that the coordinates of the deuterium atoms could be close to $(\frac{1}{4},0,0)$ [73-76] or $(\frac{1}{4}, \frac{1}{4}, 0)$ [77] It has recently been shown for Mg_2NiH_4 that the reliability factor of the neutron data analysis cannot decisively rule out either of the two possible hydrogen locations. The neutron diffraction analysis of Didisheim *et al.* [76] for Mg_2FeH_6 indicate that the compound crystallizes with the K_2PtCl_6 structure so that the hydrogen atoms form an actahedral cage around each iron atom with short Fe-H bond length. In this structure the hydrogen (deuterium) atoms are located close to the ($\frac{1}{4}$, 0,0) positions, which is consistent with the Mössbauer, Raman and IR measurements [76]. In our work on Mg_2NiH_4, however, we also considered the structure found by Darriet *et al.* [77].

We shall discuss in what follows the theoretical results obtained for Mg_2FeH_6 and Mg_2NiH_4 and interpret several experimental results. The energy bands of Mg_2FeH_6 are plotted in Fig. 13 along several high symmetry directions of the fcc BZ. In order to understand these results we should keep in mind the following essential features of the structural data. The M'-M' distances are large (d_{Fe-Fe} = 4.56 Å in Mg_2FeH_6 while d_{Fe-Fe} = 2.48 Å in b.c.c. iron). Therefore we expect a weak (M'd)-(M'd) interaction and narrow M' d bands. The M'-H distances are short (d_{Fe-H} = 1.54 Å in Mg_2FeH_6) and therefore strong (M'd)-H interactions

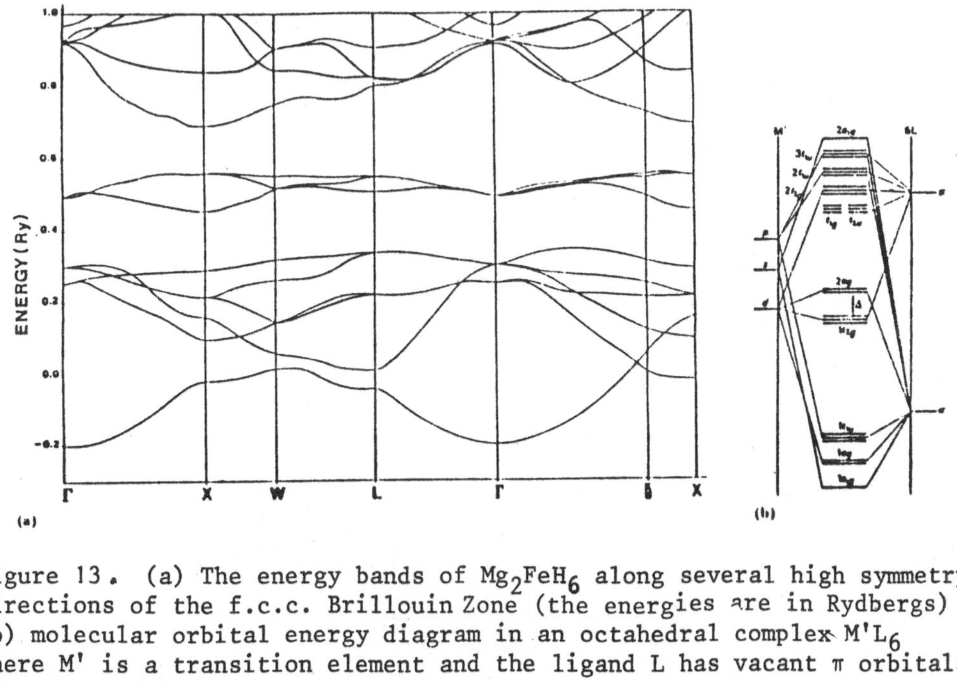

Figure 13. (a) The energy bands of Mg_2FeH_6 along several high symmetry directions of the f.c.c. Brillouin Zone (the energies are in Rydbergs) ; (b) molecular orbital energy diagram in an octahedral complex $M'L_6$ where M' is a transition element and the ligand L has vacant π orbitals from Ref. 22-d.

and thus large ligand field splittings are expected. The M-H distances are much larger because the hydrogen atoms are located close to the centres of the faces of the M cubes (d_{Mg-H} = 2.28 Å in Mg_2FeH_6). We can thus expect the electronic properties to be strongly dominated by the characteristic features of the octahedral transition metal complex $M'H_6$

since much weaker M-H and M'M interaction are expected. In order to illustrate this point we have reproduced in Fig. 13 the typical energy diagram expected for a free octahedral $M'L_6$ complex where M' is a transition element and the ligand L has occupied s and empty p orbitals.

In the present case, where the octahedral complex is in a periodic array of metal atoms, we expect that the M' s and p states are partially filled and thus are located below the M' d orbitals. Nevertheless we observe strong similarities between the bands plotted in Fig. 13 and the schematic diagram of Fig. 13. At the Brillouin zone center Γ we observe, in increasing order of energy, (i) the Γ_1 state which is a hybridized (M's)-M(s) state lowered by the H s interactions corresponds to the $1a_{1g}$ molecular orbital level , (ii) a doubly degenerate Γ_{12} state which is a bonding combination of the H s states and the $M'd_{z^2}$ and $d_{x^2-y^2}$ orbitals whose lobes point towards the hydrogen sites it corresponds to the $1e_g$ molecular levels and (iii) a triply degenerate H s state which is nonbonding with respect to the transition metal atom it corresponds to the $1t_{1u}$ molecular degenerate levels .

The total DOS of Mg_2FeH_6 is plotted in Fig. 14. The structure observed at low energy corresponds to the 6 low-lying bands which, as

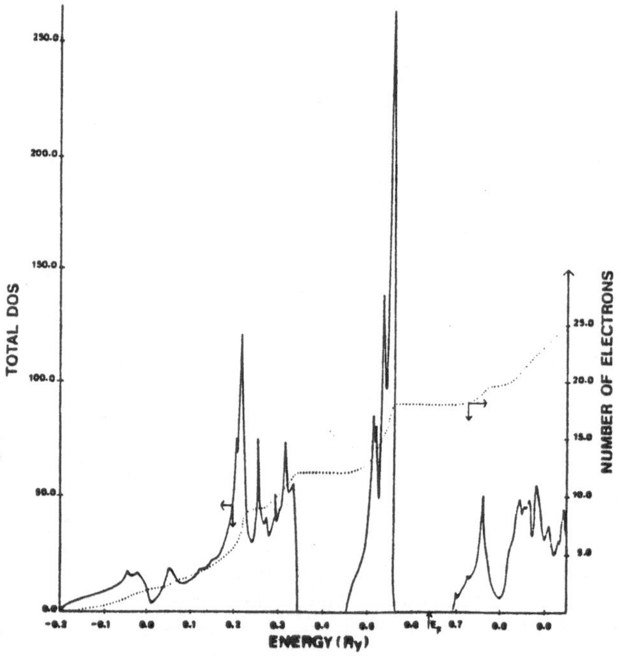

Figure 14 . The total DOS of Mg_2FeH_6 (——, left-hand scale) and the number of electrons (....., right-hand scale) (the arrow indicates the position of the Fermi energy E_F and the units are states of both spin per Rydberg per unit cell), from Ref. 28-d .

explained above result from strong (M'd)-(Hs) and H-H interactions and are also hybridized with M s-p and M's states. A gap separates the first

six bands from the next three-band complex which is formed by the narrow M' d (t_{2g}) states. Another energy gap of 1.8 eV separates the M'd(t_{2g}) states from the higher states in which we find the antibonding combination of the doubly degenerate d_{z^2} and $d_{x^2-y^2}$ orbitals with the H s states. These states are empty since the compound has 18 valence electrons which fill the 9 first bands. Mg_2FeH_6 is thus found to be a semiconductor. The origin of the gap between the filled non-bonding t_{2g} levels and the empty antibonding (M' d(e_g) -H s) states is due to the large ligand field splitting caused by the strong interactions between the transition metal elements and their octahedral environment of hydrogen atoms, the M'-H distances being rather short.
The DOS of Mg_2NiH_4 obtained with the $K_2PtC\ell_6$ type structure[75] is plotted in Fig. 15. We observe similar features as in Mg_2FeH_6 besides the expected changes originating from differences in the lattice constants and the atomic species. We find however that since the number of hydrogen atoms is only 4, the M' d(e_g)-(H s) and (H s)-(H s) interactions

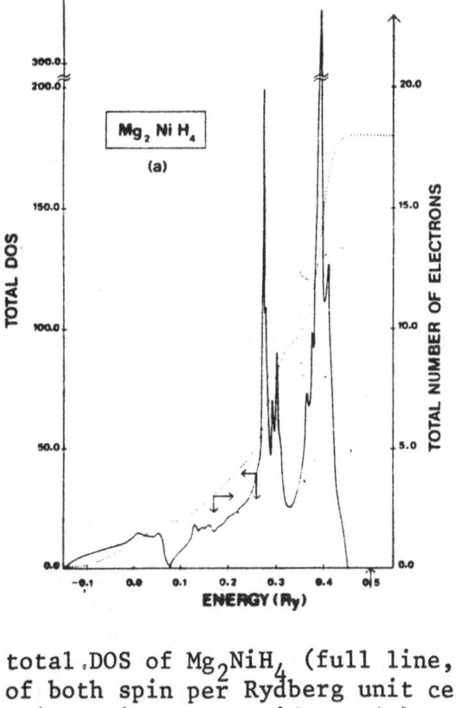

Figure 15. The total DOS of Mg_2NiH_4 (full line, left hand-side scale). Units are states of both spin per Rydberg unit cell. The number of electrons (dotted line, right hand-side scale).

are weaker than in Mg_2FeH_6 ; thus the splitting of the bonding d (e_g) from the non-bonding d (t_{2g}) states is not as strong. However, the energy gap which appears in the metal d bands between the filled d (t_{2g}) and the empty antibonding M' d($t2_g$)-H (s) states has the same origin as in Mg_2FeH_6. Thus, Mg_2NiH_4 is also found to be a semiconductor. The theoretical value of the energy gap is 1.36 eV.

Despite the absence of precise experimental determinations of the energy gaps, the compounds belonging to this family are known to be semiconductors from the preliminary resistivity data of Moyer and coworkers (73) and the unpublished resistivity data of Genossar (78) in which the conductivities of both the low and the high temperature phases of Mg_2NiH_4 have been found to increase exponentially with temperature. The conductivities of the Ca-Ir-H, Sr-Ir-H, Ca-Ru-H systems range from $6.4X10^{-8}$ to $2.6X10^{-7}$ Ω^{-1} cm^{-1} ; the conductivities of the Ca-Rh-H and Sr-Rh-H systems are also small but are much higher than those of the other systems since they are of the order of 10^{-1} $\Omega^{-1}cm^{-1}$ [5]. The colours of these hydrides are olive green for Mg_2FeH_6 and reddish for Mg_2NiH_4 ; apart from the Fe-H vibrations, Mg_2FeH_6 is transparent in the IR down to 400 cm^{-1} (76).

The present results suggest that the filling of the antibonding M' $d(e_g)$ -H (s) states is energetically unfavourable owing to the existence of an energy gap and could explain why the maximum hydrogen content corresponds to 18 valence electrons which can be accommodated in the states below the energy gap. Mössbauer, Raman and IR data have been obtained for Mg_2FeH_6 [8] and have been shown to be consistent with the presence of octahedral low spin iron (II) $[FeH_6]^{4-}$ ions. Our results show that the M' $d(t_{2g})$ states are filled, leaving the antibonding M' $d(e_g)$ -H(s) states empty. This picture confirms the experimental interpretation (76) made in terms of a low spin d^6 configuration. However, the ionic picture proposed (76) is certainly too schematic. We have emphasized in our partial DOS analysis that the low-lying bands are not just hydrogen-derived states. They contain a strong $M'd(e_g)$ bonding contribution together with a M s-p component.

For Mg_2NiH_4 a partial DOS analysis (28-e) has been performed using two different possible positions of the H atoms in the unit cell (73,77). Our results have been compared to the densities of occupied states obtained by X-ray emission spectroscopy (28-e) corresponding to the (Ni2p3/2) ←(Ni3d) and (Mg 1s) ← (Mg valence band) transitions. We conclude that the octahedral type of arrangement of the H atoms around the transition metal appears to be favoured.

6. CONCLUSIONS

We have shown that the theoretical band structure work on stoichiometric metal hydrides has clarified the nature of the chemical bond in these compounds and has been very successful in the interpretation of a large variety of experimental data. In our view it is important for fundamental as well as for applied research to further extend these studies to hydrides of intermetallic compounds, and to further examine the interesting problem of superconductivity of metallic hydrides.

REFERENCES

1. G.G. LIBOWITZ, "Solid State Chemistry of Binary Metal Hydrides", ed. W.A. BENJAMIN, Inc., New York (1965).
2. W.M. MUELLER, J.P. BLACKLEDGE and G.G. LIBOWITZ eds., "Metal Hydrides", Academic Press, New York (1968).
3. 'Topics in Applied Physics, Vols 28-29, Hydrogen in Metals' ed. by G. ALEFELD and J. VÖLKL, Springer-Verlag (1978).
4. ed. by L. SCHLAPBACH, Spinger-Verlag, in print.
5. J. FRIEDEL, Phil. Mag. 43, 153 (1952).
6. H. HOHENBERG and W. KOHN, Phys. Rev. 1313, B864 (1964) ; W. KOHN and L.J. SHAM, Phys. Rev. 140, A1133 (1965).
7. Z.D. POPOVIC and M.J.STOTT, Phys. Rev. Lett. 33, 1164 (1974) ; Z.D. POPOVIC, M.J. STOTT, J.P. CARBOTTE and G.R. PIERCY, Phys. Rev. B 13, 590 (1976).
8. C.O. ALMBLADH, U. VON BARTH, Z.D. POPOVIC and M.J. STOTT, Phys. Rev. B 14, 2250 (1976).
9. M. MANNINEN, P. HAUTOJÄRVI and M. NIEMINEN, Solid State Commun. 23, 795 (1977) ; M. MANNINEN and M. NIEMINEN, J. Phys. F. 9, 1333 (1979).
10. J.K. NORSKOV, Solid State Commun. 24, 691 (1977) ; ibid. Phys. Rev. B 20, 446 (1979).
11. P. JENA and K.S. SINGWI, Phys. Rev. B 17, 3518 (1978).
12. L.M. KAHN, F. PERROT and M. RASOLT, Phys. Rev. B 21, 5594 (1980) ; F. PERROT and M. RASOLT, Phys. Rev. B 23, 6534 (1981).
13. J. Khalifeh, Thèse d'Etat, Strasbourg, France (1982), unpublished ; C. DEMANGEAT, M.A. KAHN, G. MORAITIS and J.C. PARLEBAS, J. de Physique 41, 1001 (1980).
14. A.C. SWITENDICK, Ber. Bunsenges. Phys. Chem. 76, 535 (1972).
15. R.P. GUPTA, J. Less Common Met. 88, 299 (1982)
16. R.P. MESSMER, S.K. KNUDSON, K.H. JOHNSON, J.B. DIAMOND and C.Y. YANG, Phys. Rev. B 13, 1396 (1976).
17. D.E. ELLIS and G.S. PAINTER, Phys. Rev. B 2, 2887 (1970).
18. R.W. SIMPSON, N.F. LANE and R.C. CHANEY, J. of N. Mat. 69 and 70, 581 (1978).
19. R. PODLOUCKY, R. ZELLER and P. DEDERICHS, Phys. Rev. B 22, 5777 (1981).
20. B.M. KLEIN and W.E. PICKETT, J. Less Common Met.
21. a) A. C. SWITENDICK, Solid State Commun. 8, 1463 (1970).
    b) ibid, Int. J. Quantum Chem. 5, 459 (1971).
    ibid, J. of Less-Common Met. 49, 283 (1976).
    c) ibid, in "Transition Metal Hydrides, Advances in Chemistry Series" No 167 ed. Robert Bau, (The Americal Physical Society 1978) p. 264.
    d) ibid. in "Topics in Applied Physics", Vol. 28, Hydrogen in Metals, ed. by G. ALFEFELD and J. VÖLKL, Springer-Verlag 1978, p. 101.
22. a) Michele GUPTA, Solid State Commun. 27, 1355 (1978).
    b) M. GUPTA and A.J. FREEMAN, Phys. Rev. B 17, 3029 (1978).
    c) M. GUPTA, Solid State Commun. 29, 47 (1979).
    d) ibid, J. Phys. F. 10, 2649 (1980).

e) ibid, Phys. Rev. B 22, 6074 (1980).
f) ibid, J. of the Less-Common Met. 73, 321 (1980).
g) ibid, in Metal Hydrides ed. by G. BAMBAKIDIS, Plenum Press (1981) p. 255.
h) ibid, Phys. Rev. B 24, 7099 (1982).
i) ibid, Phys. Rev. B 25, 1027 (1982).
j) ibid in 'Electronic Structure and Properties of Hydrogen in Metals' ed. by P. JENA and C.B. SATTERTHWAITE, NATO Conference Series VI, Plenum Press., p. 321 (1983).

23. a) D.A. PAPACONSTANTOPOULOS and B.M. KLEIN, Phys. Rev. Lett. 35, 110 (1975).
b) B.M. KLEIN, E.N. ECONOMOU and D.A. PAPACONSTANTOPOULOS, Phys. Rev. Lett. 39, 574 (1977).
c) D.A. PAPACONSTANTOPOULOS, B.M. KLEIN, E.N. ECONOMOU and L.L. BOYER, Phys. Rev. B 17, 141 (1978).
d) D.A. PAPACONSTANTOPOULOS in Metal Hydrides, ed. by G. BAMBAKIDIS, Plenum Press (1981).

24. C.D. GELATT, J.A. WEISS and H. EHRENREICH, Phys. Rev. B 17, 1940 (1978).

25. a) D.J. PETERMAN, B.N. HARMON, J. MARCHIANDO and J.H. WEAVER, Phys. Rev. B 19, 4867 (1979).
b) D.J. PETERMAN and B.N. HARMON, unpublished.

26. N.I. KULIKOV, V.N. BORZUNOV and D.A. ZVONKOV, Phys. Stat. Sol. (b) 86, 83 (1978) ; N.I. KULIKOV, Phys. Stat. Sol. (b) 91, 753 (1979).

27. A. FUJIMORI, F. MINAMI and N. TSUDA, Phys. Rev. B 22, 3573 (1980) ; A. FUJIMORI and N. TSUDA, Solid State Commun. 41, 491 (1982).

28. a) Michele GUPTA, J. Phys. F 12, 57 (1982)
b) ibid., J. Less-Common Met. 88, 221 (1983)
c) Solid State Commun. 42, 501 (1982)
d) ibid., J. Less-Common Met. 103, 325 (1984)
e) Michele GUPTA, E. BELIN and L. SCHLAPBACH, J. Less-Common Met. 103, 389 (1984).

29. J.C. SLATER, Phys. Rev. 51, 846 (1937).

30. G. LEHMANN and M. TAUT, Phys. Stat. Sol. (b) 54, 469 (1972) ; O. JEPSEN AND O.K. ANDERSON, Solid State Commun. 9, 1763 (1971).

31. W.L. McMILLAN, Phys. Rev. 167, 331 (1968).

32. G.D. GASPARI and B.L. GYORFFY, Phys. Rev. Lett. 29 801 (1972); I.R. GOMMERSALL and B.L. GYORFFY, J. Phys. F 4, 1204 (1974).

33. D.E. EASTMAN, J.K. CASHION and A.C. SWITENDICK, Phys. Rev. Lett. 27, 35 (1971) ; L. SCHLAPBACH and J.P. BUIGER, J. de Physique 43, L273 (1982) ; P. BENNETT and J.C. FUGGLE, Phys. Rev. B, to be published.

34. E. BELIN, L. SCHLAPACH and Michèle GUPTA J. Phys. F. 13, L193 (1983).

35. C.A. MACKLIET, D.J. GILLESPIE and A.I. SCHINDLER, Solid State Commun. 15, 207 (1974).

36. F.M. MUELLER, A.J. FREEMAN, J.O. DIMMOCK and A.M. FURDYNA, Phys. Rev. B 1, 4617 (1970).

37. F.A. LEWIS, "The Palladium Hydrogen System" Academic Press,

New York (1967).
38. A. TRAVERSE, H. BERNAS, L. DUMOULIN and Michèle GUPTA, Solid State Commun. 40, 725 (1981).
39. D.J. PETERMAN, J.H. WEAVER and D.T. PETERSON, Phys. Rev. B 23, 3906 (1981).
40. L. SCHLAPBACH, J. OSTERWALDER and H.C. SIEGMANN J. of The Less Common Met. 88, 291 (1982).
41. W.E. PICKETT, A.J. FREEMAN and D.D. KOELLING, Phys. Rev. B 22, 2695 (1980).
42. Z. BIEGANSKI and B. STALINSKI, Phys. Status Solidi A2, K161 (1970) ; Z. BIEGANSKI and M. DRULIS, Phys. Status Solidi A44, 91 (1977).
43. W.E. WALLACE and K.H. MADER, J. Chem. Phys. 48, 84 (1968) ; B. STALINSKI, Bull. Acad. Pol. Sci. C1 35, 997 (1957).
44. H. SHAKED, J. FABER Jr., M.H. MUELLER and D.G. WESTLAKE, Phys. Rev. B 16, 340 (1977).
45. B. STALINSKI, Bull. Acad. Pol. Sc. C1 35, 1001 (1957); J.N. DAOU, C.R. Acad. Sci. 250, 3165 (1960) ; J.N. DAOU and J. BONNET, C.R. Acad. Sci. 261, 1675 (1965).
46. D.L. JOHNSON and D.K. FINNEMORE, Phys. Rev. 158, 376 (1967).
47. Z. BIEGANSKI, D. GONZALEZ-ALVAREZ and F.W. KLAAYSEN, Physica 37, 153 (1967).
48. W.E. WALLACE, Ber Bunsenges. Phys. Chem. 76, 832 (1972) ; G.K. SHENOY, B.D. DUNLAP, D.G. WESTLAKE and A.E. DWIGHT, Phys. Rev. B 14, 41 (1976).
49. B.W. VEAL, D.J. LAM and D.G. WESTLAKE, Phys. Rev. B 19, 2856 (1979).
50. C.B. SATTERTHWAITE and I.L. TOEPKE, Phys. Rev. Lett. 25, 741 (1970)
51. T. SKOSKIEWICZ, Phys. Stat.Sol. (a) 11, K123 (1972) ; B. STRITZKER and W. BÜCKEL, J. Phys. 257, 1 (1972).
52. B. STRITZKER and H. WÜHL, in "Topics in Applied Physics, Vol. 28, Hydrogen in Metals", ed. G. ALEFELD and J. VÖLKL, Springer-Verlag, Berlin, Heidelberg, New York 1978, p. 243.
53. A.M. LAMOISE, J. CHAUMONT, F. MEUNIER and H. BERNAS, J. Phys. Lett. 36, L271 (1975).
54. C.B. SATTERTHWAITE and D.T. PETERSON, J. Less Common. Met. 26, 361 (1972).
55. M.F. MERRIAM and D.S. SCHREIBER, J. Phys.Chem. Solids 24, 1375 (1963).
56. D.S. McLACHLAN, I. PAPADOPOULOS and T.B. DOYLE, J. de Phys. Paris C6, 430 (1978).
57. J.M. ROWE, J.J. RUSH, H.G. SMITH, M. MOSTOLLER and H.E. FLOTOW, Phys. Rev. Lett. 33, 1297 (1974).
58. A. EICHLER, H. WÜHL and B. STRITZKER, Solid State Commun. 17, 213 (1975).
59. B.N. GANGULY, Z. Phys. 265, 433 (1975).
60. T. SPRINGER, in "Topics in Applied Physics, vol. 28, Hydrogen in Metals", Ed. G. ALEFELD and J. VÖLKL, Springer-Verlag, Berlin, New York (1978) p. 75 and references therein.
61. S. BARISIC, J. LABBE and J. FRIEDEL, Phys. Rev. Lett. 25,

919 (1970).
62. W.H. BUTLER, Phys. Rev. B 15, 5267 (1977).
63. J.M. ROWE, N. VAGELATOS, J.J. RUSH and H.E. FLOTOW, Phys. Rev. B 12, 2959 (1975).
64. J.J. REILLY and R.H. WISWALL Jr., Inorg. Chem. 13, 218 (1974).
65. P. THOMPSON, M.A. PICK, F. REIDINGER, L.M. CORLISS, J.M. HASTINGS and J.J. REILLY, J. Phys. F. 8, (1978) L75 ; P. FISSCHER, W. HÄLG, L. SCHLAPBACH, F. STUCKI and A.F. ANDRESEN, Mat. Res. Bull. 13 ; (1978) 931 ; D. FRUCHART, M. COMMANDRE, D. SAUVAGE, A. ROUAULT and R. TELLGREN, Journal of the Less-Comm. Met. 74 (1980) 55 ; W. SCHÄFER, G. WILL and T. SCHOBER, Mat. Res. Bull. 15 (1980) 627 ; P. THOMPSON, J.J. REILLY, F. REIDINGER, J.M. HASTINGS and L.M. CORLISS, J. Phys. F. Metal Phys. 9 (1979) L61.
66. Y. YAMASHITA and S. ASANO, Progr. Theor. Phys. 48, 2119 (1972) ; D.A. PAPACONSTANTOPOULOS, Phys. Rev. B 11, 4801 (1975).
67. R. HEMPELMANN, O. OHLENDORF and E. WICKE, Proc. Int. Symp. on Hydrides for Energy Storage, Geilo Norway (1977), Pergamon Press .
68. E.A. STARKE, C.H. CHEN and P.A. BECK, Phys. Rev. 126, 1746(1962).
69. F. STUCKI, Ph. D. Dissertation number 6835, E.T.H. Zurich (1981).
70. L.J. SWARTZENDRUKER, L.H. BENNETT and R.E. WATSON, J. Phys. F 6, L331 (1976).
71. A. HEIDEMAN, G. KAINDL, D. SALOMON, H. WIPF and G. WORTMANN, Phys. Rev. Lett. 36, 213 (1976) ; J.S. CARLOW and R.E. MEADS, J. Phys. F 2, 982 (1972) ; G.K. WERTHEIM and D.N.E. BUCHANAN, J. Phys. Chem. Solids 28, 225 (1967).
72. J.J. REILLY and R.H. WISWALL, Inorg. Chem. 7, 2254 (1968).
73. R.O. MOYER, Jr., C. STANITZKI, J. TANAKA, M.L. KAY and R. KLEINBERG, J. Solid State Chem., 3 (1971) 541.
J.S. THOMPSON, R.O. MOYER, Jr., and R LINDSAY, Inorg. Chem. 14 (1975) 1866.
R. LINDSAY, R.O. MOYER, Jr., J.S. THOMPSON and D. KUHN, Inorg. Chem. 15 (1976) 3050.
74. Z. GAVRA, M.H. MINTZ, G. KIMMEL and Z. HADRI, Inorg. Chem., 18 (1979) 3595. J. GENOSSAR and P.S. RUDMAN, J. Phys. Chem. Solids, 42 (1981) 611. S. ONO, H. HAYAKAWA, A. SUZUKI, K. NOMURA, N. NISHIMIYA and T. TABATA, J. Less-Common Met.,88 (1983) 63.
75. K. YVON, J. SCHEFER and F. STUCKI, Inorg. Chem., 20 (1981) 2776. D. NOREUS and L. G. OLSSON, J. Chem. Phys., 78 (1983) 2419.
76. J.J. DIDISHEIM, P. ZOLLIKER, K. YVON, P. FISCHER, J. SCHEFER, M. GUBELMANN and A.F. WILLIAMS, Inorg. Chem. to be published.
77. B. DARRIET, J.L. SOUBEYROUX, D. FRUCHART and M. PEZAT, Proc. Int. Symp. on the Properties and Applications of Metal Hydrides IV, Eilat, April 9-13, 1984, in J. Less-Common Met., 103 (1984) 153.
78. J. GENOSSAR, private communication.

ERRATA TO VOLUME I

NUCLEAR MAGNETIC RESONANCE AND MOLECULAR MOTIONS IN LYOTROPIC LIQUID CRYSTALS

C. CHACHATY, J.P. QUAEGEBEUR and B. PERLY

(Structure and Dynamics of Molecular Systems vol. I, pp. 187-201, 1985)

Page 189, paragraph 3 line 7, read :

with $S_{ii} = \frac{1}{2} < 3l_i^2 - 1 >$ where l_i is a direction cosine of the director.

Page 189, equation (2) should be read :

$$S_{ij} = \frac{1}{2} < 3\, l_i l_j - \delta_{ij} >$$

Page 190, line 8 from the bottom, read :

$$S_{11} = \frac{1}{2} P_2(\cos\beta')(3\sin^2\beta \cos^2\gamma - 1)$$

Page 190 and 191, in equation (4) read :

$\frac{\eta}{2} \sin^2\beta \cos 2\gamma$ instead of $\frac{\eta}{2} \sin^2\beta \cos^2\gamma$

Page 191, last line, read :

with $F'_k = \sum_{k'} D_{kk'}(0,\beta,\gamma) F_{k'}$

Page 192, line 9, read :

$F^2_{\pm 2} = \frac{3}{8} < \sin^2\beta\ r^{-3} >^2$ instead of $\frac{3}{4} \sin^2\beta\ r^{-3} >^2$

INDEX OF SUBJECTS

INDEX OF NAMES